BOTANICAL RESEARCH AND PRACTICES

HORTICULTURE IN THE 21ST CENTURY

Botanical Research and Practices

Additional books in this series can be found on Nova's website under the Series tab.

Additional E-books in this series can be found on Nova's website under the E-books tab.

BOTANICAL RESEARCH AND PRACTICES

HORTICULTURE IN THE 21ST CENTURY

ABIGAIL N. SAMPSON
EDITOR

Nova Science Publishers, Inc.
New York

For permission to use material from this book please contact us:
Telephone 631-231-7269; Fax 631-231-8175
Web Site: http://www.novapublishers.com

LIBRARY OF CONGRESS CATALOGING-IN-PUBLICATION DATA
Horticulture in the 21st century / editor: Abigail N. Sampson.
p. cm.
Includes index.
ISBN 978-1-61668-582-9 (hardcover)
1. Horticulture. 2. Agricultural innovations. I. Sampson, Abigail N. II.
Title: Horticulture in the twenty-first century.
SB318.H67 2010
635--dc22

Published by Nova Science Publishers, Inc. ✣ New York

CONTENTS

Preface		**vii**
Chapter 1	Mechanisms of Cadmium Toxicity in Crops: A Review *Conceição Santos, Marta Monteiro and Maria Celeste Dias*	**1**
Chapter 2	Breeding Sweet Corn: Eating Quality vs. Agronomic Performance *Bernardo Ordás, Amando Ordás and Pedro Revilla*	**41**
Chapter 3	Grafting: An Environmentally Friendly Technique to Overcome Soil-Borne Diseases and Improve the Out of Season Production of Watermelon, Cucumber and Melon *Fotios A. Bletsos and Harold C. Passam*	**81**
Chapter 4	Application of Biotechnology Tools in the Conservation and Characterization of Critically Endangered Ancient Moroccan Species Argania Spinosa (Sapotaceae) *Pedro Martínez-Gómez, Abdellatif Hafidi, Abel Piqueras, and Khalid Majourhat*	**121**
Chapter 5	Irrigating Ornamental Horticultural Crops with Saline and Alternative Water Sources *Genhua Niu and Raul I. Cabrera*	**143**
Chapter 6	Deficit Irrigation Strategies: Use of Stress Physiology Knowledge to Increase Water Use Efficency in Tomato and Potato *Radmila Stikić, Sladjana Savić, Zorica Jovanović, Sven-Erik Jacobsen, Fulai Liu and Christian R. Jensen*	**161**
Chapter 7	High Temperature Stress Tolerance and the Changes in Antioxidative Ability in Mycorrhizal Strawberry Plants *Yoichi Matsubara*	**179**
Chapter 8	*In Vitro* Propagation of the Bishop's Hat, *Epimedium alpinum* L. *Snježana Mihaljević*	**193**

Chapter 9	Development of Safe New Cultivation System for Apples Based on *S*-Rnase and SSR Information *Shogo Matsumoto*	**203**
Chapter 10	Heritability and Correlations Among Ornamental Traits of Pansy (Viola Wittrokiana) *Jian Wang and Man Zhu Bao*	**213**
Chapter 11	Developments in Applying Biotechnology Tools in Environmental Engineering *Bing-Gang Ma, Xiao-Yan Lu, Zhip-Ping Wu and Hu-Ping Zhang*	**223**
Index		**257**

PREFACE

Horticulture is the industry and science of plant cultivation, including the process of preparing soil for the planting of seeds, tubers, or cuttings. Horticulturists work and conduct research in the disciplines of plant propagation and cultivation, crop production, plant breeding and genetic engineering, plant biochemistry, and plant physiology. Horticulture usually refers to gardening on a smaller scale, while agriculture refers to the large-scale cultivation of crops. This book presents topical data on horticulture including: the mechanisms of cadmium toxicity in crops, using grafting techniques at the seedling stage to overcome soil-borne diseases, the application of biotechnology tools in the conservation of critically endangered plant species, salinity tolerance of ornamental horticultural species, and others.

Chapter 1- Soil pollution by metals is a major environmental challenge. Metals are transmitted from plants to humans *via* the food chain. In particular, cadmium (Cd), a non essential metal, is listed as a "priority substance" by EC (2455/2001/EC, 2001) and by Environmental Protection Agency (EPA, USA), and is released to environment mainly due by anthropogenic activities.

Remediation strategies claim a better knowledge on Cd phytotoxicity and detoxification mechanisms. Cadmium toxicity involves, for example, growth reduction, wilting, photosynthesis and consequently carbon metabolism decreases. At the cell level, recent data support that Cd may have cytotoxic effects, in particular leading to oxidative stress, with impairment of antioxidative enzymes. Cd induced-genotoxicity is not so well studied, but some studies in crops suggest that Cd may induce DNA degradation (clastogenicity), may decrease DNA content and may have mitogenic effects. Moreover, molecular techniques (e.g. microsatellites, AFLP, RAPDs) showed to be powerful tools to assay Cd genotoxicity, with several studies supporting Cd induced point mutations.

This chapter revises some of the most recent advances concerning Cd induced effects at cytotoxicological and genotoxicological levels in several crops (e.g. *Lactuca* sp. and *Brassica* sp.) and in other model species as *Thlaspi* sp. and *Arabidopsis* sp.. Particular emphasis is also given to the whole plant performance, concerning this metal toxicity, including gas-exchange and chlorophyll a fluorescence parameters, carbohydrates, Calvin cycle enzymes and pigments content. The chapter also revises some important aspects concerning Cd accumulation and distribution in plants and their relevance to the trophic transfer of Cd.

Finally, the potential use of new techniques as flow cytometry (quantitative cytometry) and molecular markers are critically explored in these kinds of toxicological assays, and their potential use to develop new endpoints in ecotoxicology is also revised.

Chapter 2- In many parts of the world, cultivars of maize (*Zea mays* L.) developed and grown primarily for grain are harvested early as a vegetable, the kernels being eaten when they are immature (Tracy 1999). However, the term "sweet corn" is reserved for a specific type of maize that differs from normal maize by one or more recessive alleles that alter the carbohydrate pattern of the endosperm and confer some desirable characteristics to the kernel from the point of view of the consumer.

Historically, the first mutant used since the XIX century for sweet corn has been the *sugary1* (*su1*) allele. In the 50's, the mutant *shrunken2* (*sh2*) was introduced and in the 70's the gene *sugary enhancer1* (*se1*) appeared. These three mutants are the backbone of the sweet corn industry in temperate zones. Later on, we will briefly give a glance to other genes.

Chapter 3- Due to the limited availability of suitable land and the high consumer demand for out of season produce, growers tend to cultivate Cucurbit crops (watermelon, cucumber and melon) repeatedly on the same soil and often under unfavourable environmental conditions (e.g. low temperatures, insufficient light, the presence of soil-borne pathogens etc.), resulting in reduced yield and quality. However, many of these constraints may be largely overcome by grafting plants at the seedling stage on to compatible rootstocks. Moreover, this beneficial practice has the additional advantage of being 'friendly' towards the environment. Suitable rootstocks are those that exhibit resistance to soil-borne pathogens, tolerance to low temperatures and improve the growth vigour of the scion without adversely affecting fruit quality. Hence, by grafting, growers can cultivate sensitive cultivars with desirable yield and quality characteristics out of season and in pathogen-infested soil. Grafting is carried out in modern, commercial nurseries either by hand, or by semi- or fully-automated robots. To produce uniform, robust, grafted plants at an economically acceptable level, nurseries must employ advanced management technology to ensure seedling uniformity, precision in grafting and subsequent care of the grafted seedlings. The cost of grafting is more than compensated for by the resulting higher and earlier yield than that of non-grafted plants. Additionally, because it is an 'environmentally friendly' technique, grafting is particularly applicable to the production of organic produce, for which today there is increasing consumer demand.

Chapter 4- Argan tree [*Argania spinosa* (L.) Skeel] is an endemic species from Southwest Morocco being the unique representative of the tropical Sapotaceae in this area. This species plays a great socio-economical and ecological role in these arid zones due to the prized oil from its kernels and its foliage to feed cheep and goat herds. However, a great zoo-anthropologic action led to the absence of natural regeneration of this species. In this work, the application of recent biotechnology tools for conservation and characterization of these critically endangered species is exposed. Promising propagation methods included *in-vitro* protocols well adapted to this species that allowed the introduction, micropropagation and rooting of plant material. On the other hand, a standard karyotyping protocol has been described as preliminary tool to start molecular studies. In addition, molecular markers based on PCR amplification of the DNA have also become an essential tool for the characterization and conservation of argan tree. Regarding this PCR amplification of the DNA, three main strategies, RAPD (random amplification of the DNA), SSR (using sequences from related species) and SCAR (sequence-characterized amplified regions) markers are being assayed to date. These molecular markers have been applied in the genetic characterization of argan germplasm, the establishment of genetic relationships between morphotypes, and the future construction of genetic maps. In addition, these initially DNA research open the door to the future DNA sequencing of this species.

Chapter 5- Water shortages and poor water quality are critical issues in many regions of the world. With a rapidly increasing population and diminishing water supplies, the competition for fresh water among agriculture, industry, urban and recreational users has become intense. Use of alternative water sources, such as municipal reclaimed water and other poor-quality, non-potable saline waters, for irrigating horticultural crops and urban landscapes may be inevitable in the 21st century. The major concern of using these alternative water sources is their elevated salinity, which causes injury to tissues, growth or yield reductions on sensitive plants, and also causes salt accumulation in root zones if irrigation is not managed properly. Many horticultural commodities are high-value crops and are traditionally irrigated with high quality water. Recent research results have shown promising results on irrigating a variety of ornamental horticultural plants and crops with saline water at moderate salinity levels with little negative effects on plant growth and quality. This chapter presents and reviews the latest research results on salt tolerance for a number of ornamental horticultural species. Salinity tolerance of these plants varied widely with species, environmental conditions, substrate or soil type, and irrigation management. Tolerant plants are able to resist salt stress by limiting or reducing uptake of sodium (Na^{+}) and chloride (Cl^{-}) ions, confining these ions in roots or reducing transport of the ions accumulated in roots to shoots, selective ion uptake, and tolerating high osmotic stress.

Chapter 6- Many vegetables, including tomato and potato, have high water requirements, so that supplemental irrigation is necessary for successful production. The predictions are that the demand for irrigation will increase considerably in years to come to alleviate the consequences of climate change and more frequent and severe droughts. Water is expected to become the main limiting factor in agricultural production. However, as a consequence of global climate changes and environmental pollution, water available for agriculture is often reduced. Emphasis therefore must be placed on crop physiology and crop management under dry conditions in order to make plants more efficient water users. Recent results have demonstrated that regulated deficit irrigation (RDI) and alternate partial root drying (ARD) are new deficit irrigation strategies which may decrease demand for agricultural use of water. The application of these techniques to crops such as tomato and potato has demonstrated benefits in terms of improved water-use efficiency and a stable yield, in addition to an increased yield quality. Both methods are developed on what is known on plant responses to drought stress and to the mechanism of root-to-shoot drought stress signalling. In this paper we review the research on the responses of tomato and potato to ARD and RDI irrigation strategies, with special emphasis on drought stress signalling mechanisms. Comparison of the beneficial effects of ARD and RDI were done with the aim to suggest which of these strategies may be used to get maximum tomato and potato water productivity and economic returns in dry environments.

Chapter 7- In strawberry cultivation, high temperature in summer often restricts rooting, root growth and induces browning of roots and shoots during nursery production resulting in withering and causing great losses of plants. Then, it is necessary to establish stable technique for nursery production in strawberry cultivation during the high temperature period in summer including the problems of global warming. Arbuscular mycorrhizal fungus (AMF) has the effect of promoting host plant growth mainly by enhancing phosphorus uptake through symbiosis. On the other hand, promotion of active oxygen synthesis in plants occurs under several stress conditions including temperature stress, and antioxidative ability has relation with the resolution. In this study, the changes in activities of antioxidative ability through symbiosis with AMF in strawberry plants under high temperature stress were investigated. The strawberry (*Fragaria×ananassa* Duch., cv. Nohime) runner plants were treated with split root system method. Ten weeks after AMF (*Glomus mosseae*) inoculation, plants were transferred under 35°C-constant condition using growth chamber. Three weeks after high temperature treatment, dry weight of shoots and roots became greater in AMF plants than in non-AMF, resulting that growth promotion appeared in mycorrhizal plants. The severity of browning in shoots and roots caused by high temperature stress was eased in AMF plants than in non-AMF. In this case, the alleviation of root browning appeared in both AMF- and AMF+ root system, so that induced tolerance to high temperature stress was recognized in mycorrhizal plants. AMF plots showed higher activities in SOD and APX than non-AMF; the effect differs with plant portions and investigated periods. In addition, some AMF plots showed higher values in followed items; chlorophyll content, protein content, polyphenol content, ascorbic acid content. These effects occurred especially after high temperature stress treatment. From these findings, tolerance to high temperature stress and the increase in antioxidative ability occurred in mycorrhizal strawberry plants, so that the activation of antioxidative ability is supposed to be closely associated with the high temperature stress tolerance.

Chapter 8- The genus *Epimedium* (Bishop's Hat, Barrenwort) in the family *Berberidaceae* comprises about 50 species of herbaceous perennials. A number of species, hybrids and cultivars are grown as ornamentals in woodland gardens and naturalized areas. Certain species are also in use in traditional Asian medicine. The plant can be propagated by division of its long rhizomes or by seeds. The difficulties associated with gathering the seeds and with slow growth make bishop's hat a suitable species for propagation by tissue culture. *In vitro* propagation of *Epimedium alpinum* L. was carried out using organogenic callus induced from immature seeds transiently exposed (48 hours) to 20 μM CPPU or 80 μM TDZ followed by culture on hormone-free woody-plant medium (WPM). To promote adventitious shoot induction, the effects of different concentrations of 6-benzyladenine (BA) in combination with 2,4-dichlorophenoxyacetic acid (2,4-D) and gibberellic acid (GA_3) were investigated. WPM containing 2.2 μM BA and 1.1 μM 2,4-D was optimal for callus proliferation while shoot induction and elongation was enhanced on media with 44 μM BA and 0.55 μM 2,4-D. Histological examination of the cultures revealed that the regenerated plants were derived from *de novo* developed shoots. Shoots were successfully rooted on hormone-free WPM and well-developed plantlets were transferred to soil. Somatic chromosome number of the regenerated plants was found to be the same as the source plants.

Chapter 9- Apple production without agrochemicals is not easy, and consumers demand safe edible fruit. Reducing the amount and frequency of spraying with agrochemicals is important not only for safe fruit production, but also for the protection of the environment. Until now, as the apple has gametophytic self- and cross-incompatibility, growers have planted several genetically distinct cultivars in an orchard for stable fruit production. In such cases, agrochemicals sprayed on cultivars for harvest at late season may well drift onto the cultivars at harvest time. To avoid this, we developed a new apple cultivation system consisting of one cultivar with suitable pollinizers. We selected 'Maypole' and 'Dolgo' as pollinizers for 'Fuji' cultivar by S-RNase allele analyses of more than 430 cultivars and species. We also determined the arrangement of the pollinizers and the cultivar with the pollinator, Osmia cornifrons, based on S-RNase and SSR analyses of seeds in fruits pollinated mainly by O. cornifrons and pollen loaves collected by O. cornifrons.

Chapter 10- The heritabilities and the correlations of some ornamental traits of six pansy (*Viola wittrokiana*) inbred lines, such as size of flower (SF), number of flowers per plant (NF), length of flower stalk (LFS), number of branches per plant (NB), height of plant (HP), width of plant (WP), flower index (FI), value of HP/WP (H/W), length of leaf (LL), width of leaf (WL), leaf index (LI), number of nodes (NN) and area of leaf (AL), were calculated. 9 out of 13 traits were significantly different and therefore the broad sense heritability, genetic advances, relative genetic advances, coefficient of genetic variation and coefficient of correlations of these traits were calculated. The results showed that the broad sense heritability values of most traits were high (>80%) except for FI (0.73). The coefficients of genetic variation of SF, FI, WP, HP, LI and NN were not high (0.20-0.35). Genotypic correlations among these traits were generally high. Strong negative genotypic correlations among SF with NF, NB, HP, WP and NN could be observed, and NF was strongly positively correlated to NB, HP, WP and NN. Significant SF×FI positive correlation (r=0.911) and NF×FI negative correlation (r=-0.927) indicated that the flower size should be the more important trait than the number of flower for the breeding of pansy.

Abbreviations: size of flower (SF), number of flowers per plant (NF), length of flower stalk (LFS), number of branches per plant (NB), height of plant (HP), width of plant (WP), flower index (FI), value of HP/WP (H/W), length of leaf (LL), width of *leaf (WL), leaf index (LI), number of nodes (NN) and area of leaf (AL).*

Chapter 11- The explosion in biological knowledge is touching almost all areas of science and engineering. Improved understanding can lead to improved design, operation, and maintenance of environmental engineered systems. The integration of molecular biology tools in environmental engineering is a challenge. We discuss our views on the following two critical issues: The opening section provides an informative introduction to Environmental Biotechnology. Like white biotechnology, environmental biotechnology, often referred to as "grey biotechnology", also focuses on sustainability. For instance, environmental biotechnology deals with the treatment of sewage water, the purification of exhaust gas or the decontamination of soils or ground water using specific microorganisms. Environmental technologies are likely to play a more important role as they involve the use of environment-friendly mechanisms. This means starting environmental protection from the very beginning of production. In 1992, at the 21st Agenda of the UN Conference on Environment and Development in Rio de Janeiro, environmental biotechnology was, for the first time, seen to have an important role to play in tackling environmental problems. Traditional biotechnological processes are an integral part of classical "end of pipe" technologies. One could also refer to this as "first generation environmental biotechnology". Traditional environmental biotechnology therefore relies on the metabolic by-products of naturally occurring organisms. The rapid progress in biotechnology, molecular biology and gene technology broadens the application spectrum of environmental biotechnology: Desired metabolic products can be provided in a targeted and inexpensive way. Apart from genetically optimized organisms, isolated enzymes or biological products are available in nearly all amounts desired. Second generation environmental biotechnology is more and more based on high-tech methods and processes. This section provides a critical review of recent biotechnological approaches to environment protection. The second section focuses on the application of biotechnology in the bioremediation of environment pollution. Bioremediation is an alternative to traditional remediation technologies such as land filling or incineration. Bioremediation works by either transforming or degrading contaminants to non-hazardous or less hazardous chemicals. A number of bioremediation strategies have been developed to treat contaminated wastes and sites. Several strategies are currently being developed on biological wastewater treatment including aquatic plants, membrane and microorganisms. Bioremediation is also applied in removing pollutants from contaminated soil and petroleum industry. Bioremediation has become one of the most rapidly developing fields of environmental restoration.

In: Horticulture in the 21st Century
Editors: Abigail N. Sampson, pp. 1-40

ISBN 978-1-61668-582-9

Chapter 1

MECHANISMS OF CADMIUM TOXICITY IN CROPS: A REVIEW

Conceição Santos, Marta Monteiro and Maria Celeste Dias[1]
Laboratory of Plant Biotechnology and Cytometry
Department of Biology and CESAM, University Aveiro 3810 193, Portugal

ABSTRACT

Soil pollution by metals is a major environmental challenge. Metals are transmitted from plants to humans *via* the food chain. In particular, cadmium (Cd), a non essential metal, is listed as a "priority substance" by EC (2455/2001/EC, 2001) and by Environmental Protection Agency (EPA, USA), and is released to environment mainly due by anthropogenic activities.

Remediation strategies claim a better knowledge on Cd phytotoxicity and detoxification mechanisms. Cadmium toxicity involves, for example, growth reduction, wilting, photosynthesis and consequently carbon metabolism decreases. At the cell level, recent data support that Cd may have cytotoxic effects, in particular leading to oxidative stress, with impairment of antioxidative enzymes. Cd induced-genotoxicity is not so well studied, but some studies in crops suggest that Cd may induce DNA degradation (clastogenicity), may decrease DNA content and may have mitogenic effects. Moreover, molecular techniques (e.g. microsatellites, AFLP, RAPDs) showed to be powerful tools to assay Cd genotoxicity, with several studies supporting Cd induced point mutations.

This chapter revises some of the most recent advances concerning Cd induced effects at cytotoxicological and genotoxicological levels in several crops (e.g. *Lactuca* sp. and *Brassica* sp.) and in other model species as *Thlaspi* sp. and *Arabidopsis* sp.. Particular emphasis is also given to the whole plant performance, concerning this metal toxicity, including gas-exchange and chlorophyll a fluorescence parameters, carbohydrates, Calvin cycle enzymes and pigments content. The chapter also revises some important aspects concerning Cd accumulation and distribution in plants and their relevance to the trophic transfer of Cd.

Finally, the potential use of new techniques as flow cytometry (quantitative cytometry) and molecular markers are critically explored in these kinds of toxicological assays, and their potential use to develop new endpoints in ecotoxicology is also revised.

---

1 Corresponding author: celeste.dias@ua.pt; phone: + 351234370766.

INTRODUCTION

Metals are among the most persistent environmental pollutants, and although some are essential plant micronutrients (e.g. copper or iron) others such as cadmium (Cd) have no known biological functions, except for the recently described Cd carbonic anhydrase in marine diatoms (Lane and Morel 2000; Lane et al. 2005). Cadmium is of worldwide concern because it is a toxic trace pollutant for humans, animals and plants. It enters the environment mainly from industrial processes and phosphate fertilizers, and is then transferred to the food chain. Due to its known toxicity, Cd is listed as "priority substance" by EC (2455/2001/EC 2001) and by Environmental Protection Agency (EPAUSA).

Cadmium occurs naturally in soils (e.g. in complexes), but it has been estimated that the anthropogenic emissions of Cd are in the range of 30 000 t per year (Shutzendubel and Polle 2002), mostly due to mining activities, burning of fossil fuels (ATSDR 1999), or metallurgical industry (Robards and Worsfold 1991). In agricultural soils, treated sewage sludge ("biosolids") and phosphate fertilizers (Speir et al. 2003; McLaughlin et al. 2006; Singh and Agrawal 2007) are also important sources of Cd contamination. Unlike other metals (as mercury (Hg) or lead (Pb)), Cd has relative mobility in soils and does not bind strongly to organic matter (Nelson and Campbell 1991). Besides its long environmental persistence, Cd has an extremely long biological half-life that essentially accounts for its bioaccumulation in individuals (e.g. John and Leventhal 1996). Cadmium has been largely studied and it was demonstrated to be cytotoxic, mutagenic and/or carcinogenic (e.g. Filipic et al. 2006) in animal cells. Despite less studied in plants, available data also support Cd has a cytotoxic and genotoxic environmental pollutant. We will explore in this review some of the recent research on Cd phytotoxicity (mostly directed to cytotoxicity and genotoxicity), and critically evaluate the use of new techniques used to Cd cytotoxicity and/or genotoxicity. Furthermore, the implications of metal accumulation in plants to the trophic transfer of this metal are highlighted.

1. CADMIUM IN SOILS AND UPTAKE BY PLANTS

Cadmium has relative mobility in the plant-soil system. Despite the fact that Cd concentrations in non-polluted soils are variable and dependent on sources of minerals and organic material, it was estimated that on average they contain 0.04 to 0.32 μM of Cd (Wagner 1993). Soil solutions with Cd concentrations from 0.32 to 1 μM are considered moderately polluted (Sanitá di Toppi and Gabbrielli 1999).

Cadmium toxicity has been disclosed as early as 1955 in Japan as Itai–itai disease, and for the first time, Cd pollution was shown to have severe consequences on human health (Bertin and Averbeck 2006). Cadmium levels up to 800 mg/kg have been reported for soils in polluted areas (IARC, 1993). In a Korean mining area, Jung and Thornton (1996) found Cd concentrations up to 40 mg/kg in surface soils. Also, Cd concentrations of 0.5 mg/kg or more have been found in rice grown in Cd-polluted areas of Japan (Nogawa et al. 1989) and China (Cai et al. 1990). Furthermore, in a recent field study in Europe performed by Peris et al. (2007) the Cd content in edible parts of vegetables such as lettuce were found to be above the maximum levels established by the Commission Regulation no. 466/2001 for horticultural crops (466/2001/EC, 2001).

Plants may accumulate Cd mostly by root uptake, but also by foliar uptake. The degree to which plants are able to uptake Cd, is conditioned by its concentration in the soil, and its bioavailability, modulated by the presence of organic matter, pH, redox potential, temperature and concentrations of other elements (Benavides et al. 2005). Cadmium uptake by plants grown in contaminated soils has been studied, mostly in sludge-amended soils (e.g. McLaughlin et al. 2006, Singh and Agrawal 2007) and in soils treated with Cd-enriched phosphate fertilizers (e.g. Huang et al. 2003). For example, Cd bioavailability increases with decreasing pH in soil, but Cd uptake may be reduced at low pH because of competition with H^+ ions at root uptake sites (Greger 1999). Acid rain may acidify soil and surface waters increasing the geochemical mobility of Cd (Campbell 2006). Contrarily, by binding Cd ions, organic matter may decrease Cd uptake (He and Singh 1993, Prasad 1995). Chloride levels may also affect Cd availability as soil sodium chloride has an antagonistic effect on metal toxicity (Bhartia and Singh 1994). Finally, the rhizosphere may have a pH different from the soil directly around it, and therefore can significantly affect the uptake of most plant nutrients.

Cadmium co-existence in soil solution with other essential and non-essential metals leads to various synergic and antagonistic interactions conditioning plant uptake and tissue content (Palóve-Balang et al. 2006). Cadmium ions may compete with macro and micronutrients and, such as calcium (Ca) and zinc (Zn), for the same transmembrane carriers (Sanitá di Toppi and Gabbrielli 1999), putatively leading to nutrient deficiencies (Krupa et al. 2002).

Once inside the root, and similarly to other mineral ions, Cd moves across the root through the symplast and/or the apoplast (Sanitá di Toppi and Gabbrielli 1999) until it reaches the endoderm. The endodermal cell layer acts then as a barrier for apoplastic diffusion into the vascular system, forcing most ions to enter the symplastic pathway prior entering the xylem (Jacoby and Moran 2002, McLaughlin 2002). Subsequently, three processes govern the movement of metals from outer root regions to xylem: sequestration on metals inside the cells, symplast transport into the stele and release into the xylem (Benavides et al. 2005). Cell membranes play, therefore, a key role in metal (and particularly Cd) homeostasis, preventing or reducing entry into the cell. On other hand, the apoplast continuum of the root epidermis and cortex is readily permeable to solutes.

The mechanisms regulating metal transport across the plasma membrane to the stele are still not fully understood (McLaughlin 2002). For cationic metals, such as Cd, the main route for uptake across the plasma membrane is the large negative electrochemical potential produced as a result of the membrane H^+ translocating adenosine triphosphatase (ATPases) (McLaughlin 2002). For example, in lettuce plants exposed to increasing Cd concentrations, it was shown that high amounts of Cd in roots were correlated with high contributions from H^+-ATPase in the active process of Cd uptake (Costa and Morel 1994). Other studies suggest that the main route for uptake of divalent metals is via ion channels, such as Cd^{2+} and Mg^{2+} channels (McLaughlin 2002).

Subsequent to metal uptake into the root symplasm, three processes govern the movement of metals from the root into the xylem: sequestration of metals inside root cells, symplastic transport into the stele and release into the xylem (Clemens et al. 2002). During their transport through the plant, metals become bound to cell walls, which can explain why normally Cd^{2+} ions are mainly retained in the roots, and only small amounts are translocated to the shoots (Cataldo et al. 1983, Greger 1999). But once loaded in the xylem sap, Cd is translocated to the aerial parts of plants through the transpiration stream, where they might be present as a divalent ion (Greger 1999) or complexed by several ligands, such as amino acids, organic acids and/or, perhaps, phytochelatins (PC) (e.g. Sanitá di Toppi and Gabbrielli 1999, Gong et al. 2003).

2. Cadmium Phytotoxicity

Plants represent unique models for *in situ* monitoring of soil contamination and can potentially be used as biomonitors of environmental quality. Moreover, the contamination of crops with metals (such as Cd), by soil or irrigation with contaminated water may raise the problem of trophic transfer and ultimately human contamination.

In some highly metal polluted regions, Cd phytotoxicity may lead to a decrease in agricultural crop productivity (Vassilev and Yordanov 1997). Different studies involving different plant species, Cd compounds and concentrations, demonstrated that Cd is strongly phytotoxic and causes in general growth inhibition and even plant death, although the mechanisms involved in its toxicity are still not completely understood. In fact, it is well demonstrated that Cd interferes at several physiological levels (e.g. it affects photosynthesis, respiration and nitrogen metabolism, it induces oxidative stress and genotoxicity) culminating in poor growth and low biomass production (Sanitá di Toppi and Gabbrielli 1999, Fodor 2002), and eventual senescence and death. Understanding Cd uptake/allocation and plant cytological and physiological responses is therefore a requisite for the long-term safety and conservation of agricultural resources and ecosystems.

Apart from the recently discovered Cd-dependent enzyme in diatoms stated above, Cd is considered a non-essential and toxic element without any other known biological role (Palóve-Balang et al. 2006), that can affect plants on various organizational and functional levels. Plants respond to metal accumulation by expressing various manifestations of toxicity that can be detected and analyzed at various levels of organization ranging from gross morphology to cellular, biochemical or molecular levels, and thus can be useful to monitor as well as assess environmental metal pollution. Moreover, the sedentary nature of plants is a major advantage of plant-based assays for monitoring toxic chemicals in the environment. As for other stresses, the type and/or degree of response(s) to metal stress depend not only on stress intensity and duration but also on the tissue type and the age of the plant.

2.1. Cadmium Cytotoxicity

Cadmium stress usually leads to a battery of unspecific symptoms that include chlorosis, necrotic lesions, wilting, reddish coloration and growth reduction (e.g. Sanitá di Toppi and Gabbrielli 1999, Azevedo et al. 2005a-c). Disturbances in plant water relations are also widely known as one of the first effects of Cd toxicity. It was even suggested that water stress caused by Cd may lead to a cascade of physiological and metabolic processes, including photosynthesis impairment (Barceló and Poschenrieder 1990).

Among the several metabolic pathways that have shown to be highly affected by Cd toxicity, carbon metabolism and antioxidant system are particularly sensitive and are of particular importance. The first one is involved in supplying energy to the plant and the cells have to deal with the multiple direct/indirect effects of Cd in the photosynthetic apparatus, and the second pathway is crucial as the Cd-exposed cells have to develop efficient strategies to overcome the increase of oxidative stress. Below, we'll explore in more detail these two metabolic aspects.

Carbon Metabolism

The photosynthetic apparatus is particularly susceptible to Cd toxicity and the reduction in photosynthetic rate is a common response in plants exposed to Cd (Mysliwa-Kurdziel et al. 2002, Burzynski and Klobu 2004). Inhibition of photosynthetic rate could be due to several structural and functional disorders in the different parts of this process. Photosynthesis can be inhibited at several levels: photosynthetic pigments synthesis, electron transport, enzymes of the Calvin cycle, structural disorganization of chloroplasts and uptake of nutrients (Larbi et al. 2002, Mysliwa-Kurdziel et al. 2002, Vassilev et al. 2002, Mallick and Mohn 2003, Burzynski and Klobu 2004, Balakhnina et al. 2005, Han et al. 2006, López-Millán et al. 2009).

Photosynthetic pigments have been shown as one of the main targets of the toxic Cd action. One of the most usual symptoms of Cd stress is chlorosis of the leaves due to an impairment of photosynthetic pigment biosynthetic pathways (Baryla et al. 2001, Mysliwa-Kurdziel 2002, Carrier et al. 2003), but also to a strong interaction between Cd and Fe that reduces uptake of Fe and causes Fe deficiency in leaves as well as to enzymatic degradation (Krupa et al. 2002, Vassilev et al. 2002). Cadmium can alter chlorophyll biosynthesis by inhibiting protochlorophyllide reductase. Moreover, Cd^{2+}, like other metals, can interfere with photosynthetic pigments through the substitution of the Mg^{2+} ion in the chlorophyll molecules by Cd^{2+} (Mysliwa-Kurdziel 2002). These substituted chlorophylls have much lower fluorescence quantum yields when compared to Mg-chlorophylls (Krupa et al. 2002).

Several authors have reported decreased levels of chlorophyll pigments in different plant species treated with Cd (e.g. Vassilev et al. 2002, Pietrini et al. 2003, López-Millán et al. 2009). A recent report of Fagioni and Zolla (2009) revealed that spinach plants exposed to Cd develops chlorosis and this is prevalently localized in the basal leaves. The reduced chlorophyll a synthesis and photosynthesis leading to senescence and cell death of the spinach plants. In *Phaseolus vulgaris*, Cd induced decreases in leaf chlorophyll content and other photosynthetic pigment. This effect was more marked in plants exposed to 10 μM Cd than to 100 μM (Han et al. 2006). A similar effect was observed in sugar beet, and attributed to a Cd-induced Fe deficiency at low Cd treatments (Larbi et al. 2002). Also in *Lycopersicon esculentum* (Delpérée and Lutts 2008), *Zea* mays (Ekmekci et al. 2008), *Brassica napus* (Ben et al. 2009) and *Lactuca sativa* (Monteiro et al. 2009b) Cd toxicity induced a decrease in photosynthetic pigments. Since the chlorophyll content may directly influence the functioning of the photosynthetic apparatus and thus affect overall plant metabolism, it is considered a key factor when assessing the impact of Cd stress (Fodor 2002).

The ratio Chl *a*/Chl *b* is another related endpoint relevant for Cd toxicity assessment. Although, there is no known direct influence of metal ions on the process of transformation of Chl *a* to Chl *b*, changes on Chl *a*/Chl *b* ratio are commonly reported on metal stressed plants (Mysliwa-Kurdziel and Strzalka 2002). Both increases and decreases in this ratio have been found in plants treated with Cd^{2+} (Mysliwa-Kurdziel and Strzalka 2002 and references therein). In *Brassica juncea*, the excess levels of Cd decreased the contents of chlorophylls but increased the ratio of chlorophyll a/b (Singh and Tewari 2003).

Several investigations on the effects of Cd toxicity on plant photochemical processes have been performed. It is generally accepted that water oxidizing system of PSII (photosystem II) is affected by Cd by replacing the Mn^{2+} ions, thereby inhibiting the reaction of PSII (Krantev et al. 2008). Cadmium exerts multiple effects on both donor and acceptor sites of PSII. On the donor site, the presence of Cd inhibits the oxygen evolving cycle and, consequently, oxygen evolution; on the acceptor site, it inhibits electron transfer from Q^-_A to Q^-_B (Ekmekci et al. 2008). The maximum photochemical efficiency of PSII was found to be reduced in different plant species exposed to Cd (Burzynski and Klobus 2004, Balakhnina et al. 2005, He et al. 2008, Bi et al. 2009, Monteiro et al. 2009b). PSII is extremely sensitive to Cd and its function is inhibited to a much greater extent than that of PSI (Mallick and Mohn 2003, Ekmekci et al. 2008). Pal'ove-Balang et al. (2006) reported that PSI activity is only slightly affected by Cd. However a recent study of Fagioni and Zolla (2009) in spinach plants reports a high sensitivity of PSI to Cd. According to Siedlecka and Baszynski (1993), the disturbed functional activity of PSI is due to Cd-induced Fe deficiency that limits the level of ferredoxin and $NADP^+$-oxydoreduction. A report of Siedlecka and Krupa (1996) on the interactions between Cd and Fe in bean plants also confirms this suggestion.

Cd reduces the photochemical quenching while the non-photochemical quenching increased (Krupa et al. 1993, Linger et al. 2005, Ekmekci et al. 2008). Di Cagno et al. (2001) also confirm this findings in Cd treated sunflower plants.

Limitations of photosynthesis in Cd treated plants are mainly due to non-mesophyll factors (Vassilev et al. 2002, Zhu et al. 2005, He et al. 2008) and inhibition of RuBisCO (ribulose 1,5-bisphosphate carboxylase/ oxygenase) activity is considered the main target of Cd toxicity (Siedlecka et al. 1997, Krantev et al. 2008). Cd^{2+} ions interfere with RuBisCO activation, lower its activity and damage its structure by substituting for Mg^{2+} ions, which are important cofactors of carboxylation reactions, and may also shift RuBisCO activity towards oxygenation reactions (Siedlecka et al. 1997, Pietrini et al. 2003). Cadmium toxicity causes an irreversible dissociation of the large and small subunits of RuBisCO, thus leading to total inhibition of the enzyme (Malik et al. 1992).

Photosynthesis impairment due to RuBisCO inhibition was described for Cd-treated bean plants (Siedlecka et al. 1997), barley (Vassilev et al. 1997), sunflower (Di Cagno et al. 2001) and *L. esculentum* (Chaffei et al. 2004). In maize Cd-treated plants the photosynthesis impairment was correlated with a drop in the activities of RuBisCO and also PEPCase (phosphoenol-pyruvate carboxylase) (Krantev et al. 2008). Moreover, in pigeon pea and wheat plants, Cd seems to inhibit the activity of all enzymes of the Calvin's cycle (Malik et al. 1992).

The negative effects of Cd can also be observed in carbohydrates metabolism (Delpérée and Lutts 2008, Dunwei et al. 2009) due to the inhibition of several enzymes such as RuBisCO (Mobin and Khan 2007, Krantev et al. 2008), fructose-6-phosphate kinase (Malik et al. 1992), fructose-1,6-bisphosphatase (Sheoran et al. 1990), $NADP^+$-glyceraldehyde-3-phosphate dehydrogenase (Sheoran et al. 1990), PEPCase (Krantev et al. 2008), aldolase (Sheoran et al. 1990), and carbonic anhydrase (Mobin and Khan 2007).

Profound anatomical changes in leaves and structural disorganization of chloroplasts are also in the basis of the inhibition of photosynthesis (Mysliwa-Kurdziel et al. 2002). Cadmium induces a disorder in chloroplast ultrastructure (Bi et al. 2009), reduction on its size, decrease the number of grana and tylakoids (Pietrini et al. 2003, Vassilev et al. 2002) and change the shape and internal organization of the thylakoid membranes and stroma (Gratão et al. 2009). According to Skorzinska and Baszynski (1993) stroma thylakoids degradation is more severe than grana degradation. In leaves of *Pisum sativum*, Cd produced a significant inhibition of growth as well as a reduction in the transpiration and photosynthesis rate, chlorophyll content and an alteration in the nutrient status. The ultrastructural analysis of *P. sativum* grown with 50 μM $CdCl_2$, showed cell disturbances characterized by an increase of mesophyll cell size, and a reduction of intercellular spaces, as well as severe disturbances in chloroplast structure (Sandalio et al. 2001). Also on oilseed rape (*B. napus*), leaf chlorosis was attributable to a marked decrease in the chloroplast density caused by a reduction in the number of chloroplasts per cell and a change in cell size (Baryla et al. 2001).

Another unfavourable effects of toxic metals on plants are the inhibition of the normal uptake and utilization of mineral nutrients (Sandalio et al. 2001, Fodor 2002, Vassilev et al. 2002). One of the crucial factors of Cd^{2+} influence on plant metabolism and physiological processes is its relationship with other mineral nutrients. As mentioned above, Cd^{2+} transport across cell membranes is most likely facilitated by metal transporters that normally act to mobilize essential metals. Thus, by substituting for essential divalent cations, Cd^{2+} limits their uptake. Alternatively, Cd^{2+} may bind to specific groups of proteins and lipids or channel proteins of membranes, thereby inhibiting transport and disturbing the uptake of many macro and micronutrients. Furthermore, destruction of the cell membranes can also alter the ratio of essential elements and cause the decrease in their content, thereby inducing nutrient deficiencies (Cseh 2002). One of the most important mechanisms for impairment of the uptake of nutrients by Cd is via the inhibition of Fe transport into the shoot, which has a pronounced effect on many aspects of the structure and function of the photosynthetic apparatus (Krupa et al. 2002). The induced iron shoot deficiency reduces the pool of Fe-containing electron carriers in the photosynthetic electron transport chain, causes disorganization of the chloroplast structure and even reduces RuBisCO content (Siedlecka and Krupa 1998). Cadmium is also known to cause other important disturbances in nutrient levels that can severely affect normal plant metabolism. Specifically it can decrease the levels of magnesium (Mg), potassium, phosphate, Ca and Zn, and increase manganese content (Krupa et al. 2002).

Plant growth inhibition is a classical parameter commonly used in the assessment of Cd toxicity to plants (Linger et al. 2005, Monteiro et al. 2009b) and a recommended endpoint in standard tests for toxicity assessment (OECD, 2006). Apart from being an important indicator of toxicity at an individual level, growth inhibition is a non-specific manifestation of alterations at a biochemical level that are produced as a more specific response of plants to the particular stress. Cadmium induces a decrease and/or inhibition of plant growth as a consequence of CO_2 assimilation decline and/or inhibition (Larbi et al. 2002, Krantev et al. 2008, López-Millán et al. 2009), and of translocation of photosynthetic products (Linger et al. 2005) and cell division inhibition (Dalla Vecchia et al. 2005, Linger et al. 2005). Excessive concentrations of Cd (100 and 1000 μM) affected photosynthesis and consequently plant growth in *Medicago sativa*, *Raphanus sativus*, and *L. sativa* (Benzarti et al. 2008). Similar results were obtained by Monteiro et al. (2009b) in *L. sativa* with 100 μM $Cd(NO_3)_2$.

Oxidative Stress

Exposure to abiotic and biotic stress results, in general, at some stage of exposure, in an increase in reactive oxygen species (ROS) (e.g. Schützendübel and Polle 2002). In particular, metals stimulate the formation of ROS, either by direct electron transfer involving metal cations, or as a consequence of metal-mediated inhibition of metabolic reactions. Reactive oxygen species are generated in plant cells during normal metabolic processes, such as respiration and photosynthesis (Mittler 2002). The ROS intermediates are partially reduced forms of O_2 and typically result from the excitation of O_2 to form a singlet ($^1O_2$), or from the transfer of one to three electrons to O_2 to form, respectively, O_2^-, H_2O_2 or HO^- (Mittler 2002). Oxidative stress in plants occurs when the delicate balance between antioxidants and ROS is tilted in favour of ROS (Razinger et al. 2009).

Some ROS may function as important signalling molecules (e.g., by altering gene expression and modulating the activity of specific defence proteins), but they can be extremely harmful at high concentrations (Apel and Hirt 2004). Reactive oxygen species may lead to the oxidation of proteins, lipids and nucleic acids, and often culminating with, e.g., membrane damage, mutagenesis and inactivation of enzymes, and thus affecting cell viability (Apel and Hirt 2004). As a consequence, tissues injured by oxidative stress generally contain increased concentrations of carbonylated proteins and malondialdehyde (MDA) (e.g. Azevedo et al. 2005c).

Schützendübel and Polle (2002) distinguished three different molecular mechanisms of metal toxicity: (a) ROS production by auto-oxidation and Fenton reaction; (b) blocking of essential functional groups in biomolecules, a reaction that has been mainly been reported for non-redox-reactive metals such as Cd; (c) displacement of essential metal ions from biomolecules. However, metals without redox capacity such as Cd do not seem to act directly on the production of ROS via Fenton and/or Haber Weiss reactions (Dietz et al. 1999). Instead, Cd can expedite ROS production indirectly by, for example, substituting an essential micronutrient, or binding unspecifically to thiol groups disturbing various metabolic processes (Razinger et al. 2009). For example, Cd may disrupt the photosynthetic electron chain, leading to ROS production of O_2^- and $^1O_2$ (Dietz et al. 1999).

The ROS formed in response to Cd may differ in time, nature and localisation. Tobacco cells exposed to $CdCl_2$ developed successive waves of ROS differing in nature and subcellular localization: the first one consisted in the transient NADPH oxidase-dependent accumulation of H_2O_2 followed by the accumulation of O_2^- in mitochondria. A third wave of ROS consisted in fatty acid hydroperoxide accumulation and was concomitant with cell death (Garnier et al. 2006).

One of the plant responses to ROS production is the increase in anti-oxidant enzyme activities providing protection from oxidative damage induced by several environmental stress (Apel and Hirt 2004). A variety of proteins function as scavengers of superoxide and H_2O_2. Among the major ROS-scavenging enzymes in plants are catalase (CAT), peroxidase (POX) and superoxide dismutase (SOD) (Mittler 2002). The superoxide released by processes such as oxidative phosphorylation is first converted to H_2O_2 and then reduced originating water, in a detoxification pathway that involves multiple enzymes, with SOD catalysing the first step (superoxide conversion to H_2O_2) and then CAT and various POX removing H_2O_2 from the cell. This type of response seems however to be species, time and stress-condition dependent. For example, CAT activity has been shown to be suppressed in diverse plant species exposed to Cd (Chaoui et al. 1997, Chaoui and El Ferjani 2005). In sensitive barley cultivars, activities of the enzymes involved in ROS detoxification were markedly enhanced by increasing Cd supply, with the exception of CAT, while in the tolerant ones, there was either only a slight increase or no change in the antioxidative enzymes activities, and the differential Cd tolerance was not related to uptake or accumulation of Cd (Tiryakioglu et al. 2006).

In the newly discovered hyperaccumulator *Sedum alfredii* Hance, POX played an important role during Cd hyperaccumulation (Zhang and Qiu 2007). In *Arabidopsis thaliana*, Cd elevated the activities of FeSOD and MnSOD, but CuZnSOD activity was diminished in comparison with control (Drazkiewicz et al. 2007). Also in alfalfa Cd exposure led only to a mild increase of POX (and other antioxidant molecules) (Ortega-Villasante et al. 2007), while in transgenic tall fescue plants, Cd led to an overexpression of the CuZnSOD and POX genes, which are utilized in scavenging ROS. For CAT, Cd induced a CATA3 transcript level, but this effect was reverted by ascorbate (Lee et al. 2007).

The anti-oxidative battery also includes non-protein scavengers, including ascorbate (ASC) and glutathione (GSH) (Mittler 2002) that play a pivotal role in the response of plant cells to abiotic stress (Horemans et al. 2007). In fact, these compounds act in a network in complement with anti-oxidative enzymes. Ascorbate peroxidase (APX) and glutathione reductase (GR), as well as GSH, are important components of the ascorbate–glutathione cycle responsible for the removal of H_2O_2 in different cellular compartments (Jiménez-Ambriz et al. 2007). Glutathione is also the substrate for the biosynthesis of PC, which are involved in metal detoxification (Zenk 1996).

Comparing the hyperaccumulator *Arabidopsis halleri* with the non accumulator *A. thaliana*, Chiang et al. (2006) found that, among others, APX and MDAR4 (in the ascorbate-glutathione pathway) were expressed at higher levels in *A. halleri*. These changes were supported by higher enzymatic activities APX in the hyperaccumulator species. Also, in *A. thaliana*, Cd decreased GSH/GSSG ratio ((Drazkiewicz et al. 2007). In *Pisum sativum*, Cd reduced GSH and ASC contents, and CAT, GR, POX and CuZnSOD activities, while MnSOD activity increased. Comparatively, CAT and CuZnSOD were down-regulated at transcriptional level, while MnSOD, FeSOD and GR were up-regulated (Rodriguez-Serrano et al. 2006). The same authors also reported that Cd induced ethylene and ROS, while nitric oxid (NO) decreased, which correlated with senescence processes induced by Cd (Rodriguez-Serrano et al. 2006). In *Coffea arabica* L. suspension cultures, Cd induced activities of SOD, CAT and GR (ensuring the availability of reduced GSH) while POX activity seemed inhibited (Gomes-Junior et al. 2006). Comparing the *Arabidopsis* mutant rax1-1 with the mutant Cd hypersensitive 2 plants, Ball et al. (2004) showed a direct link between GSH biosynthesis and stress defence. Other enzymes, such as lipoxygenase, have their activities stimulated by Cd exposure (Tamas et al. 2009).

Cadmium produces alterations in the functionality of membranes, often related with induced changes in lipid composition (Ouariti et al. 1997), and by affecting the enzymatic activities associated with membranes, such as the H^+-ATPase (Fodor 2002). Tommasini et al. (1998) analysed an ATP-binding-cassette (ABC) transporter of *Arabidopsis thaliana* with high sequence similarity to the human (MRP1) and yeast (YCF1) glutathione-conjugate transporters and in the ycf1 mutant, Cd resistance was partially restored and glutathione-conjugate transport activity was observed. In the same species, tolerance to Cd of transgenic plants is correlated with decreased accumulation of lipid peroxidation-derived reactive aldehydes compared to wild-type plant, and this is probably related with an increased activity of Ath-ALDH3 that limits aldehyde accumulation and oxidative stress (Sunkar et al. 2003). Moreover, Cd uptake is apparently a prerequisite for the change in dehydroascorbate transport activity, and this seems to be independent of the Cd-induced H_2O_2 production (Horemans et al. 2007). Transcript profiles of roots of *A. thaliana* and *T. caerulescens* plants exposed to Cd and Zn showed differences in gene expression between both species, mostly genes involved in lignin, GSH and sulphate metabolism (van Mortel et al. 2008).

Many plant metal transporters remain to be identified at the molecular level. Thomine et al. (2000) isolated AtNramp cDNAs from *Arabidopsis*, which expresses in both roots and aerial parts. AtNramp3 gene leads to slightly enhanced Cd resistance of root growth, and overexpression of AtNramp3 results in Cd hypersensitivity of *Arabidopsis* root growth (Thomine et al. 2000). Shigaki et al. (2005) identified in an *Arabidopsis* mutant a vacuolar transporter CAX1 with higher transport for Cd. Also, screening Cd-responsive genes in *Arabidopsis,* Susuki et al. (2002) identified a gene encoding a putative metal binding protein CdI19, which, upon introduction into yeast cells, conferred marked toleration of Cd exposure. The authors proposed that CdI19 plays an important role in the maintenance of metal homeostasis and/or in detoxification by endowing plasma membranes with the capacity to serve as an initial barrier against inflow of free metal ions into cells. Using a functional cloning strategy, Li et al. (2002) described for *A. thaliana* a detoxifying carrier AtDTX1 localised in the plasma membrane that mediates the efflux of plant derived or exogenous toxic compounds from the cytoplasm and that was capable of detoxifying Cd. Clemens et al. (1998) tested putative plant cation transporters for Cd uptake activity and suggested that LCT1 may represent a plant cDNA encoding a plant Ca cation uptake or an organellar Ca transport pathway in plants contributing to transport of Cd^{2+} across membranes.

The interaction of Cd with other nutrients in the antioxidative response is being largely demonstrated: a model was proposed by *Rodríguez-Serrano et al. (*2009), for the cellular response to long-term Cd exposure involving Ca, ROS and NO. Using pea (*P. sativum*), these authors demonstrated that Cd affected differentially the expression of SOD isozymes at both transcriptional and posttranscriptional levels, leading to SOD activity reduction. Calcium down regulated CuZnSOD, while NO synthase-dependent production was depressed by Cd exposure. Also Aravind et al. (2009) showed that Zn supplementation inhibited the oxidative products of proteins. Also Zn protection against Cd-induced ROS was confirmed by the reduced levels of oxidative products of proteins (Aravind et al. 2009). Moreover, the inhibition of ROS in Cd production in isolated plasma membranes was reversed by Ca and Mg (Heyno et al. 2008). In *B. napus*, selenium applied separately or in combination with Cd did not affect the activity of antioxidative enzymes in roots, but diminished it in the shoot (Filek et al. 2008). Finally, available data suggest that Cd, when not detoxified rapidly enough, may trigger, via the disturbance of the redox control of the cell, a sequence of reactions leading to growth inhibition, stimulation of secondary metabolism, lignification, and subsequent cell death (Schützendübel and Polle 2002).

The role of organelles in the metabolic pathways related with oxidative stress (e.g. fatty acid beta-oxidation, photorespiration, and metabolism of ROS and RNS) induced by Cd have been studied. It was described for *Arabidopsis* that ROS appeared first in the mitochondria and subsequently in the chloroplast (Bi et al. 2009). The same authors also described that previous treatment with ascorbic acid or CAT decreased the production of ROS and prevented inhibition of photosynthesis and organelle changes (Bi et al. 2009). Cadmium influenced differently ROS production in isolated plasma membranes and in mitochondria, in these last ones with increased levels of H_2O_2. *In vivo*, Cd stimulated ROS production in the mitochondrial electron transfer chain and inhibition of NADPH oxidase activity in the plasma membrane (Heyno et al. 2008). *Rodríguez-Serrano et al.* (2009) reported that treatment of plants with $CdCl_2$ significantly increased peroxissome speed, which was dependent on endogenous ROS and Ca, but was not related to actin cytoskeleton modifications. The same authors also proposed that this increase in peroxisomal motility may protect the cell against the Cd-imposed oxidative stress (*Rodríguez-Serrano et al.* 2009). Another example of Cd toxicity in cell components, is, besides the nucleus (and DNA content that will be explored below), the effect at the microtubular cytoskeleton, as microtubule alterations appeared after treatment with low Cd concentration, pointing to microtubules or microtubule-associated proteins, among the main targets of Cd (Fusconi et al. 2007).

The above described responses are merely a fraction of the multiple responses of plants to Cd stress, still yet to unfold. The dimension of the pathways putatively involved in response to Cd stress is evident in the still few works on genes responses. For example, in tolerant genotypes of backcrosses (BC1) between A. *halleri* ssp. *halleri* and the non-tolerant relative *A. lyrata* ssp. *petraea*, a hundred and thirty-four genes that expressed more than sensitive genotypes were identified, and most belonged to diverse functional classes, including ROS, detoxification, cellular repair, metal sequestration, water transport, signal transduction, transcription regulation, and protein degradation (Craciun et al. 2006). Chiang et al. (2006) used an *Arabidopsis* cDNA microarray to compare the gene expression of the Zn/Cd hyperaccumulator *A. halleri* and a non-hyperaccumulator, *A. thaliana* for better understanding the hyperaccumulating mechanism, and proposed genes that could be candidates for the future engineering of plants with large biomass for use in phytoremediation. Also, transcriptome analyses of the antioxidative enzymes in leaves of pea plants grown in the presence of Cd and treated with some modulators of the signal transduction cascade suggested the existence of cross-talk between these elements and ROS metabolism during Cd stress. At least Ca channels, phosphorylation/dephosphorylation processes, NO, cGMP, salicylic acid (SA) and H_2O_2 were involved in some steps between the Cd signal and transcript expression of CuZnSOD, CAT and monodehydroascorbate reductase (Romero-Puertas et al. 2007). Finally, a recent proteomic analysis of Cd exposed *Spinacea oleracea* plants showed differently modified profiles in apical and basal leaves (Fagioni and Zolla 2009). Also, metabolic analyses of Cd exposed spinach (*Spinacea oleracea*) plants showed reinforced 2-C-methyl-D-erythritol 2,4-cyclodiphosphate (MEcDP) accumulation, an intermediate of the methylerythritol 4-phosphate (MEP) pathway, which was blocked (Rivasseau et al. 2009). Transcriptional responses of a *Chlamydomonas reinhardtii* cell wall-deficient mutant to metal stress was investigated and authors identified, sequenced, and quantified the induction of a number of transcripts that are up-regulated by cadmium chloride (Rubinelli et al. 2002).

2.2. Cadmium Genotoxicity

The phenotype analysis of Cd exposed plants is of unquestionable importance, but in toxicology studies, phenotype studies should be complemented with analyses of the genome. This genome approach may allow a precocious detection of Cd effects and plant cell responses, allowing to survey putative occurrence of toxic effects, e.g. genotoxicity induced by the metal. Moreover, the combined phenotypic and molecular survey of Cd-exposed plants will allow its integration in prevention/intervention programs in the target agro-ecossystem.

Molecular and (cyto)genetic analyses are particularly important to assess putative genotoxicity as, like other metals, Cd can damage the genome or DNA of plants and/or induce mutations. There are different types of genotoxic effects: mutagenesis, which is a permanent change in DNA sequence within a gene; clastogenesis that refers to a damage in chromosome structure, usually resulting in a gain, loss or rearrangement of chromosome pieces within the genome; aneugenesis, which refers to the gain or loss of one or more chromosomes (aneuploidy) or to a complete haploid set of chromosomes (euploidy) (e.g. Panda and Panda 2002).

Genotoxicity induced by Cd has been extensively studied in mammals and particularly in humans. Moreover, exposure to this metal has been linked to several types of cancer, such as lung, prostate and renal cancer, and has been shown to induce tumours in experimental animals and exposed human cell lines (Waalkes 2003) and to induce large deletion mutations in mammalian cells (Filipic et al. 2006). Therefore, Cd and its compounds are classified as Category 1 human carcinogens by the International Agency for Research on Cancer (IARC 1993). However, the molecular mechanisms underlying the genotoxic and carcinogenic potential in organisms are still not well understood.

Cadmium may induce genotoxicity by inducing damages in DNA, and/or by leading to inactivation of mechanisms of DNA repair (Hartwig 1994). In this last model, Cd may interfere with DNA repair acting as a mutagen by direct inhibition of an essential DNA mismatch repair, resulting in a high level of genetic instability (Hartwig 1994, Jin et al. 2003, Slebos et al. 2006). Alternatively, genotoxicity may be induced indirectly by promoting the production of ROS, which may then damage nucleic acids (Hartwig 1994, Valverde et al. 2001, Apel and Hirt 2004). Furthermore other authors have shown that Cd cation can directly damage DNA, binding to DNA, possibly at guanine, adenine and thymine centres (Hossain and Huq 2002). Moreover, Cd may affect mitotic activity (often evaluated by the mitotic index (MI) and percentage of DNA synthesising cells (Fusconi et al. 2007).

Evaluation in plants of the Cd induced mutagenicity and/or genetic instability is of the utmost importance because, among several reasons, it may contribute to the inherited change of many phenotypic traits in the progeny of exposed plants. Furthermore, plants have been shown to provide ideal models for genotoxicity assays for screening as well as monitoring of environmental mutagens or genotoxins (Grant 1994, Knasmuller et al. 1998, Grant 1999). Different techniques have been used in plant bioassays for the detection of environmental metal pollution. A brief compilation of relevant works on molecular/genetic markers and their main achievements in the study of Cd genotoxicity in plants is presented in Table 1. Most of the studies on Cd genotoxicity used model species as *Arabidopsis* sp., *Vicia* sp., *Allium* sp. or *Oryza* sp. For example, Kovalchuk et al. (2001) developed an efficient and very sensitive system of transgenic *A. thaliana* that when exposed to several metals, one being Cd, exhibited a pronounced uptake-dependent increase in the frequencies of both somatic intrachromosomal recombination and point mutation.

Among the molecular markers, random-amplified polymorphic DNAs (RAPDs), amplified fragment length polymorphism (AFLP) and simple sequence repeats or microsatellite markers (SSR), open enormous advantages and need to be more explored in environmental risk assessments. Tanhuanpää et al (2007) compared four markers associated with grain Cd concentration in *Avena sativa*: 2 RAPDs, 1 REMAP (retrotransposon-microsatellite amplified polymorphism) and 1 SRAP (sequence-related amplified polymorphism), and as result suggested the use of molecular markers linked to differentiated Cd accumulation as an alternative to phenotyping selection.

Random Amplified Polymorphic DNA

These markers use random oligonucleotide (10-20) as initiators; have the advantage of being simple and non expensive, and if technical issues are adequately controlled, they may provide sensitive information on Cd induced genotoxicity. However, RAPD results are often difficult to reproduce around different laboratories (eg. Shasany et al. 2005). The low reproducibility is in fact considered the main disadvantage of this technique, namely when compared to others such as AFLPs.

Penner et al. (1995) identified RAPDs markers linked to a gene governing Cd uptake in durum wheat (*Triticum turgidum* L. var. *durum*). The genotoxicity induced by metals, including Cd, in kidney bean (*P. vulgaris*) seedlings was analysed by RAPDs, and revealed polymorphisms based on the presence and/or absence of DNA fragments in treated samples compared with untreated ones (Enan 2006). DNA polymorphisms detected by RAPDs analysis were also found in barley (*Hordeum vulgare*) (Liu et al. 2005 and 2009) and in rice (*Oryza sativa*) (Liu et al. 2007) exposed to Cd pollution, supporting the use of these markers in environmental toxicology.

Microsatellite or Single Sequence Repeat

SSRs are sequences of in general 1-6 nucleotide units repeated in tandem and randomly spread in plant genomes, usually in regions of low selective pressure. Deletions or duplications of those repeats through e.g. replication slippage, result in microsatellite instability (MSI) relative to other genomic loci. The variation in the repeat number is in nature so frequent that SSRs are very polymorphic and thus extremely useful for fine-scale genetic analysis, being used for genotyping and forensic analysis (GuhaMajumdar et al. 2008) and in the detection of genomic DNA damage and/or mutational events (e.g. deletions, insertions, point mutations). SSRs are likely to be one of the most reproducible techniques, especially when compared to RAPDs, which has the main disadvantage of low reproducibility with a consequent inconsistency of results (Powell et al. 1996, Jones et al. 1997). Due to these advantages, SSRs have been used to study genotoxic effects in several animal species (e.g. Zienolddiny et al. 2000, Jin et al. 2003, Ohshima 2003, Slebos et al. 2006). Despite providing robust information on MSI (putative mutations in the selected SSR), these markers are less used than other molecular markers, in particular when compared to RAPDs, in assessing metal induced genotoxicity. Some metals have been found to induce MSI in animal models: nickel (Ni) has been reported to promote genetic instability in hamster (Ohshima 2003) and in, e.g., human lung cancer cell lines(Zienolddiny et al. 2000), and exposure of human cell lines to environmentally relevant quantities of Cd led to statistically significant increases in MSI (e.g. Slebos et al. 2006).

In plant research, SSRs are already a powerful tool in e.g., environmental population genetics focusing on the relationships between environmental selective agents (stressors) and genotypic variability of plant natural populations (van Rossum et al. 2004). As for animals, despite SSRs have the potential to be used in the surveying of plant genomic DNA for evidence of genetic instability as in a genotoxic bioassay for the detection of DNA damage induced by environmental contaminants, up to moment, their use is limited and restricted to few reports. For instance, chloroplast SSR markers were efficient for elucidating the pattern of genetic differentiation in metal-tolerant populations in *Silene paradoxa* (Mengoni et al. 2001). For genomic DNA, using a population of 91 wheat (*T. aestivum* L.) recombinant inbred lines segregating for Al tolerance, Milla and Gustafson (2001) provided a genetic linkage map of the chromosome arm 4DL based on RFLP, SSR and AFLP markers. In another approach, van Rossum (2004) used five SSR loci to investigate the genetic structure in a metallicolous population of *A. halleri*. Also, Ma et al. (2004) identified a SSR marker (Bmag353) associated with both Al resistance and citrate secretion, which provides a valuable tool for marker-assisted selection of Al-resistant lines. Several markers, among them SSR, were also used to map rye chromosome 7R and a new Alt gene (related with Al tolerance) was found to be located in this chromosome (Matos et al. 2005). In *T. caerulescens*, Basic and Besnard (2006) suggested the use of SSR markers for population genetic or mapping studies between and within several *Brassicaceae*, in particularly for genes involved in traits such as metal tolerance and/or hyperaccumulation. Similarly, Jimenez-Ambriz et al. (2007) compared neighbouring metallicolous and nonmetallicolous *T. caerulescens* populations growing in contaminated soils, using phenotypic and SSR markers. Recently, Monteiro et al. (2009a) demonstrated the occurrence of MSI and therefore genotoxicity in lettuce (*L. sativa*) plants exposed to Cd during 28 days since germination. Mutagenic effects of Cd were evaluated on nine SSR loci. No MSI was found in leaves, but a 2-bp deletion in one lettuce root SSR was detected among the SSRs that were analysed. However, this effect depends on the age/status of the plants and period of exposure, as the same authors also demonstrated that when five week old plants (not germinated in the presence of Cd) were exposed to Cd during 14 days under similar conditions as above, no genotoxicity was observed (Monteiro et al. 2005).

Amplified Fragment Length Polymorphism

This approach is based on DNA or cDNA (cDNA-AFLP) giving valuable information on putative point mutations in the genomes and/or functional information on genes controlling/associated with some traits, such as metal resistance. For the hyperaccumulator *A. halleri* ssp. *halleri*, Dräger et al. (2005) reported 128 transcript-derived sequence fragments (TDFs) identified in a cDNA-AFLP approach designed to identify metal-regulated transcripts in roots. In addition, the authors showed that in *A. halleri* roots the transcript levels of AhPDR11, encoding an ABC transport protein are induced in response to metal exposure. Fusco et al. (2005) also used cDNA-AFLP analysis to identify genes that exhibited a modulated expression following Cd treatment in *B. juncea*. Using AFLP markers and 13 co-dominant expressed sequence tags (EST), Deniau et al. (2006) constructed a map in *T. caerulescens*, used for mapping quantitative trit locus (QTL) for Zn and Cd concentrations in roots and shoots. Also, in the backcross segregating population produced from the crosses of *A. halleri* ssp. *halleri* (tolerant) and *A. lyrata* ssp. *petraea* (non tolerant), the analysis of the transcriptome by cDNA-amplified fragment length polymorphism (cDNA-AFLP) showed differently expressed transcripts according to different sensitivities to Cd (Cracium et al. 2006).

Most plant assays for genotoxicity focus however, cytogenetic techniques, in particular Comet assay and micronucleus (MNC), using model species as *Allium cepa*, *Vicia faba*, *Hordeum vulgare*, *Tradescantia sp.*, etc. These techniques, despite being highly valuable, are very time consuming, and their sensitivity may be hampered by technical constrain.

Comet assay

This is a general but sensitive assay for genotoxicity, that measures DNA damage as percentage tail DNA (% tDNA), by using neutral or alkaline conditions that allow to detect single/double strand breakage. This technique has been used with success in several biological organisms (e.g. Juhel et al. 2007, Kuzmick et al. 2007) and even in human cells (e.g. Palus et al. 2003, Bakare et al. 2007). For instance, in plants, DNA damage measured by the Comet assay was performed in association with a battery of markers of oxidative stress (Lin et al. 2007), showing that oxidative stress induced by Cd accumulation in *Vicia faba* contributed to DNA damage.

Koppen et al. (1996) exposed *V. faba* roots to several toxic compounds, including Cd, and demonstrated that the Comet test was a suitable marker to assess genotoxicity, despite the heterogeneity in the extent of DNA migration reported. The Comet assay was evaluated as a potential tool for the assessment of ecogenotoxicity in a wetland plant, *Bacopa monnieri* L., exposed to Cd and the authors reported that this metal induced DNA damage, with more severe effects in roots than leaves (Vajpayee et al. 2006). Gichner et al. (2004) also compared Cd induced DNA damage in tobacco roots vs. leaves, showing through Comets that genotoxicity was only displayed in roots. Also, Zhang et al. (2006) assessed different types of DNA damage induced by Cd in leaves of *V. faba*. These authors found that Cd induced DNA damage (especially double strand breakages) together with apoptosis (detected by DAPI staining), revealing a possible relationship between apoptosis and DNA damage (Zhang et al. 2006).

Micronucleus Test

Due to mitotic anomalies and/or chromosome fragmentation, some chromosomes-fragments may not be incorporated into the daughter nucleus at cell division and lead to the formation of single or multiple micronuclei in the cytoplasm. These tests may be combined with a cytogenetic technique, the fluorescent in situ hybridization (FISH) that localizes (presence or absence) of specific chromosomic DNA sequences. For example, using a probe labeling the (peri-)centromeric region, FISH assay discriminates micronuclei containing a whole chromosome (labelling positive for the centromere) from micronuclei with acentric chromosome fragments (labelling negative for the centromere). Micronuclei test is however, time consuming and often subjective, and require cell division. Some criteria have to be followed, such as: the main nuclei and the micronuclei must be clearly separated, the diameter of the micronucleus should be approximately one third of the main nucleus, and should stain similarly.

Despite being time consuming, micronuclei tests have been largely used in toxicological assays for the last twenty years. Micronuclei frequency increased in water hyacinth (*Eichhornia crassipes*) exposed to Cd (Rosas et al. 1984). This data was confirmed by Mishra et al. (2007) who demonstrated that Cd levels affected the frequency of micronuclei and MI in root meristem of this species. Also it was observed an induction of micronuclei in *V. faba* root tips treated with Cd and Cr (De Marco et al. 1988). In *H. vulgare* roots, exposure to Cd increased mitotic irregularities comprising c-mitoses, anaphase bridges, breaks, stickiness, lagging and vagrant chromosomes and micronuclei (Zhang and Yang 1994). Also, in the same species, micronuclei increase was reported for *H. vulgare* roots exposed to Cd (Zhang and Xiao 1998). Similar results were described for *A. cepa* roots exposed to maleic hydrazide and methyl mercuric chloride, where the adaptive response after Cd treatment was also analysed (Panda et al. 1997). Dovgaliuk et al. (2001) showed for *A. cepa* apical meristem cells that Cd (and other metals) induced both clastogenic and aneugenic effects (including mitosis and cytokinesis disturbances). The toxicity of Cd in *V. faba* root tip cell was supported by the increase of micronuclei frequency, and this toxicity of Cd was increased by its combination with surfactant or acid rain (Liu et al. 2004). The increase of micronuclei induction, often concomitant with increases of oxidative stress parameters, was proposed to be interpreted as a consequence of oxidative stress, upholding the view that Cd-induced DNA damage is, to some extent, via generation of ROS (Rosa et al. 2003). Also, in *V. faba* exposed to Cd the population of micronuclei increased and a putative correlation with oxidative stress was proposed by the authors (Rosa et al. 2003). Ünyayar et al. (2006) tested different concentrations of Cd in two crop species, *Allium sativum* and *V. faba*, and demonstrated that the frequency of micronuclei increased considerably for the concentration of 10 μM. Cadmium also increased micronuclei frequency in the medicinal species *Adhatoda vasica* (Jahangir et al. 2006). The exposure of *A. cepa* roots to several metals, including Cd, lowered the MI value and changed the proportion of mitotic phases (mainly prophases and telophases) in MI value. Mitotic disturbances (c-metaphases, sticky and lagging chromosomes, chromosome bridges, binucleate cells, micronuclei) were also observed (Glinska et al. 2007). Cadmium genotoxicity was assessed by micronuclei in *A. sativum*, and an induced frequency of micronuclei that exhibited a dose-dependent increase in Cd treatments was observed (Celik et al. 2008). Furthermore, this effect was suggested to be controlled by giberellic acid (Celik et al. 2008).

Flow Cytometry (FCM) Analyses

Initially, flow cytometry (FCM) has been used in health and biological research for many different purposes and appeared as a relatively rapid test applicable to any organism or tissue from which cellular or nuclear suspensions can be obtained. FCM has been less used in plant sciences, and most uses are essentially on nuclear DNA content (nDNA), ploidy and cell cycle analyses. FCM principle is based on the excitation by a laser of particles "cytomes" passing individually through a flow chamber in the flow cytometer. In response to this excitation each particle (cytome) scatters light (forward and side scatter) and/or emits fluorescence. For FCM fluorescence analyses, specific fluorochromes are used. For example for the analysis of DNA, propidium iodide (IP, intercalating) or DAPI (specific for AT bases) are used, allowing the analysis of the relative fluorescence of stained isolated nuclei. In plants, FCM analyses of relative nDNA content yields histograms showing a dominant peak (nuclei at the G_0/G_1 phase) and a minor peak (G_2 nuclei), separated by a low number of nuclei in S phase. To estimate ploidy levels, the position of the G_1 peak in a histogram of an unknown sample is compared to that of a reference plant with known ploidy. FCM presents several important advantages when compared to chromosome counting, namely: convenience (sample preparation is easy), rapid output (dozens of samples can be prepared and analysed in one working day), non requirement of dividing cells and it needs only few milligrams of tissue (e.g. Loureiro et al. 2006).

Flow cytometric assays detect minute differences in nDNA content and chromosomal damage produced by clastogenic agents through the quantification of the increase of the coefficient of variation (CV) of the G_0/G_1 peak (Otto and Oldiges 1980). Flow cytometry measurement of the dispersion in the nDNA content as induced by the interactions of DNA with environmental agents, emerged then as a powerful tool in cytogenetic investigations and in genotoxicity testing (Otto et al. 1981). It is largely used in animal ecotoxicological studies, for genotoxicity assessment (e.g. Bickham et al. 1998).

However, FCM utilization in plant ecotoxicological studies remain less explored. Changes in nDNA content were observed in *Zea mays* plants exposed to coal fly ash (McMurphy and Rayburn 1993) and to the fungicides captan (Rayburn et al. 1993) and triticonazole (Biradar et al. 1994). Moreover, the mean CV of G_1 peaks increased in *Z. mays* with coal fly ash treatments (McMurphy and Rayburn 1993), indicating DNA degradation. Using *Trifolium repens,* Citterio et al. (2002) proposed a new biomonitoring methodology for soil genotoxicity assessment, combining FCM with AFLP, and showed that DNA content changed (expressed as a decrease in DNA index) with increasing concentrations of metals, and increased the debris background at the highest concentrations of Cd and Cr tested. For exposure to polycyclic aromatic hydrocarbons, Aina et al. (2007) used the same method but did not found any differences in DNA content in exposed vs. control *T. repens* plants.

Monteiro et al. (2004 and 2007) exposed *in vivo* 5-week-old lettuce (*L. sativa*) plants during 14 days to a relatively high concentration of Cd (100 μM Cd) and assessed Cd induced genotoxicity by FCM. The authors found no changes in nDNA content or in the mean CV of the G_0/G_1 peak (a putative measure of DNA degradation). The same group also assessed genotoxic and mitogenic effects of Cd on the hyperaccumulator *T. caerulescens* vs. *Thlaspi arvense* L. (field pennycress) which is a non-accumulator plant, and found differences only for some organs. More recently, exposure of *P. sativum* plants to Cr (VI) revealed significant changes by the appearance of aneuploidies and polyploidization in the higher concentrations (Rodriguez et al. 2010). Moreover, rice or barley roots (but not leaves) from plants exposed to Al also showed changes in CV, and mitogenic effects (with accumulation of cells in one stage of the cell cycle).

3. Plant Metal Accumulation and Hyperaccumulation

3.1. Accumulator Species

While the growth of most species is limited when growing on contaminated sites, some grow on contaminated sites and some are tolerant to toxic levels. When exposed to a highly contaminated environment, plants respond by excluding the metals (maintaining low and, within possible, constant metal concentration in their shoots up to a critical soil value) or accumulate the metal (accumulators) presenting therefore high levels of the metal inside the organs. As a particular case, indicator species have internal metal concentrations that reflect the external metal levels.

Although Cd is not an essential or beneficial element for plants, (hyper)accumulators generally exhibit measurable Cd concentrations, particularly in roots, but also in leaves, most probably as a result of inadvertent uptake and translocation (Assunção et al. 2003). A Cd foliar concentration above 100 μg/g DW (0.01%) is considered exceptional and it is used as a threshold value for Cd hyperaccumulation (100 mg/Kg DW) (Reeves and Baker 2000). The metal hyperaccumulation characteristic is not common in higher terrestrial plants and less than 0.2% of all angiosperms have been identified as metal hyperaccumulators (Reeves and Baker 2000).

Screening hyperaccumulators and accumulators is a key step, for both fundamental research on functional pathways involved in metal uptake-transport-accumulation processes, and for more applied studies concerning their use in metal contaminated soil phytoremediation.

Some species have been mostly studied in metal toxicology and phytoremediation studies, most of them due to their accumulation properties. Lettuce, an important human food crop, has been largely studied for the profile of Cd-distribution and cytotoxicity. Within the Brassicaceae (for review see Memon and Schroder 2009), in the genus *Thlaspi*, the alpine pennycress (*T. caerulescens*, Ganges ecotype) is a Zn, Cd and Ni hyperaccumulator plant commonly used as a model in metal transport and accumulation studies (e.g. Zhao et al. 2003).

T. caerulescens plants have been found by Reeves and Baker (2000) to contain more than 100 mg/Kg Cd frequently, and more than 1000 mg/Kg Cd occasionally, with very large variations between sites and populations, and considerable intrasite variability. Several studies have shown that *T. caerulescens* ecotype from metalliferous soils of a Zn/Pb mine spoil in the southern France (Ganges ecotype) is far superior in Cd accumulation than other ecotypes (e.g. Prayon from Belgium); in hydroponic conditions it was able to accumulate >10,000 mg/kg Cd in the shoots without showing any symptoms of phytotoxicity (Lombi et al. 2000). Recently it was proposed that Cd could play a physiological role in *T. caerulecens* by enhancing carbonic anhydrase activity (Liu et al. 2008); also the related non-accumulator, the field pennycress (*T. arvense* L.), is often used for comparative purposes.

B. juncea L. is also a well known Zn/Cd accumulator, in which the metal efflux transporter BjCET2 was demonstrated to play important roles in metal tolerance of this species (Xu et al. 2009). The high levels of Cd accumulation in its seeds (posing risks to human health) require studies on Cd redistribution in this species (e.g. Sankaran and Ebbs 2008).

Also field assays demonstrated that *Amaranthus retroflexus*, *Polygonum aviculare*, *Gundelia tournefortii*, *Noea mucronata* and *Scariola orientalis* accumulated metals, and could be used in phytorremediation strategies (Chehregani et al. 2009). Other species are known to accumulate metals, such as *Melastoma malabathricum* (Watanabe et al. 2008) and *Cassia siamea* (Jambhulkar et al. 2009).

Metal-Binding Ligands

Several plant species, namely accumulator and tolerant species, have evolved mechanisms to respond to metal uptake and accumulation that may include the chelation and sequestration of metals by particular ligands and, in some cases, the subsequent compartmentalization of the ligand-metal complex in vacuoles, achieving cellular metal homeostasis and consequent detoxification (e.g. Sung-Hyun et al. 2009).

There are many metal-binding biomolecules, most with role(s) in sequestering, transporting or storing the accumulated metal. Only a small fraction of the metal content in plants is present in the form of free aqua ions, and plants produce a number of possible ligands, including organic acids (e.g. carboxilic acids are present at high levels in plant vacuoles and play a role in hyperaccumulation, probably by sequestering them in the vacuole), nicotianamine (e.g. Gendre et al. 2007), amino acids (such as histidine), peptides and proteins (Damien et al. 2009).

Metallothioneins (MT) and PCs are particularly important in plant cell responses to metal exposure. MT are proteins of low molecular weight (approximately up 14000 Da). As MT are rich in cysteine, through the thiol group of its cysteine residues, they have the capacity to bind metals (either nutrient such as Zn, Cu) or xenobiotic (such as Cd or Hg). Concerning PCs, these are a family of enzymatically synthesized cysteine-rich peptides (Cobbett and Goldsbrough 2002). Damien et al. (2009) proposed a distinction between normal and hyperaccumulating plants, as nonaccumulating plants coordinate the majority of Cd and As using PCs and hyperaccumulators do not. Cadystins, Cd peptides, or γ-glutamyl peptides are nonprotein metal-binding polypeptides that possess the structure of (γGlu-Cys)nGly (Sung-Hyun et al. 2009).

Cadmium is recognised as a strong primary inducer of some of these metal binding ligands. Recently, Sung-Hyun et al. (2009) exposed *Echinochloa crusgalli* var. *frumentacea* to high Cd levels and purified a Cd-binding ligand (Cd-BL) from the roots and characterized its amino acid composition.

Plant vacuoles play crucial roles in the metal cellular homeostasis (Barkla and Pantoja 1996). The tonoplast functions as an effective and selective metal diffusion barrier. Vacuolar compartmentalization prevents the free circulation of Cd ions in the cytosol and forces them into a limited area (Sanitá di Toppi and Gabbrielli, 1999). Most progress on the role of the vacuole in metal transport, homeostasis and detoxification has been made with the micronutrients Zn and Fe (Martinoia et al. 2007). However, several studies also demonstrate the importance of the vacuole as a site of accumulation of a number of metals including Cd (Ma et al. 2005, Ueno et al. 2005). One example is the accumulation of Cd and PCs in the vacuole involving an ABC transporter (Hall 2002). Oat root tonoplast vesicles were found to accumulate Cd cation by a $2H^+$/ion antiport mechanism (Salt and Wagner 1993). Also, several transporters such as MTPs, CAXs and ABC transporters were already described as acting as Cd or Cd-chelate transporters (Martinoia et al. 2007).

Plant vacuoles of metal exposed plants accumulate PCs-Cd complexes in the vacuole (Cobbett and Goldsbrough 2002). This was evidenced, for example, by the preferential accumulation of Cd and/or PCs confined to the vacuole in tobacco mesophyll protoplasts derived from Cd-treated plants (Vogeli-Lange and Wagner 1990). Also, detoxification mechanisms were also described for *L. sativa* and *T. arvense* plants in which PCs play an important role (Ebbs et al. 2002, Maier et al. 2003). In *T. caerulescens* a main storage of Cd cation (as electron-dense granules) was found inside vacuoles by means of complexation with malate, an organic acid (Ma et al. 2005, Ueno et al. 2005).

4. Trophic Transfer, Bioaccumulation and Biomagnification of Cd

Plants have developed mechanisms for sequestering metals in their systems limiting metal phytotoxicity, but these plants may still be a threat to animal consumers and therefore a risk to ecological and human food chains (McLaughlin 2002). Animals and plants may amplify the Cd content relatively to their environment (Robards and Worsfold 1991, Greger 1999). Therefore, uptake of topsoil Cd (usually more contaminated that subsoil), and trophic transfer is likely the most important route for human exposure to Cd (IARC 1993). Cadmium bioaccumulation is affected by several factors such as its physico-chemical form, presence of other metals, pH, salinity, temperature, season, cation-exchange capacity of soils and the species taking up the Cd (e.g. Robards and Worsfold 1991). Several studies have demonstrated the transfer of Cd from producers, including plants: Xie et al. (2009) showed Cd bioaccumulation in mayfly fed on periphyton biofilms; Yu and Fleeger (2006) suggested that eutrophication by nitrate enrichment could enhance Cd trophic transfer from microalgae to suspension-feeding benthic invertebrates and Monteiro et al. (2008) demonstrated Cd accumulation in isopods fed with Cd biologically contaminated plants, including lettuce.

Metals biomagnification (progressive accumulation of a metal with increasing trophic levels) during transfer from contaminated soils may result in predatory exposure to toxic metal concentrations (Grenn and Tibbert 2008). Croteau and co-workers (2005) have demonstrated that Cd was progressively enriched among trophic levels in two discrete epiphyte-based food webs composed of macrophyte-dwelling invertebrates or fishes. Mathews and Fisher (2008) have also shown biomagnification of Cd (and of MeHg and Po) from phytoplankton to zooplankton.

However, biomagnification process still remains a matter of controversy: trophic barriers were found along the soil-plant-herbivorous insect pathway reducing relative uptake of Cd (Merrington et al. 2001). Zhuang et al. (2009) demonstrated that transfer of metals (Pb, Zn Cu and Cd) to chicken from plant and insect was limited. Alonso et al. (2009) reported that a significant Cd transfer through a food chain (where *T. aestivum* was used as model plant in the chain) was found but not with the occurrence of biomagnification in the predator species (*Chrysoperla carnea*). Also, in a soil-plant arthropod system, Cd was biomagnified, contrarily to Zn (Green et al. 2006), in a soil-pea plant-aphid system, Zn was biomagnified while Cd was biominimized (Green and Tibbert 2008).

Factors Affecting Trophic Transfer of Metals

The bioaccumulation of metals is known to differ among species and metals because of differences in uptake and loss rates, exposure pathways and influences of environmental parameters (Fisher and Reinfelder 1995, Wang and Fisher 1999). However, less is known about the influence of these factors in the internal storage and detoxification of accumulated metal and subsequent impacts on trophic transfer. Since the ingestion of metal-contaminated food can serve as a source of metals to consumers and can result in sub-lethal toxicity (e.g. Fisher and Hook 2002), understanding the mechanisms that influence metal trophic transfer is a critical step in the management of metal contaminated ecosystems. In general, to completely understand metal cycling through trophic levels, several factors which control the bioavailability of tissue-bound metals to predators must be considered and understood (e.g. tissue metal distributions and concentrations, duration of exposure, nutritional status and exposure history of predator). Different species will accumulate and partition metals in varying ways depending on the detoxification mechanisms employed. The subsequent bioavailability of those partitioned metals to a consumer will be dictated by digestive and assimilative mechanisms of its digestive tract and gut passage time (Wang and Fisher 1999). Added to this complexity is the varying ability of consumers to discriminate between different foods and contaminants, their nutritional status at the time of consumption, the degree of exposure, and the exposure history for the metal in question, all of which can influence the degree of metal assimilation (Wang and Fisher 1999).

Subcellular Partition of Metals

The internal distribution and detoxification of metals within an organism can be used to explain trophic transfer of metals but also to predict metal toxicity for the organism itself. The internal metal sequestration strategies of different species are complex and variable and the determination of the metal concentrations in different compartments can be used to understand the complex relationship between metal accumulation and toxicity.

Metals can be present in various chemical forms in an organism, including the following: (a) free ionic form or complexed ion species (e.g., $CdCl_2$, $CdCl^+$, $CdCl_3^-$); (b) bound in the active center of functional proteins and enzymes; (c) bound to low molecular weight organic acids (e.g., citrate, malate); (d) bound to sequestration proteins (MT and PCs); (e) bound in vesicles of the lysosomal system, as intracellular granules; (f) precipitated in extracellular granules, mineral deposits, residual bodies, and exoskeletons; (g) bound to cellular constituents potentially causing dysfunction (e.g. DNA) (Vijver et al. 2004).

The various internal metal fractions all have their own binding capacity for metals, which has implications for food-chain transfer to higher trophic levels. A study on the relationship between subcellular Cd distribution in an oligochaete and its trophic transfer to a predatory shrimp showed that only metal present in the soluble fraction (organelles and protein fraction) of prey is available for the predator (Wallace et al. 1998). Factors influencing the subcellular distribution in the prey will directly alter trophic transfer to predators. Wallace et al. (1998) showed that differences in subcellular distribution of Cd between resistant and non-resistant worms directly affected Cd availability for the predatory shrimp. When fed resistant worms, shrimp absorbed about 4 times less Cd than when fed non-resistant worms (Wallace et al. 1998). Similar conclusions were found in a study using bivalves as prey, where the metal partitioning to organelles, denaturated cytosolic proteins, and MT comprise a subcellular compartment of that was considered as trophically available metal to predators (Wallace et al. 2003. Furthermore, Wallace and Luoma (2003) building on previous studies (Wallace and Lopez 1997, Wallace et al. 1998), postulated that Cd associated with the subcellular fractions, organelles, heat-denatured proteins, and heat-stable proteins of prey was trophically available metal and was assimilated at an efficiency of approximately 100% by the predator, while Cd bound to metal-rich granules was less bioavailable to predators. Such subcellular partitioning is dynamic in response to metal exposure and other environmental conditions, and is metal- and organism-specific. The different metal pools are not equally bioavailable to predators; thus, the determination of the metal concentration in the different subcellular compartments and the differences in its assimilation by consumers can be a useful tool to understand metal transfer to higher trophic levels.

Other applications of this approach have been proposed. Recent studies in aquatic organisms have revealed that the subcellular partitioning model may provide an improved method to predict Cd toxicity. As intracellular metal accumulation and the subsequent subcellular distribution of the metal in the cells are directly related to metal toxicity, it is likely that the metal concentration in a particular subcellular fraction will serve as a better toxicity predictor than the activity of the free metal ion in bulk solution (Wang and Rainbow 2006). Rainbow (2002) proposed that when accumulated metal destined for storage in a detoxified form (e.g. by MT and granules) exceeds the detoxified binding capacity, the metals are subsequently bound with other (metabolically available) forms, with the potential to cause toxicity to the organism. The significance of the subcellular distribution of accumulated metals in toxicity assessments is now receiving increasing attention among aquatic (e.g.

Cheung et al. 2006, Perceval et al. 2006, Steen Redeker et al. 2007) and terrestrial organisms (Vijver et al. 2006, Vijver et al. 2007). Recently, Monteiro et al. (2008) demonstrated that Cd bound to heat-stable proteins was the least bioavailable fraction to isopods while Cd bound to heat-denatured proteins was the most trophically available to isopods, pointing out the ecological importance of the plant subcellular Cd distribution to the trophic transfer of Cd.

5. Conclusions and Future Prospects

Cadmium is one of the most highly toxic trace pollutants for humans, animals and plants. However, the major source of Cd contamination is due to anthropogenic activities. Several studies demonstrated its cytotoxic, mutagenic and/or carcinogenic effects in animal cells. In plants, available also data support that Cd leads to cytotoxic and genotoxic effects.

Cd interaction with plants and its toxicological effects at the plant physiological level, as well as factors conditioning its uptake by plants, are well described. In particular, several studies concern to Cd concentration in soil and its bioavailability is modulated by the presence of organic matter, pH, redox potential, temperature and concentrations of other elements. Once inside the plant, Cd usually leads to a battery of unspecific symptoms that include chlorosis, necrotic lesions, wilting, reddish coloration, growth reduction and disturbances in plant water relations. However, particular physiological aspects of Cd/plant interaction still remain obscure. At the carbon metabolism level, Cd inhibits photosynthetic pigments synthesis, electron transport, uptake of nutrients, CO_2 assimilation and enzymes of the Calvin cycle, and also induce a structural disorganization of chloroplasts. However, at what extent, directly or indirectly, this metal affects photosynthetic proteins and genes regulation is an open field for research. In this issue functional proteomics and, mostly, metabolomics (by detecting products) may contribute to elucidate Cd regulation of some related photosynthetic pathways. For example, the effects of cadmium on the transcription rate of chloroplast genes and/or the level of their transcripts remains poorly understood until now, although this is very important for deciphering the molecular mechanisms of cadmium damaging.

For animal cells it is well demonstrated that cadmium affects cellular processes such as cell-cycle progression, proliferation, differentiation, DNA replication and repair, and apoptosis, but the dimension of these effects in plant cells still remain unclear. It is known that Cd ions stimulate the free oxygen radical production and increased the activity of several antioxidant enzymes. Under severe and extreme stress conditions, the capacity of the antioxidant system may not be sufficient to minimize the harmful effect of oxidative injury. Plant survival under these conditions depends on its ability to perceive the stimulus, generate and transmit signals and induce biochemical changes. Research on signal molecules that mediate stress tolerance could be an important step towards a better understanding the capacity of plants to cope with these environments.

Together with other –omics, plant quantitative cytometry (also called plant cytomics) will bring new tools to better understand oxidative stress responses and how it may regulate, for example, gene expression of oxidative defence of other pathways such as cell cycle. Also, the expanding research using genomic technology, in particular DNA microarrays, allows a more comprehensive analysis of gene expression profiles of plants (from sensitive genomic model

Arabidopsis thaliana to the hyperaccumulator *Thlaspi caerulescens*) exposed to Cd. This will be a powerful tool to understand the expression of genes related to cellular protection and damage control mechanisms (e.g. genes coding for metallothioneins or antioxidant proteins) under Cd stress. In humans, as a whole, a number of genes appear to be coordinately regulated toward survival from Cd toxicity, and the evaluation of equivalent coordination networks in plants is still to clarify.

Concerning Cd induced-genotoxicity, most studies focus on genetic markers such as micronuclei or Comets. Molecular markers, despite less used, have in the last years demonstrated to be reliable markers in toxicity evaluations. Genetic methodologies (such as microsatellites, AFLP, RAPDs, micronucleus and flow cytometry) showed to be powerful tools to assay Cd genotoxicity. As specific markers of the genome characteristics/regions, one should emphasise the need to combine these markers for more comprehensive information. Moreover, the efficacy of these markers to verify genotoxic effects within an ERA perspective requires highly controlled experimental conditions and coordinated practices within the laboratories. Comparisons among these markers require controlled conditions (plant age, cadmium concentration and chemical formulation).

Accumulator and tolerant species have evolved mechanisms to respond to metal uptake and accumulation that include the chelation and sequestration of metals by particular ligands and, in some cases, the subsequent compartmentalization of the ligand-metal complex in vacuoles, achieving cellular metal homeostasis and consequent detoxification. Accumulator's species generally exhibit measurable Cd concentrations, particularly in roots, but also in leaves. Screening hyperaccumulators and accumulators could be the key step, for both fundamental research on functional pathways involved in metal uptake-transport-accumulation processes, to provide model systems to study the mechanisms of Cd tolerance and to use these species in phytoremediation strategies.

The use of soils and water contaminated with Cd in agriculture and the consume of contaminated products is especially important since it may represent a high risk to human food chain. Therefore, one concern of Cd pollution, like other heavy metals, is its trophic transfer, bioaccumulation and biomagnification. Bioaccumulation of metals differs among species and metals because of differences in uptake and loss rates, exposure pathways and influences of environmental parameter. Future investigation on the influence of these factors in the internal storage and detoxification of the accumulated metal and subsequent impacts on trophic transfer will improve the knowledge of metal bioaccumulation.

Table 1. A brief review of genotoxicity studies in Cd exposed plants

Species studied	Method used	Genotoxicity (organ)	References
Allium cepa Vicia faba and Tradescantia	Micronucleus	Yes (root tips and pollen mother cells)	Steinkellner et al. (1998)
Bacopa monnieri	Comet assay	Yes (+ roots and - leaves)	Vaipayee et al. (2006)
Allium cepa	Chromosome aberrations	Yes (roots)	Borboa and de La Torre (1996)
Helianthus annuus	SSRs[3]	No (root and leaves)	Gomes et al. (2005)
Helianthus annuus	Flow cytometry	No (roots and leaves)	Azevedo (pers. comm.)
Hordeum vulgare	RAPDs[1]	Yes (root tips)	Liu et al. (2005)

Table 1. (Continued)

Species studied	Method used	Genotoxicity (organ)	References
Nicotiana tabacum	Comet assay	Yes (roots); No (leaves)	Gichner et al. (2004)
Oryza sativa	AFLP[2]	Yes (roots)	Aina et al. (2007)
Oryza sativa	RAPDs[1]	Yes (root tips)	Liu et al. (2007)
Phaseolus vulgaris	RAPDs[1]	Yes (seedlings)	Enan (2006)
Pisum sativum	Flow cytometry	Yes (roots)	Fusconi et al. (2006)
Thlaspi caerulescens and Thlaspi arvense	SSRs[3]	No (roots)	Paiva (2008)
Vicia faba	Micronucleus	Yes (root tips)	Beraud et al. (2007)
Vicia faba	Comet assay	Yes (leaves)	Lin et al. (2007)
Raphanus sativus L		Yeas roots (leaves and roots)	Villatoro-Pulido et al. (2009)
Allium sativum	Micronucleus	Yes (root tips)	Celik et al. 2008
Vicia faba	Micronucleus	Yes (root tips exposed to Leachates of municipal solid waste containing several metals)	Feng et al. 2009
Vicia faba	Comet assay	Yes (leaves)	Zhang et al. 2006
Bacopa monnieri L	Alkaline Comet assay	Yes (leaves and roots)	Vajpayee et al. 2006
Allium sativum and Vicia faba	Micronuclei	Yes (root tips)	Unyayar et al. 2006
Vicia faba	Comet assay	Yes (root tip)	Lin et al. 2005
Vicia faba	Micronucleus	Yes (root tips, exposed to municipoal solid waste with low concentrations of Cd)	Radetski et al. 2004
Tradescantia	Micronuclous	Yes (used as positive control)	Gong et al. 2003
Allium cepa	Chromosome lagging of, multipolar anaphases, C-mitoses, etc	Yes (roots)	Dovgaliuk et al. 2001
Arabidopsis thaliana	Intrachromosomal recombination	Yes	Kovalchuk et al. 2001
Allium cepa	Anaphase-telophase chronmosome aberration	Yes (root)	Rank et al. 1998

[1]RAPDs - Random Amplified Polymorphic DNA.

[2]AFLP - Amplified Fragment Length Polymorphism.

[3]SSRs - Microsatellites or Simple Sequence Repeats.

[4]REMAP - retrotransposon-microsatellite amplified polymorphism)

References

2455/2001/EC European Commission Regulation (EC) no. 2455/2001 of 20 November 2001 established a list of priority substances in the field of water policy and amending Directive 2000/60/EC. *OJ L 331, p. 1–5.*

466/2001/EC, 2001. European Commission Regulation (EC) no. 466/2001 of 8 March 2001 setting maximum levels for certain contaminants in foodstuffs. 77: 1-13.

Aina, R., Labra, M., Fumagalli, P., Vannini, C., Marsoni, M., Cucchi, U., Bracale, M., Sgorbati, S., Citterio, S., 2007. Thiol-peptide level and proteomic changes in response to cadmium toxicity in *Oryza sativa* L. roots. Environmental and Experimental Botany 59: 381-392.

Alonso, E., González-Núñez, M., Carbonell, G., Fernández, C., Tarazona, J.V., 2009. Bioaccumulation assessment via an adapted multi-species soil system (MS.3) and its application using cadmium. Ecotoxicololy and Environmental Safety 72(4): 1038-44.

Apel, K., Hirt, H., 2004. Reactive oxygen species: metabolism, oxidative stress, and signal transduction. Annual Review of Plant Biology 55: 373-399.

Aravind, P., Prasad, M.N., Malec, P., Waloszek, A., Strzałka, K., 2009. Zinc protects *Ceratophyllum demersum* L. (free-floating hydrophyte) against reactive oxygen species induced by cadmium. Journal of Trace Elements in Medicine and Biology 23(1):50-60.

Assunção, A.G.L., Schat, H., Aarts, M.G.M., 2003. *Thlaspi caerulescens*, an attractive model species to study heavy metal hyperaccumulation in plants. New Phytologist 159: 351-360.

ATSDR, 1999. Toxicological profile for cadmium. In: Agency for toxic substances and disease registry, draft for public comment. Public health service, U.S. department of health and human services, Atlanta, GA, pp. 1-397.

Azevedo, H., Pinto, G., Fernandes, J., Loureiro, J., Santos, C., 2005a. Cadmium effects on sunflower growth and photosynthesis. Journal of Plant Nutrition 28, 2211-2220.

Azevedo, H., Pinto, G., Santos, C., 2005b. Cadmium effects in sunflower: Membrane permeability and changes in catalase and peroxidase activity in leaves and calluses. Journal of Plant Nutrition 28: 2233-2241.

Azevedo, H., Pinto, G., Santos, C., 2005c. Cadmium effects in sunflower: Nutritional imbalances in plants and calluses. Journal of Plant Nutrition 28: 2221-2231.

Bakare, A.A., Pandey, A.K., Bajpayee, M., Bhargav, D., Chowdhuri, D.K., Singh, K.P., Murthy, R.C., Dhawan, A., 2007. DNA damage induced in human peripheral blood lymphocytes by industrial solid waste and municipal sludge leachates. Environmental and Molecular Mutagenesis 48(1): 30-7.

Balakhnina, T. Y., Kosobryukhov, A. A., Ivanov, A. A., Kreslavskii, V. D., 2005. The effect of cadmium on CO_2 exchange, variable fluorescence of chlorophyll, and the level of antioxidant enzymes in pea leaves. Russian Journal of Plant Physiology 52(1): 15-20.

Ball, L., Accotto, G.P., Bechtold, U., Creissen, G., Funck, D., Jimenez, A., Kular, B., Leyland, N., Mejia-Carranza, J., Reynolds, H., Karpinski, S., Mullineaux, P.M., 2004. Evidence for a direct link between glutathione biosynthesis and stress defence gene expression in *Arabidopsis*. The Plant Cell 16(9): 2448-62.

Barcelo, J., Poschenrieder, C., 1990. Plant water relations as affected by heavy metal stress: a review. Journal of Plant Nutrition 13: 1-37.

Barkla, B.J., Pantoja, O., 1996. Physiology of ion transport across the tonoplast of higher plants. Annual Review of Plant Physiology and Plant Molecular Biology 47: 159-184.

Baryla, A., Carrier, P., Franck, F., Coulomb, C., Sahut, C., Havaux, M., 2001. Leaf chlorosis in oilseed rape plants (*Brassica napus*) grown on cadmium-polluted soil: causes and consequences for photosynthesis and growth. Planta 212: 696–709.

Basic, N., Besnard ,G., 2006. Gene polymorphisms for elucidating the genetic structure of the heavy-metal hyperaccumulating trait in *Thlaspi caerulescens* and their cross-genera amplification in *Brassicaceae*. Journal of Plant Research 119(5): 479-87.

Ben, A., Charles, G., Hourmant, A., Ben, J., Branchard, M., 2009. Physiological behaviour of four rapeseed cultivar (*Brassica napus L.*) submitted to metal stress. C R Biology 332(4): 363-70.

Benavides, M.P., Gallego, S.M. Tomaro, M.L., 2005. Cadmium toxicity in plants. Brazilian Journal of Plant Physiology 17: 21-34.

Benzarti, S., Mohri, S., Ono, Y., 2008. Plant response to heavy metal toxicity: comparative study between the hyperaccumulator *Thlaspi caerulescens* (ecotype Ganges) and on accumulator plants: lettuce, radish, and alfalfa. Environmental Toxicology 23(5): 607-16.

Beraud, E., Cotelle, S., Leroy, P. Ferard, J.F., 2007. Genotoxic effects and induction of phytochelatins in the presence of cadmium in *Vicia faba* roots. Mutation Research 633: 112-116.

Bertin, G., Averbeck, D., 2006. Cadmium: cellular effects, modifications of biomolecules, modulation of DNA repair and genotoxic consequences (a review). Biochimie 88:1549–1559.

Bhartia, N., Singh, R.P., 1994. Antagonistic effect of sodium chloride to differential heavy metal toxicity regarding biomass accumulation and nitrate assimilation in *Sesamum indicum* seedlings. Phytochemistry 35: 1157-1161.

Bi, Y., Chen, W., Zhang, W., Zhou, Q., Yun, L., Xing, D., 2009. Production of reactive oxygen species, impairment of photosynthetic function and dynamic changes in mitochondria are early events in cadmium-induced cell death in *Arabidopsis thaliana*. Biology of the Cell 101(11): 629-43.

Bickham, J.W., Mazet, J.A., Blake, J., Smolen, M.J., Ballachey, B.E., 1998. Flow cytometric determination of genotoxic effects of exposure to petroleum in mink and sea otters. Ecotoxicology 7: 191-199.

Biradar, D.P., Pedersen, W.L. Rayburn, A.L., 1994. Nuclear DNA analysis of maize seedlings treated with the triazole fungicide, triticonazole. Pesticides Science 41: 291-295.

Borboa, L., de La Torre, C., 1996. The genotoxicity of Zn(II) and Cd(II) in *Allium cepa* root meristematic cells. New Phytologist 134: 481-486.

Burzyński, M., Kłobus, G., 2004. Changes of photosynthetic parameters in cucumber leaves under Cu, Cd, and Pb stress. Photosynthetica 42 (4): 505-510.

Cai, S.W., Yue, L., Hu, Z.N., Zhong, X.Z., Ye, Z.L., Xu, H.D., Liu, Y.R., Ji, R.D., Zhang, W.H. Zhang, F.Y., 1990. Cadmium exposure and health effects among residents in an irrigation area with ore dressing wastewater. Science of Total Environment 90: 67-73.

Campbell, P.G.C., 2006. Cadmium - A priority pollutant. Environmental Chemistry 3: 387-388.

Carrier, P., Baryla, A., Havaux, M., 2003. Cadmium distribution and microlocalization in oilseed rape (*Brassica napus*) after long-term growth on cadmium-contaminated soil. Planta 216: 939–950.

Cataldo, D.A., Garland, T.R. Wildung, R.E., 1983. Cadmium uptake kinetics in intact soybean plants. Plant Physiology 73: 844-848.

Celik, A., Unyayar, S., Cekiç, F.O., Güzel, A., 2008. Micronucleus frequency and lipid peroxidation in *Allium sativum* root tip cells treated with gibberellic acid and cadmium. Cell Biology and Toxicology 24(2): 159-64.

Chaffei, C., Pageau, K., Suzuki, A., Gouia, H., Ghorbel, H.M., Mascalaux-Daubresse, C., 2004. Cadmium toxicity induced changes in nitrogen management in *Lycopersicon esculentum* leading to a metabolic safeguard through an amino acid storage strategy. Plant and Cell Physiology 45: 1681–1693.

Chaoui, A., El Ferjani, E., 2005. Effects of cadmium and copper on antioxidant capacities, lignification and auxin degradation in leave\s of pea (*Pisum sativum* L.) seedlings. C R Biology 328: 23-31.

Chaoui, A., Mazhoudi, S., Ghorbal, M.H., El Ferjani, E., 1997. Cadmium and zinc induction of lipid peroxidation and effects on antioxidant enzyme activities in bean (*Phaseolus vulgaris* L.). Plant Science 127: 139-147.

Chehregani, A., Noori, M., Yazdi, H.L., 2009. Phytoremediation of heavy-metal-polluted soils: screening for new accumulator plants in Angouran mine (Iran) and evaluation of removal ability. Ecotoxicology and Environmental Safety 72(5): 1349-53.

Cheung, M.-S., Fok, E.M.W., Ng, T.Y.-T., Yen, Y.-F. Wang, W.-X., 2006. Subcellular cadmium distribution, accumulation, and toxicity in a predatory gastropod, *Thais clavigera*, fed different prey. Environmental Toxicology and Chemistry 25:174-181.

Chiang, H.C., Lo, J.C., Yeh, K.C., 2006. Genes associated with heavy metal tolerance and accumulation in Zn/Cd hyperaccumulator *Arabidopsis halleri*: a genomic survey with cDNA microarray. Environmental Science Technology 40(21): 6792-8.

Citterio, S., Aina, R., Labra, M., Ghiani, A., Fumagalli, P., Sgorbati, S. Santagostino, A., 2002. Soil genotoxicity assessment: a new strategy based on biomolecular tools and plant bioindicators. Environmental Science Technology 36: 2748-2753.

Clemens, S., Antosiewicz, D.M., Ward, J.M., Schachtman, D.P., Schroeder, J.I., 1998. The plant cDNA LCT1 mediates the uptake of calcium and cadmium in yeast. Proceedings of the National Academy of Sciences 95(20): 12043-8.

Clemens, S., Palmgren, M.G. Kramer, U., 2002. A long way ahead: understanding and engineering plant metal accumulation. Trends in Plant Science 7: 309-315.

Cobbett, C., Goldsbrough, P., 2002. Phytocelatins and metallothioneins: roles in heavy metal detoxification and homeostasis. Annual Review of Plant Biology 53: 159-182.

Costa, G., Morel, J.L., 1994. Efficiency of H^+-ATPase activity on cadmium uptake by 4 cultivars of lettuce. Journal of Plant Nutrition 17: 627-637.

Craciun, A.R., Courbot, M., Bourgis, F., Salis, P., Saumitou-Laprade, P., Verbruggen, N., 2006. Comparative cDNA-AFLP analysis of Cd-tolerant and -sensitive genotypes derived from crosses between the Cd hyperaccumulator *Arabidopsis halleri* and *Arabidopsis lyrata* ssp. petraea. Journal of Experimental Botany 57(12): 2967-83.

Croteau, M.-N., Luoma, S.N. Stewart, A.R., 2005. Trophic transfer of metals along freshwater food webs: Evidence of cadmium biomagnification in nature Limnol. Oceanography 50: 1511–1519.

Cseh, E., 2002. Metal permeability, transport and efflux in plants. In: Prasad, M. N. V., Strzalka, K., (Eds.). Physiology and Biochemistry of Metal Toxicity and Tolerance in Plants. Kluwer Academic Publishers, Dordrecht, Netherlands. pp. 1-36.

Dalla, Vecchia, F., La Rocca, N., Moro, I., de Faveri, S., Andreoli, C., Rascio, N., 2005. Morphogenetic, ultrastructural and physiological damages suffered by submerged leaves of *Elodea canadensis* exposed to cadmium. Plant Science 168(2): 329–338.

Damien, L., Callahan, Alan, J. M., Baker, Spas, D., Kolev, Anthony, G., 2006. Wedd metal ion ligands in hyperaccumulating plants. Journal of Biological Inorganic Chemistry 11: 2–12.

Delpérée, C., Lutts, L., 2008. Growth Inhibition Occurs Independently of Cell Mortality in Tomato (*Solanum lycopersicum*) Exposed to High Cadmium Concentrations. Journal of Integrative Plant Biology 50(3): 300-310.

De Marco, A., Paglialunga, S., Rizzoni, M., Testa, A., Trinca, S., 1988. Induction of micronuclei in *Vicia faba* root tips treated with heavy metals (cadmium and chromium) in the presence of NTA. Mutation Research 206(3): 311-5.

Deniau, A.X., Pieper, B., Ten, Bookum, W.M., Lindhout, P., Aarts, M.G., Schat, H., 2006. QTL analysis of cadmium and zinc accumulation in the heavy metal hyperaccumulator *Thlaspi caerulescens*. Theoretical and Applied Genetics 113(5): 907-20.

Di Cagno, R., Guidi, L., De Gara, L., Soldatini, G., 2001. Combined cadmium and ozone treatments affect photosynthesis and ascorbate-dependent defences in sunflower. New Phytologist 151 (3): 627-636(10).

Dietz, K.-J., Baier, M., Krämer, U., 1999. Free radicals and reactive oxygen species as mediators of heavy metal toxicity in plants. In: Prasad, M. N. V., Hagemeyer, J., (Eds.). Heavy metal stress in plants. Springer-Verlag, Berlin, Germany.

Dovgaliuk, A.I., Kaliniak, T.B., Blium, I.B., 2001. Cytogenetic effects of toxic metal salts on apical meristem cells of *Allium cepa* L. seed roots. Toxicology Genetic 35(2):3-10.

Dräger, D.B., Voigt, K., Krämer, U., 2005. Short transcript-derived fragments from the metal hyperaccumulator model species *Arabidopsis halleri*. Zeitschrift für Naturforschung C. 60(3-4): 172-8.

Drazkiewicz, M., Skórzyńska-Polit, E., Krupa, Z., 2007. The redox state and activity of superoxide dismutase classes in *Arabidopsis thaliana* under cadmium or copper stress. Chemosphere 67(1): 188-93.

Dunwei, C., Jiang, D., Dai, T., Jing, Q., Cao, W., 2009. Effects of cadmium on plant growth and physiological traits in contrast wheat recombinant inbred lines differing in cadmium tolerance. Chemosphere.

Ebbs, S., Lau, I., Ahner, B., Kochian, L., 2002. Phytochelatin synthesis is not responsible for Cd tolerance in the Zn/Cd hyperaccumulator *Thlaspi caerulescens* (J. and C. Presl). Planta 214: 635-640.

Ekmekçi, Y., Tanyolaç, D., Ayhan, B., 2008. Effects of cadmium on antioxidant enzyme and photosynthetic activities in leaves of two maize cultivars. Journal of Plant Physiology 165: 600-611.

Enan, M.R., 2006. Application of random amplified polymorphic DNA (RAPD) to detect the genotoxic effect of heavy metals. Biotechnology Applied Biochemistry 43: 147-154.

Fagioni, M., Zolla, L., 2009. Does the different proteomic profile found in apical and basal leaves of spinach reveal a strategy of this plant toward cadmium pollution response? Journal of Proteomic Research 8(5): 2519-29.

Feng, S., Wang, X., Wei, G., Peng, P., Yang, Y., Cao, Z., 2009. Leachates of municipal solid waste incineration bottom ash from Macao: heavy metal concentrations and genotoxicity. Chemosphere 67(6):1133-7.

Filek, M., Keskinen, R., Hartikainen, H., Szarejko, I., Janiak, A., Miszalski, Z., Golda, A., 2008. The protective role of selenium in rape seedlings subjected to cadmium stress. Journal of Plant Physiology 165(8): 833-44

Filipic, M., Fatur, T. Vudrag, M., 2006. Molecular mechanisms of cadmium induced mutagenicity. Human Experimental Toxicology 25: 67-77.

Fisher, N.S. Hook, S.E., 2002. Toxicology tests with aquatic animals need to consider the trophic transfer of metals. Toxicology 181: 531-536.

Fisher, N.S., Reinfelder, J.R., 1995. The trophic transfer of metals in marine systems. In: Tessier, A., Turner, D. R., (Eds.). Metal Speciation and Bioavailability in Aquatic Systems. John Wiley and Sons Ltd, London. pp. 363-406.

Fodor, F., 2002. Physiological responses of vascular plants to heavy metals. In: Prasad, M. N. V., Strzalka, K., (Eds.). Physiology and Biochemistry of Metal Toxicity and Tolerance in Plants. Kluwer Academic Publishers, Dordrecht, Netherlands. pp. 149-177.

Fusco, N., Micheletto, L., Dal Corso, G., Borgato, L., Furini, A., 2005. Identification of cadmium-regulated genes by cDNA-AFLP in the heavy metal accumulator *Brassica juncea* L. Journal of Experimental Botany 56(421): 3017-27.

Fusconi, A., Gallo, C., Camusso, W., 2007. Effects of cadmium on root apical meristems of *Pisum sativum* L.: cell viability, cell proliferation and microtubule pattern as suitable markers for assessment of stress pollution. Mutation Research. 632(1-2): 9-19.

Fusconi, A., Repetto, O., Bona, E., Massa, N., Gallo, C., Dumas-Gaudot, E. Berta, G., 2006. Effects of cadmium on meristem activity and nucleus ploidy in roots of *Pisum sativum* L. cv. Frisson seedlings. Environmental and Experimental Botany 58: 253-260.

Garnier, L., Simon-Plas, F., Thuleau, P., Agnel, J.P., Blein, J.P., Ranjeva, R., Montillet, J.L., 2006. Cadmium affects tobacco cells by a series of three waves of reactive oxygen species that contribute to cytotoxicity. Plant Cell and Environment 29(10): 1956-69.

Gendre, D., Czernic, P., Conéjéro, G., Pianelli, K., Briat, J.F., Lebrun, M., Mari S., 2007. TcYSL3, a member of the YSL gene family from the hyperaccumulator *Thlaspi caerulescens*, encodes a nicotianamine-Ni/Fe transporter. The Plant Journal 49(1):1-15.

Gichner, T., Patková, Z., Száková, J., Demnerová, K., 2004. Cadmium induces DNA damage in tobacco roots, but no DNA damage, somatic mutations or homologous recombination in tobacco leaves. Mutation Research 559(1-2): 49-57.

Glińska, S., Bartczak, M., Oleksiak, S., Wolska, A., Gabara, B., Posmyk, M., Janas, K., 2007. Effects of anthocyanin-rich extract from red cabbage leaves on meristematic cells of *Allium cepa* L. roots treated with heavy metals. Ecotoxicology and Environmental Safety 68(3): 343-50.

Gomes, A., Azevedo, H., Santos, C. Lopes, T., 2005. Microsatellite analysis of *Helianthus annuus* L genome exposed to cadmium and lead. 42nd Congress of Toxicology EUROTOX, Cracow, Poland.

Gomes-Junior, RA., Moldes, C.A., Delite, F.S., Pompeu, G.B., Gratão, P.L., Mazzafera, P., Lea, P.J., Azevedo, R.A., 2006. Antioxidant metabolism of coffee cell suspension cultures in response to cadmium. Chemosphere 65:1330-1337.

Gong, P., Kuperman, R.G., Sunahara, G.I., 2003. Genotoxicity of 2,4- and 2,6-dinitrotoluene as measured by the Tradescantia micronucleus (Trad-MCN) bioassay. Mutation Research 538(1-2):13-8.

Gong, J.-M., Lee, D.A., Schroeder, J.I., 2003. Long-distance root-to-shoot transport of phytochelatins and cadmium in *Arabidopsis*. Proceedings of the National Academy of Sciences U S A 100: 10118-10123.

Grant, W.F., 1994. The present status of higher-plant bioassays for the detection of environmental mutagens. Mutation Research 310: 175-185.

Grant, W.F., 1999. Higher plant assays for the detection of chromosomal aberrations and gene mutations - a brief historical background on their use for screening and monitoring environmental chemicals. Mutation Research 426: 107-112.

Gratão, P.L., Monteiro, C.C., Rossi, M.L., Martinelli, A.P., Peres, L.E., Medici, L.O., Lea, P.J., Azevedo, R.A., 2009. Differential ultrastructural changes in tomato hormonal mutants exposed to cadmium. Environmental and Experimental Botany 67 (2): 387-394.

Green, I.D., Jeffries, C., Diaz, A., Tibbett, M. 2006. Contrasting behaviour of cadmium and zinc in a soil-plant-arthropod system. Chemosphere 64(7): 1115-21.

Green, I.D., Tibbett, M., 2008. Differential uptake, partitioning and transfer of Cd and Zn in the soil-pea plant-aphid system. Environmental Science and Technology 42(2): 450-5.

Greger, M., 1999. Metal availability and bioconcentration in plants. In: Prasad, M. N. V., Hagemeyer, J., (Eds.). Heavy metal stress in plants - from molecules to ecosystems. Springer-Verlag, Berlin, Germany. pp. 1-27.

GuhaMajumdar, M., Dawson-Baglien, E., Sears, B.B., 2008. Creation of a chloroplast microsatellite reporter for detection of replication slippage in *Chlamydomonas reinhardtii*. Eukaryotic Cell 7(4): 639-46.

Hall, J.L., 2002. Cellular mechanisms for heavy metal detoxification and tolerance. *Journal of Experimental Botany* 53: 1-11.

Han, S., Lee, J., Oh, C., Kim, P., 2006. Alleviation of Cd toxicity by composted sewage sludge in Cd-treated Schmidt birch (*Betula schmidtii*) seedlings. Chemosphere 65 (4): 541-546.

Hartwig, A., 1994. Role of DNA-repair inhibition in lead-induced and cadmium-induced genotoxicity - a review. Environmental Health Perspect 102: 45-50.

He, J., Ren, Y., Zhu, C., Yan, Y., Jiang, D., 2008. Effect of Cd on growth, photosynthetic gas exchange, and chlorophyll fluorescence of wild and Cd-sensitive mutant rice. Photosynthetica 46: 466-470.

He, Q.B., Singh, B.R., 1993. Effect of organic-matter on the distribution, extractability and uptake of cadmium in soils. *Journal of Soil Science* 44:641-650.

Heyno, E., Klose, C., Krieger-Liszkay, A., 2008. Origin of cadmium-induced reactive oxygen species production: mitochondrial electron transfer versus plasma membrane NADPH oxidase. New Phytologist 179(3): 687-99.

Horemans, N., Raeymaekers, T., Van Beek, K., Nowocin, A., Blust, R., Broos, K., Cuypers, A., Vangronsveld, J., Guisez, Y., 2007. Dehydroascorbate uptake is impaired in the early response of *Arabidopsis* plant cell cultures to cadmium. *Journal of Experimental Botany* 58(15-16): 4307-17.

Hossain, Z., Huq, F., 2002. Studies on the interaction between Cd^{2+} ions and nucleobases and nucleotides. *Journal of Inorganic Biochemistry* 90: 97-105.

Huang, B., Kuo, S., Bembenek, R., 2003. Cadmium uptake by lettuce from soil amended with phosphorus and trace element fertilizers. Water Air Soil Pollution 147: 109-127.

IARC, (1993). Beryllium, cadmium, mercury, and exposures in the glass manufacturing industry: IARC Monogr Eval Carcinog Risks Hum, Lyon, France

Jacoby, B., Moran, N., 2002. Mineral Nutrient Transport in Plants In. Handbook of Plant Crop Physiology, Marcel Dekker AG, Switzerland. pp 337-361.

Jahangir, T., Khan, T.H., Prasad, L., Sultana, S., 2006. Reversal of cadmium chloride-induced oxidative stress and genotoxicity by *Adhatoda vasica* extract in Swiss albino mice. Biological Trace Element Research 111(1-3): 217-28.

Jambhulkar, H.P., Juwarkar, A.A., 2009. Assessment of bioaccumulation of heavy metals by different plant species grown on fly ash dump. Ecotoxicology and Environmental Safety 72(4):1122-8.

Jiménez-Ambriz, G., Petit, C., Bourrié, I., Dubois, S., Olivieri, I., Ronce, O., 2007. Life history variation in the heavy metal tolerant plant *Thlaspi caerulescens* growing in a network of contaminated and non contaminated sites in southern France: role of gene flow, selection and phenotypic plasticity. New Phytologist 173(1):199-215.

Jin, Y.H., Clark, A.B., Slebos, R.J.C., Al-Refai, H., Taylor, J.A., Kunkel, T.A., Resnick, M.A., Gordenin, D.A., 2003. Cadmium is a mutagen that acts by inhibiting mismatch repair. Nature Genetics 34: 326-329.

John, D.A., Leventhal, J.S., 1996. Bioavailability of metals. In: du Bray, E. A., (Eds.). Preliminary compilation of descriptive geoenvironmental mineral deposit models. U.S. Department of the Interior, USA. pp. 10-18.

Jones, C.J., Edwards, K.J., Castaglione, S., Winfield, M.O., Sala, F., vandeWiel, C., Bredemeijer, G., Vosman, B., Matthes, M., Daly, A., Brettschneider, R., Bettini, P., Buiatti, M., Maestri, E., Malcevschi, A., Marmiroli, N., Aert, R., Volckaert, G., Rueda, J., Linacero, R., Vazquez, A. Karp, A., 1997. Reproducibility testing of RAPD, AFLP and SSR markers in plants by a network of European laboratories. Molecular Breeding 3: 381-390.

Juhel, G., O'Halloran, J., Culloty, S.C., O'riordan, R.M., Davenport, J., O'Brien, N.M., James, K.F., Furey, A., Allis, O. 2007. In vivo exposure to microcystins induces DNA damage in the haemocytes of the zebra mussel, *Dreissena polymorpha,* as measured with the comet assay. Environmental and Molecular Mutagenesis 48(1): 22-9.

Jung, M.C., Thornton, I., 1996. Heavy metal contamination of soils and plants in the vicinity of a lead-zinc mine, Korea. Applied Geochemistry 11: 53-59.

Knasmuller, S., Gottmann, E., Steinkellner, H., Fomin, A., Pickl, C., Paschke, A., God, R., Kundi, M., 1998. Detection of genotoxic effects of heavy metal contaminated soils with plant bioassays. Mutation Research 420: 37-48.

Koppen, G., Verschaeve, L., 1996. The alkaline comet test on plant cells: a new genotoxicity test for DNA strand breaks in Vicia faba root cells. Mutation Research 360(3): 193-200.

Kovalchuk, O., Dubrova, Y.E., Arkhipov, A., Hohn, B.. Kovalchuk, I., 2000. Germline DNA - Wheat mutation rate after Chernobyl. Nature 407: 583-584.

Krantev, A., Yordanova, R., Janda, T., Szalai, G., Popova, L., 2008. Treatment with salicylic acid decreases the effect of cadmium on photosynthesis in maize plants. *Journal of Plant Physiology* 165: 920–931.

Krupa, Z., Moniak, M., 1998. The stage of leaf maturity implicates the response of the photosynthetic apparatus to cadmium toxicity. Plant Science 138: 149-156.

Krupa, Z., Öquist, G., Huner, N.P.A., 1993. The effects of cadmium on photosynthesis of *Phaseolus vulgaris* - a fluorescence analysis. Physiologia Plantarum 88: 626–630.

Krupa, Z., Siedlecka, A., Skórzynska-Polit Maksymiek, W., 2002. Heavy metal interactions with plant nutrients. In: Prasad, M. N. V., Strzalka, K., (Eds.). Physiology and biochemistry of metal toxicity and tolerance in plants. Kluwer Academic Publishers, Dordrecht, Netherlands. pp. 287-301.

Kuzmick, D.M., Mitchelmore, C.L., Hopkins, W.A., Rowe, C.L., 2007. Effects of coal combustion residues on survival, antioxidant potential, and genotoxicity resulting from full-lifecycle exposure of grass shrimp (*Palaemonetes pugio* Holthius). Science Total Environment 373(1): 420-30.

Lane, T.W., Morel, F.O.M.M., 2000. A biological function for cadmium in marine diatoms. Proceedings of the National Academy of Sciences U S A 97: 4627-4631.

Lane, T.W., Saito, M.A., George, G.N., Pickering, I.J., Prince, R.C. Morel, F.M.M., 2005. Biochemistry: A cadmium enzyme from a marine diatom. Nature 435: 42-42.

Larbi, A., Morales, F., Abadía, A., Gogorcena, Y., Lucena, J.J., Abadía, J., 2002. Effects of Cd and Pb in sugar beet plants grown in nutrient solution: induced Fe deficiency and growth inhibition. Functional Plant Biology 29: 1453–1464.

Lee, S., Ahsan, N., Lee, K.W., Kim, D.H., Lee, D.G., Kwak, S.S., Kwon, S.Y., Kim, T.H., Lee, B.H., 2007. Simultaneous overexpression of both CuZn superoxide dismutase and ascorbate peroxidase in transgenic tall fescue plants confers increased tolerance to a wide range of abiotic stresses. *Journal of Plant Physiology* 164(12): 1626-38.

Li, L., He, Z., Pandey, G.K., Tsuchiya, T., Luan, S., 2002. Functional cloning and characterization of a plant efflux carrier for multidrug and heavy metal detoxification. *Journal of Biology and Chemistry* 277(7):5360-8.

Lin, A.J., Zhang, X.H., Chen, M.M., Cao, Q., 2007. Oxidative stress and DNA damages induced by cadmium accumulation. *Journal of Environmental Science (China)* 19(5): 596-602.

Lin, A.J., Zhu, Y.G., Tong, Y.P., Geng, C.N., 2005. Evaluation of genotoxicity of combined pollution by cadmium and atrazine. Bulletin of Environmental Contamination and Toxicology 74(3): 589-96.

Linger, P., Ostwald, A., Haensler, J., 2005. *Cannabis sativa* L. growing on heavy metal contaminated soil: growth, cadmium uptake and photosynthesis. Biologia Plantarum 49: 567-576.

Liu, H., Liao, B., Lu, S., 2004. Toxicity of surfactant, acid rain and Cd2+ combined pollution to the nucleus of *Vicia faba* root tip cells. Ying Yong Sheng Tai Xue Bao 15(3):493-6.

Liu, J., Qian, M., Cai, G., Zhu, Q., Wong, M.H., 2007. Variations between rice cultivars in root secretion of organic acids and the relationship with plant cadmium uptake. Environmental Geochemistry and Health 29(3):189-95.

Liu, M.Q., Yanai, J., Jiang, R.F., Zhang, F., McGrath, S.P., Zhao, F.J., 2008. Does cadmium play a physiological role in the hyperaccumulator *Thlaspi caerulescens*? Chemosphere 71(7):1276-83.

Liu, W., Li, P.J., Qi, X.M., Zhou, Q.X., Zheng, L., Sun, T.H., Yang, Y.S., 2005. DNA changes in barley (*Hordeum vulgare*) seedlings induced by cadmium pollution using RAPD analysis. Chemosphere 61(2):158-67.

Liu, W., Yang, Y.S., Li, P.J., Zhou, Q.X., Xie, L.J., Han, Y.P. 2009. Risk assessment of cadmium-contaminated soil on plant DNA damage using RAPD and physiological indices. *Journal of Hazardous Material* 161(2-3):878-83.

Lombi, E., Zhao, F.J., Dunham, S.J. McGrath, S.P., 2000. Cadmium accumulation in populations of *Thlaspi caerulescens* and *Thlaspi goesingense*. New Phytologist 145: 11-20.

López-Millán, A-F., Sagardoy, R., Solanas, M., Abadía, A., Abadía, J., 2009. Cadmium toxicity in tomato (*Lycopersicon esculentum*) plants grown in hydroponics. Environmental and Experimental Botany 65: 376–385.

Loureiro, J., Rodriguez, E., Doležel, J., Santos, C., 2006. Flow cytometric and microscopic analysis of the effect of tannic acid on plant nuclei and estimation of DNA content. Annals of Botany 98:515–527.

Ma, J.F., Nagao, S., Sato, K., Ito, H., Furukawa, J., Takeda, K., 2004. Molecular mapping of a gene responsible for Al-activated secretion of citrate in barley. *Journal of Experimental Botany* (401):1335-41.

Ma, J.F., Ueno, D., Zhao, F.J., McGrath, S.P., 2005. Subcellular localisation of Cd and Zn in the leaves of a Cd-hyperaccumulating ecotype of *Thlaspi caerulescens*. Planta 220: 731-736.

Maier, E.A., Matthews, R.D., McDowell, J.A., Walden, R.R., Ahner, B.A., 2003. Environmental cadmium levels increase phytochelatin and glutathione in lettuce grown in a chelator-buffered nutrient solution. *Journal of Environmental Quality* 32: 1356-1364.

Malik, D., Sheoran, I.S., Singh, R., 1992. Carbon metabolism in leaves of cadmium treated wheat seedlings. Plant Physiology and Biochemistry 30: 223-229.

Mallick, N., Mohn, F.H., 2003. Use of chlorophyll fluorescence in metal-stress research: a case study with green microalga *Scenedesmus*. Ecotoxicology Environment and Safety 55:64–9.

Martnoia, E., Maeshima, M., Neuhaus, H.E., 2007. Vacuolar transporters and their essential role in plant metabolism. Jounal of Experimental Botany 58(1): 83-102.

Mathews, T., Fisher, N.S., 2008. Evaluating the trophic transfer of cadmium, polonium, and methylmercury in an estuarine food chain. Environmental Toxicology and Chemistry 27(5):1093-101.

Matos, M., Camacho, M.V., Pérez-Flores, V., Pernaute, B., Pinto-Carnide, O., Benito, C., 2005. A new aluminum tolerance gene located on rye chromosome arm 7RS. Theoretical and Applied Genetics 111(2): 360-9.

McLaughlin, M.J., 2002. Bioavailability of metals to terrestrial plants. In: Allen, H. E., (Eds.). Bioavailability of metals in terrestrial ecosystems: importance of partitioning for bioavailability to invertebrates, microbes, and plants. Society of Environmental Toxicology and Chemistry (SETAC), Lawrence, USA. pp. 39-68.

McLaughlin, M.J., Whatmuff, M., Warne, M., Heemsbergen, D., Barry, G., Bell, M., Nash, D., Pritchard, D., 2006. A field investigation of solubility and food chain accumulation of biosolid-cadmium across diverse soil types. Environmental Chemistry 3: 428-432.

McMurphy, L.M., Rayburn, A.L., 1993. Nuclear alterations of maize plants grown in soil contaminated with coal fly ash. Archives of Environmental Contamination and Toxicology 25: 520-524.

Memon, A.R., Schröder, P., 2009. Implications of metal accumulation mechanisms to phytoremediation. Environmental Science and Pollution Research International 16(2):162-75.

Mengoni, A., Barabesi, C., Gonnelli, C., Galardi, F., Gabbrielli, R. Bazzicalupo, M., 2001. Genetic diversity of heavy metal-tolerant populations in *Silene paradoxa* L. (*Caryophyllaceae*): a chloroplast microsatellite analysis. Molecular Ecology 10: 1909-1916.

Merrington, G., Miller, D., McLaughlin, M.J., Keller, M.A., 2001. Trophic barriers to fertilizer Cd bioaccumulation through the food chain: a case study using a plant--insect predator pathway. Archives of Environmental Contamination and Toxicology 41(2):151-6.

Milla, R., Gustafson, J.P., 2001. Genetic and physical characterization of chromosome 4DL in wheat. Genome 44(5): 883-92.

Mishra, K.K., Rai, U.N., Prakash, O., 2007. Bioconcentration and phytotoxicity of Cd in *Eichhornia crassipes*. Environmental Monitoring and Assessment 130(1-3): 237-43.

Mittler, R., 2002. Oxidative stress, antioxidants and stress tolerance. Trends in Plant Science 7: 405-410.

Mobin, M., Khan, N.A., 2007. Photosynthetic activity, pigment composition and antioxidative response of two mustard (*Brassica juncea*) cultivars differing in photosynthetic capacity subjected to cadmium stress. *Journal of Plant Physiology* 164: 601-10.

Monteiro, M.S., Lopes, T., Loureiro, J., Mann, R.M., Santos, C., Soares, A.M.V.M., 2004. An assessment of genetic stability in cadmium-treated *Lactuca sativa* L. by flow cytometry and microsatellite analysis. Fourth SETAC World Congress and 25th Annual Meeting in North America, Portland, Oregon, USA.

Monteiro, M.S., Lopes, T., Mann, R.M., Paiva, C., Soares, A.M., Santos, C., 2009a. Microsatellite instability in *Lactuca sativa* chronically exposed to cadmium. Mutation Research 672(2): 90-4.

Monteiro, M., Santos, C., Mann, R.M., Soares, A.M.V.M., Lopes, T., 2007. Evaluation of cadmium genotoxicity in *Lactuca sativa* L. using nuclear microsatellites. Environmental and Experimental Botany 60: 421-427.

Monteiro, M.S., Rodriguez, E., Loureiro, J., Mann, R.M., Soares, A., Santos, C., 2005. The use of flow cytometry to assess the ploidy stability of cadmium-treated *Lactuca sativa* L.: a potential use of this technique for toxicological studies. Congresso da Sociedade Ibérica de Citometria, Porto, Portugal.

Monteiro, M.S., Santos, C., Soares, A.M., Mann, R.M., 2008. Does subcellular distribution in plants dictate the trophic bioavailability of cadmium to *Porcellio dilatatus* (Crustacea, Isopoda)? Environmental Toxicology and Chemistry 27(12): 2548-56.

Monteiro, M.S., Santos, C., Soares, A.M., Mann, R.M., 2009b. Assessment of biomarkers of cadmium stress in lettuce. Environmental Toxicology and Chemistry 72(3): 811-8.

Mysliwa-Kurdziel, B., Strzalka, K., 2002. Influence of metal on biosynthesis of photosynthetic pigments. In: Prasad, M. N. V., Strzalka, K., (Eds.). Physiology and biochemistry of metal toxicity and tolerance in plants. Kluwer Academic Publishers, Dordrecht, Netherlands. pp. 201-227.

Nelson, W.O., Campbell, P.G.C., 1991. The effects of acidification on the geochemistry of Al, Cd, Pb and Hg in freshwater environments: A literature review. Environmental Pollution 71: 91-130.

Nogawa, K., Honda, R., Kido, T., Tsuritani, I., Yamada, Y., Ishizaki, M. Yamaya, H., 1989. A dose-response analysis of cadmium in the general environment with special reference to total cadmium intake limit. Environmental Research 48: 7-16.

OECD, 2006. Terrestrial plant test: seedling emergence and seedling growth test. OECD-Organisation for Economic Co-operationand Development, Paris, pp 208–221.

Ohshima, S., 2003. Induction of genetic instability and chromo nickel sulfate in V79 Chinese hamster cells. Mutagenesis 18: 133-137.

Ortega-Villasante, C., Hernández, L.E., Rellán-Alvarez, R., Del Campo, F.F., Carpena-Ruiz, R.O., 2007. Rapid alteration of cellular redox homeostasis upon exposure to cadmium and mercury in alfalfa seedlings. New Phytologist 176(1): 96-107.

Otto, F.J., Oldiges, H., 1980. Flow cytogenetic studies in chromosomes and whole cells for the detection of clastogenic effects. Cytometry 1: 13-17.

Otto, F.J., Oldiges, H., Göhde, W., Jain, V.K., 1981. Flow cytometric measurement of nuclear DNA content variations as a potential in vivo mutagenicity test. Cytometry 2: 189 – 191.

Ouariti, O., Boussama, N., Zarrouk, M., Cherif, A., Ghorbal, M.H., 1997. Cadmium- and copper-induced changes in tomato membrane lipids. Phytochemistry 45: 1343-1350.

Paiva, C., 2008. Avaliação da genotoxicidade do cádmio em duas espécies de *Thlaspi*. MSc Thesis, Universidade de Aveiro, Aveiro, pp. 59.

Pal'ove-Balang, A., Kisová, J., Pavlovkin, I., Mistrík, 2006. Effect of manganese on cadmium toxicity in maize seedlings. Plant Soil Environment 5(4): 143–149.

Palus, J., Rydzynski, K., Dziubaltowska, E., Wyszynska, K., Natarajan, A.T., Nilsson, R., 2003. Genotoxic effects of occupational exposure to lead and cadmium. Mutation Research 540(1): 19-28.

Panda, B.B., Panda, K.K., 2002. Genotoxicity and mutagenicity of metals in plants. In: Prasad, M.N.V., Strzalka, K., (Eds.). Physiology and Biochemistry of Metal Toxicity and Tolerance in Plants. Kluwer Academic Publishers, Dordrecht, Netherlands. pp. 395-414.

Panda, K.K., Patra, J., Panda, B.B., 1997. Persistence of cadmium-induced adaptive response to genotoxicity of maleic hydrazide and methyl mercuric chloride in root meristem cells of *Allium cepa* L.: differential inhibition by cycloheximide and buthionine sulfoximine. Mutation Research 389(2-3): 129-39.

Penner, G.A., Bezte, L.J., Leisle, D., Clarke, J., 1995. Identification of RAPD markers linked to a gene governing cadmium uptake in durum wheat. Genome 38(3): 543-7.

Perceval, O., Couillard, Y., Pinel-Alloul, B. Campbell, P.G.C., 2006. Linking changes in subcellular cadmium distribution to growth and mortality rates in transplanted freshwater bivalves (*Pyganodon grandis*). Aquatic Toxicology 79: 87-98.

Peris, M., Mico, C., Recatala, L., Sanchez, R., Sanchez, J., 2007. Heavy metal contents in horticultural crops of a representative area of the European Mediterranean region. Science of Total Environment 378: 42-48.

Pietrini, F., Iannelli, M.A., Pasqualini, S., Massacci A., 2003. Interaction of Cadmium with Glutathione and Photosynthesis in Developing Leaves and Chloroplasts of *Phragmites australis* (Cav.) Trin. ex Steudel. Plant Physiology 133: 829–837.

Powell, W., Morgante, M., Andre, C., Hanafey, M., Vogel, J., Tingey, S., Rafalski, A., 1996. The comparison of RFLP, RAPD, AFLP and SSR (microsatellite) markers for germplasm analysis. Molecular Breeding 2: 225-238.

Prasad, M.N.V., 1995. Cadmium toxicity and tolerance in vascular plants. Environmental and Experimental Botany 35: 525-545.

Radetski, C.M., Ferrari, B., Cotelle, S., Masfaraud, J.F., Ferard, J.F., 2004. Evaluation of the genotoxic, mutagenic and oxidant stress potentials of municipal solid waste incinerator bottom ash leachates. Science of Total Environment 333(1-3): 209-16.

Rainbow, P.S., 2002. Trace metal concentrations in aquatic invertebrates: why and so what? Environmental Pollution 120: 497-507.

Rank, J., Nielsen, M.H., 1998 .Genotoxicity testing of wastewater sludge using the *Allium cepa* anaphase-telophase chromosome aberration assay. Mutation Research 418(2-3): 113-9.

Rayburn, A.L., Pedersen, W.L., McMurphy, L.M., 1993. The fungicide captan reduces nuclear DNA content in maize seedlings. Pesticide Science 37: 79-82.

Razinger, J., Drinovec, L., Zrimec, A., 2009. Real-time visualization of oxidative stress in a floating macrophyte *Lemna minor* L. exposed to cadmium, copper, menadione, and AAPH. Environmental Toxicology 23.

Reeves, R.D., Baker, A.J.M., 2000. Metal-accumulating plants. In Raskin, I., Ensley, B.D. (Eds.). Phytoremediation of Toxic Metals: Using Plants to Clean up the Environment. John Wiley and Sons Inc., New York, NY, USA. pp. 193–229.

Rivasseau, C., Seemann, M., Boisson, A.M., Streb, P., Gout, E., Douce, R., Rohmer, M., Bligny, R., 2009. Accumulation of 2-C-methyl-D-erythritol 2,4-cyclodiphosphate in illuminated plant leaves at supraoptimal temperatures reveals a bottleneck of the prokaryotic methylerythritol 4-phosphate pathway of isoprenoid biosynthesis. Plant Cell Environment 32(1):82-92.

Robards, K., Worsfold, P., 1991. Cadmium: toxicology and analysis - a review. Analyst 116: 549-568.

Rodriguez, E., Fernandes, P., Azevedo, R., Fernandes, J., Araújo, A., Santos, C., 2010. Genotoxic and cytotoxic evaluation of *Pisum sativum* plants exposed to Chromium. Europe 20th Annual Meeting of the Society of Environmental Toxicology and Chemistry (SETAC). Sevilla.

Rodríguez-Serrano, M., Romero-Puertas, M.C., Pazmiño, D.M., Testillano, P.S., Risueño, M.C., Del Río, L.A., Sandalio, L.M., 2009. Cellular response of pea plants to cadmium toxicity: cross talk between reactive oxygen species, nitric oxide, and calcium. Plant Physiology 150(1): 229-43.

Rodríguez-Serrano, M., Romero-Puertas, M.C., Zabalza, A., Corpas, F.J., Gómez, M., Del Río, L.A., Sandalio, L.M., 2006. Cadmium effect on oxidative metabolism of pea (*Pisum sativum* L.) roots. Imaging of reactive oxygen species and nitric oxide accumulation in vivo. Plant Cell and Environment 29(8): 1532-44.

Romero-Puertas, M.C., Corpas, F.J., Rodríguez-Serrano, M., Gómez, M., Del Río L.A., Sandalio, L.M., 2007. Differential expression and regulation of antioxidative enzymes by cadmium in pea plants. *Journal of Plant Physiology* 164(10): 1346-57.

Rosa, E.V., Valgas, C., Souza-Sierra, M.M., Corrêa, A.X., Radetski, C.M., 2003. Biomass growth, micronucleus induction, and antioxidant stress enzyme responses in *Vicia faba* exposed to cadmium in solution. Environmental Toxicology and Chemistry 22(3): 645-9.

Rosas, I., Carbajal, M.E., Gómez-Arroyo, S., Belmont, R., Villalobos-Pietrini, R., 1984. Cytogenetic effects of cadmium accumulation on water hyacinth (*Eichhornia crassipes*). Environmental Research 33(2):386-95.

Rubinelli, P., Siripornadulsil, S., Gao-Rubinelli, F., Sayre, R.T., 2002. Cadmium- and iron-stress-inducible gene expression in the green alga *Chlamydomonas reinhardtii*: evidence for H43 protein function in iron assimilation. Planta 215(1):1-13.

Salt, D.E., Wagner, G.J., 1993. Cadmium transport across tonoplast of vesicles from oat roots. Evidence for a Cd^{2+}/H^{+} antiport activity. *Journal of Biology and Chemistry* 268: 12297-12302.

Sandalio, L., Dalurzo, H., Gomes, M., Romero-Puertas, M., del Rio, L., 2001. Cadmium-induced changes in the growth and oxidative metabolism of pea plants. *Journal of Experimental Botanny* 52: 2115–2126.

Sanitá di Toppi, L., Gabbrielli, R., 1999. Response to cadmium in higher plants. Environmental and Experimental Botany 41: 105-130.

Sankaran, R.P., Ebbs, S.D., 2008. Transport of Cd and Zn to seeds of Indian mustard (*Brassica juncea*) during specific stages of plant growth and development. Physiologia Plantarum 132(1):69-78.

Schützendübel, A., Polle, A., 2002. Plant responses to abiotic stresses: heavy metal-induced oxidative stress and protection by mycorrhization. *Journal of Experimental Botany* 53: 1351-1365.

Shasany, A. K., Darokar, M. P., Dhawan, S., Gupta , A. K., Gupta, S., Shukla, A. K., Patra, N. K., Khanuja, S. P. S. 2005. Use of RAPD and AFLP Markers to Identify Inter- and Intraspecific Hybrids of *Mentha*. *Journal of Heredity Advance Access* 96: 542-549.

Sheoran, I.S., Singal, H.R., Singh, R., 1990. Effect of cadmium and nickel on photosynthesis and enzymes of the photosynthetic carbon reduction cycle in pigeon pea (*Cajanus cajan* L.). Photosynthesis Research 23: 345–351.

Shigaki, T., Barkla, B.J., Miranda-Vergara, M.C., Zhao, J., Pantoja, O., Hirschi, K.D.J., 2005. Identification of a crucial histidine involved in metal transport activity in the *Arabidopsis* cation/H+ exchanger CAX1. Chemical Biology 280(34):30136-42.

Siedlecka, A., Baszynski, T., 1993. Inhibition of electron flow around photosystem I in chloroplasts of Cd-treated maize is due to Cd-induced iron deficiency. Physiologia Plantarum 87: 199–202.

Siedlecka, A. Krupa, Z., 1996. Interaction between cadmium and iron and its effects on photosynthetic capacity of primary leaves of *Phaseolus vulgaris*. Plant Physiology and Biochemistry 34: 833-841.

Siedlecka, A., Krupa, Z., Samuelsson, G., Öquist, G., Gardeström, P., 1997. Primary carbon metabolism in *Phaseolus vulgaris* plants under Cd/Fe interaction. Plant Physiology and Biochemistry 35: 951–957.

Singh, P.K., Tewari, R.K., 2003. Cadmium toxicity induced changes in plant water relations and oxidative metabolism of *Brassica juncea* L. plants. *Journal of Environmental Biology* 24(1):107-12.

Singh, R.P., Agrawal, M., 2007. Effects of sewage sludge amendment on heavy metal accumulation and consequent responses of *Beta vulgaris* plants. Chemosphere 67: 2229-2240.

Skorzynska, E., Baszynski, T., 1993. The changes in PSII complex polypeptides under cadmium treatment – are they of direct or indirect nature? Acta Physiologia Plantarum 15(4): 263–269.

Slebos, R.J.C., Li, M., Evjen, A.N., Coffa, J., Shyr, Y. Yarbrough, W.G., 2006. Mutagenic effect of cadmium on tetranucleotide repeats in human cells. Mutation Research 602; 92-99.

Speir, T.W., Van Schaik, A.P., Percival, H.J., Close, M.E. Pang, L.P., 2003. Heavy metals in soil, plants and groundwater following high-rate sewage sludge application to land. Water Air Soil Pollution 150: 319-358.

Steen Redeker, E., van Campenhout, K., Bervoets, L., Reijnders, H. Blust, R., 2007. Subcellular distribution of Cd in the aquatic oligochaete *Tubifex tubifex*, implications for trophic availability and toxicity. Environmental Pollution 148: 166-175.

Steffens, J.C., 1990. The Heavy Metal-Binding Peptides of Plants. Annual Review of Plant Physiology and Plant Molecular Biology 41: 553-575.

Steinkellner, H., Mun-Sik, K., Helma, C., Ecker, S., Ma, T.H., Horak, O., Kundi, M. Knasmüller, S., 1998. Genotoxic effects of heavy metals: Comparative investigation with plant bioassays. Environmental and Molecular Mutagenesis 31: 183-191.

Sung-Hyun, K., Jin-Sung, P., In-Sook, L., 2009. Characterization of Cadmium-Binding Ligands from Roots of *Echinochloa crusgalli* var. frumentacea. *Journal of Plant Biology* 52:167–170.

Sunkar, R., Bartels, D., Kirch, H.H., 2003. Overexpression of a stress-inducible aldehyde dehydrogenase gene from *Arabidopsis thaliana* in transgenic plants improves stress tolerance. *Plant Journal* 35(4):452-64.

Suzuki, N., Yamaguchi, Y., Koizumi, N., Sano, H., 2002. Functional characterization of a heavy metal binding protein CdI19 from arabidopsis. *The Plant Journal* 32(2):165-73.

Tamás, L., Dudíková, J., Durceková, K., Halusková, L., Huttová, J., Mistrík, I., 2009. Effect of cadmium and temperature on the lipoxygenase activity in barley root tip. Protoplasma 235(1-4):17-25.

Tanhuanpää, P., Kalendar, R., Schulman, A.H., Kiviharju, E., 2007. A major gene for grain cadmium accumulation in oat (*Avena sativa* L.). Genome 50(6):588-94.

Thomine, S., Wang, R., Ward, J.M., Crawford, N.M., Schroeder, J.I., 2000. Cadmium and iron transport by members of a plant metal transporter family in Arabidopsis with homology to Nramp genes. Proceedings of the National Academy of Sciences 97(9):4991-6.

Tiryakioglu, M., Eker, S., Ozkutlu, F., Husted, S., Cakmak, I., 2006. Antioxidant defence system and cadmium uptake in barley genotypes differing in cadmium tolerance. *Journal of Trace Elements in Medicine and Biology* 20(3):181-9.

Tommasini, R., Vogt, E., Fromenteau, M., Hörtensteiner, S., Matile, P., Amrhein, N., Martinoia, E., 1998. An ABC-transporter of *Arabidopsis thaliana* has both glutathione-conjugate and chlorophyll catabolite transport activity. *The Plant Journ*al 13(6):773-80.

Ueno, D., Ma, J.F., Iwashita, T., Zhao, F.J., McGrath, S.P., 2005. Identification of the form of Cd in the leaves of a superior Cd-accumulating ecotype of *Thlaspi caerulescens* using Cd-113-NMR. Planta 221: 928-936.

Unyayar, S., Celik, A., Cekiç, F.O., Gözel, A., 2006. Cadmium-induced genotoxicity, cytotoxicity and lipid peroxidation in *Allium sativum* and *Vicia faba*. Mutagenesis 21(1):77-81.

Vaipayee, P., Dhawan, A., Shanker, R., 2006. Evaluation of the alkaline comet assay conducted with the wetlands plant *Bacopa monnieri* L. as a model for ecogenotoxicity assessment. Environmental and Molecular Mutagenetics 47: 483-489.

Valverde, M., Trejo, C., Rojas, E., 2001. Is the capacity of lead acetate and cadmium chloride to induce genotoxic damage due to direct DNA-metal interaction? Mutagenesis 16: 265-270.

van de Mortel, J.E., Schat, H., Moerland, P.D., Ver Loren van Themaat, E., van der Ent, S., Blankestijn, H., Ghandilyan, A., Tsiatsiani, S., Aarts, M.G., 2008. Expression differences for genes involved in lignin, glutathione and sulphate metabolism in response to cadmium in *Arabidopsis thaliana* and the related Zn/Cd-hyperaccumulator *Thlaspi caerulescens*. Plant Cell and Environment 31(3):301-24.

van Rossum, F., Bonnin, I., Fenart, S., Pauwels, M., Petit, D., Saumitou-Laprade, P., 2004. Spatial genetic structure within a metallicolous population of *Arabidopsis halleri*, a clonal, self-incompatible and heavy-metal-tolerant species. Molecular Ecology 13(10):2959-67.

Vassilev, A., Yordanov, I., 1997. Reductive analysis of factors limiting growth of cadmium-treated plants: a review. *Bulgarian Journal of Plant Physiology* 23: 114-133.

Vassilev, A., Vangronsveld, J., Yordanov, I., 2002. Cadmium phytoextraction: present state, biological backgrounds and research needs. *Bulgarian Journal of Plant Physiology* 28(3–4): 68–95.

Vijver, M.G., Koster, M., Peijnenburg, W.J.G.M., 2007. Impact of pH on Cu accumulation kinetics in earthworm cytosol. Environmental Science Technology 41: 2255-2260.

Vijver, M.G., van Gestel, C.A.M., Lanno, R.P., van Straalen, N.M. Peijnenburg, W.J.G.M., 2004. Internal metal sequestration and its ecotoxicological relevance: a review. Environmental Science Technology 38: 4705-4712.

Vijver, M.G., van Gestel, C.A.M., van Straalen, N.M., Lanno, R.P. Peijnenburg, W.J.G.M., 2006. Biological significance of metals partitioned to subcellular fractions within earthworms (*Aporrectodea caliginosa*). Environmental Toxicological Chemistry 807-814.

Villatoro-Pulido, M., Font, R., De Haro-Bravo, M.I., Romero-Jiménez, M., Anter, J., De Haro Bailón, A., Alonso-Moraga, A., Del Río-Celestino, M., 2009. Modulation of genotoxicity and cytotoxicity by radish grown in metal-contaminated soils. Mutagenesis 24(1):51-7.

Vogeli-Lange, R., Wagner, G.J., 1990. Subcellular localization of cadmium and cadmium-binding peptides in tobacco leaves: implication of a transport function for cadmium-binding peptides. Plant Physiology 92: 1086-1093.

Waalkes, M.P., 2003. Cadmium carcinogenesis. Mutation Research 533: 107-120.

Wagner, G.J., 1993. Accumulation of cadmium in crop plants and its consequences to human health. Advances in Agronomy 51: 173-212.

Wallace, W.G., Lee, B.G., Luoma, S.N., 2003. Subcellular compartmentalization of Cd and Zn in two bivalves. I. Significance of metal-sensitive fractions (MSF) and biologically detoxified metal (BDM). Marine Ecology Progress Series 249: 183-197.

Wallace, W.G., Lopez, G.R., 1997. Bioavailability of biologically sequestered cadmium and the implications of metal detoxification. Marine Ecology Progress Series 147: 149-157.

Wallace, W.G., Lopez, G.R. Levinton, J.S., 1998. Cadmium resistance in an oligochaete and its effect on cadmium trophic transfer to an omnivorous shrimp. Marine Ecology Progress Series 172: 225-237.

Wallace, W.G., Luoma, S.N., 2003. Subcellular compartmentalization of Cd and Zn in two bivalves. II. Significance of trophically available metal (TAM). Marine Ecology Progress Series 257: 125-137.

Wang, W.-X., Fisher, N.S., 1999. Assimilation efficiencies of chemical contaminants in aquatic invertebrates: a synthesis. Environmental Toxicology and Chemistry 18: 2034-2045.

Wang, W.X., Rainbow, P.S., 2006. Subcellular partitioning and the prediction of cadmium toxicity to aquatic organisms. Environmental Chemistry 3: 395-399.

Watanabe, T., Misawa, S., Hiradate, S., Osaki, M., 2008. Root mucilage enhances aluminium accumulation in *Melastoma malabathricum*, an aluminium accumulator. Plant Signal Behaviour 3(8):603-5.

Xie, L., Funk, D.H., Buchwalter, D.B., 2009. Trophic transfer of Cd from natural periphyton to the grazing mayfly *Centroptilum triangulifer* in a life cycle test. Environmental Pollution.

Xu, J., Chai, T., Zhang, Y., Lang, M., Han, L.., 2009. The cation-efflux transporter BjCET2 mediates zinc and cadmium accumulation in *Brassica juncea* L. leaves. Plant and Cell Report 28(8):1235-42.

Yu, R.Q., Fleeger, J.W., 2006. Effects of nutrient enrichment, depuration substrate, and body size on the trophic transfer of cadmium associated with microalgae to the benthic amphipod *Leptocheirus plumulosus*. Environmental and Toxicological Chemistry 25(11):3065-72.

Zenk, M.H., 1996. Heavy metal detoxification in higher plants: a review. Gene 179: 21-30.

Zhang, C., Qiu, B.S., 2007. Reactive oxygen species metabolism during the cadmium hyperaccumulation of a new hyperaccumulator *Sedum alfredii* (Crassulaceae). *Journal of Environmental Science (China)* 19(11):1311-7.

Zhang, Y., Xiao, H., 1998. Antagonistic effect of calcium, zinc and selenium against cadmium induced chromosomal aberrations and micronuclei in root cells of *Hordeum vulgare*. Mutation Research 420(1-3):1-6.

Zhang, Y., Yang, X., 1994. The toxic effects of cadmium on cell division and chromosomal morphology of *Hordeum vulgare*. Mutation Research 312(2):121-6.

Zhang, X.H., Lin, A.J., Su, Y.H., Zhu, Y.G., 2006. DNA damages and apoptosis induced by cd in the leaves of horsebean *Vicia faba*. Huan Jing Ke Xue 27(4):787-93.

Zhao, F.J., Lombi, E., McGrath, S.P., 2003. Assessing the potential for zinc and cadmium phytoremediation with the hyperaccumulator *Thlaspi caerulescens*. Plant Soil 249: 37-43.

Zhu, R., Macfie, S.M., Ding, Z., 2005 Cadmium-induced plant stress investigated by scanning electrochemical microscopy. *Journal of Experimental Botany* 56 (421), 2831–2838.

Zhuang, P., Zou, H., Shu, W., 2009. Biotransfer of heavy metals along a soil-plant-insect-chicken food chain: field study. *Journal of Environmental Science (China)* 21(6):849-53.

Zienolddiny, S., Svendsrud, D.H., Ryberg, D., Mikalsen, A.B., Haugen, A., 2000. Nickel (II) induces microsatellite mutations in human lung cancer cell lines. Mutation Research 452: 91-100.

In: Horticulture in the 21st Century
Editors: Abigail N. Sampson, pp. 41-79
ISBN: 978-1-61668-582-9

Chapter 2

BREEDING SWEET CORN: EATING QUALITY VS. AGRONOMIC PERFORMANCE

Bernardo Ordás, Amando Ordás and Pedro Revilla

1. INTRODUCTION

1.1. History and Germplasm

In many parts of the world, cultivars of maize (*Zea mays* L.) developed and grown primarily for grain are harvested early as a vegetable, the kernels being eaten when they are immature (Tracy 1999). However, the term "sweet corn" is reserved for a specific type of maize that differs from normal maize by one or more recessive alleles that alter the carbohydrate pattern of the endosperm and confer some desirable characteristics to the kernel from the point of view of the consumer.

Historically, the first mutant used since the XIX century for sweet corn has been the *sugary1* (*su1*) allele. In the 50's, the mutant *shrunken2* (*sh2*) was introduced and in the 70's the gene *sugary enhancer1* (*se1*) appeared. These three mutants are the backbone of the sweet corn industry in temperate zones. Later on, we will briefly give a glance to other genes.

The allele *su1* produces an increment in the sugar content, although, contrary to its name, the increment is not too high. What gives it really its typical flavor is the accumulation of phytoglycogen, a water soluble polysaccharide (WSP) that produces a creamy texture much appreciated by consumers. In *su1* endosperm, sugar, after its concentration reaches a maximum, is converted to WSP both preharvest and postharvest resulting in a decrease of sweetness over time so there is a continuous loss of sugar from the maximum, which is attained around 18–20 days after pollination (Marshall and Tracy 2003). Sucrose levels decline by 50% or more form the prime eating stage 28 days after pollination (Creech 1968). The time span for optimal harvest of *su1* sweet corn is short, especially in hot weather, which speeds conversion; besides, change in sugar content after harvest is even more rapid (Marshall and Tracy 2003).

In pre-Columbian times the American Indians knew the *su1* mutant and used maize homozygous for this allele, but they did not consume it as a vegetable. Presently, in some countries of South America maize homozygous for the *su1* allele is left to dry completely and then fried for human consumption. Sweet corn, as is used presently, is a development of settlers in the East of the US in the XVIII and XIX centuries.

According to legend, the origin of sweet corn dates back to 1779 when Captain Richard Bagnal, a member of Gen. Sullivan expedition against the Iroquois Indians, brought with him a sample of "papoon corn" (Smith 1955; Marshall and Tracy 2003). Very likely, this "papoon corn" was privately grown among settlers of New England and its importance at that time was small. Sweet corn as it is known today was first offered to the public in 1828 by Thorburn (Smith 1955). Interest in sweet corn grew rapidly, with named cultivars appearing in seed catalogs in the 1840s and 1850s (Marshall and Tracy 2003). At the time of the first efforts at inbreeding and crossing there were approximately 189 open-pollinated white and 50 yellow varieties. Obviously, there were many synonyms (Smith 1955). Important early cultivars included Darling's Early, Stowell's Evergreen, Crosby, and Country Gentleman.

Darling's Early was developed by Noyes Darling, a lawyer of New Haven (Connecticut), who was the first breeder to give a detailed account of how he obtained a variety in a letter he wrote in The Cultivator in 1844 (Singleton 1944). Darling's account was written two years before Robert Reid moved from Ohio to Central Illinois. Darling crossed a very early yellow maize with a sweet corn variety. In the following years he selected two types of "smooth" (that is normal maize) white maize and in the 6th year after the initial cross he found a few scattered kernels of white sweet corn upon the ears of the earliest variety of smooth maize. Those sweet kernels were carefully picked out and were the foundation of what was later named Darling's early.

Stowell's Evergreen, a high-row number variety, was produced crossing Menomony soft maize (probably a Southern Dent) with northern sweet corn (Tracy 2001). Another variety historically important is Crosby that was listed in catalogs by 1867. Its ancestry is unknown and differed from any of the types then known by having deep, tender kernels on 12 to 16 rowed ears. Country Gentleman, also known as Shoepeg, has tender white kernels on ears with no rows, as most of the flowers have two spikelets that do not abort.

By 1899 at least 63 sweet corn varieties were being offered by seed companies, most of which had white endosperm (Tracy 2001). At that time most consumers rejected yellow endosperm. Things changed at the turn of the XX century. In 1902, the Burpee Company of Philadelphia, Pennsylvania, released Golden Bantam, a variety derived from the cross between Darling's Early and an early yellow flint (Common Yellow or Canada Flint) (Gerdes *et al.* 1993; Marshall and Tracy 2003). The ears of Golden Bantam resemble Northern Flint, especially in outcrossed ears. Because of its excellent quality, Golden Bantam changed consumer preferences to yellow types and so, along many years, yellow varieties were predominant.

Of all the varieties that were in cultivation in the past, most of the inbreds that make up present day hybrids trace back to Golden Bantam and its derivatives, followed by Stowell's Evergreen and Country Gentleman (Tracy 2001). These three cultivars were the initial germplasm used by the first programs of inbreeding at the state agricultural experiment stations of land grant universities. Programs breeding yellow sweet corn for the fresh market emphasized Golden Bantam types, whereas those developing varieties for processing used also Evergreens and Country Gentleman (Marshall and Tracy 2003).

The first program of inbreeding in sweet corn was carried out by Pearl and Surface who, in 1907, undertook the production of seed of sweet corn adapted to Maine. They made experimental crosses after discarding the poorer lines, thus using the performance of the progeny as a basis for the selection of breeding stock (Smith 1955).

The main public programs of inbreeding in sweet corn in the first half of the XX century were those carried out in Connecticut, Purdue, Iowa and Illinois. In 1924 D.F. Jones started an inbreeding program at the Connecticut Agricultural Experiment Station with several objectives in mind, mainly to determine the feasibility of the pure line method, to develop early market hybrid resistant to bacterial wilt, and to develop a series of hybrids ripening at regular intervals for a period of approximately a month (Singleton 1948). As a result of this program several important inbreds (Table 1) and hybrids (Table 2) were released.

Table 1. Some historically important public inbred lines of sweet corn (adapted from Singleton and Jones, 1936; Singleton, 1948; Smith, 1955; Kerr, 1968; Kerr, 1973; Kerr, 1979a).

Inbred	Pedigree	Origin
C3	Spanish Gold	CAES†
C6	Whipple, Harris	CAES
C7	Whipple, Harris	CAES
C13	Golden Early Market	CAES
C23	Whipple, C. Ferre	CAES
C30	P39 mutation Crookham	CAES
C68	C7 outcross	CAES
I45	Bantam Evergreen	Iowa State Univ.
I453	(P39 × I45) $I45^2$	Iowa State Univ.
I2132	(TSR × I45) I4329	Iowa State Univ.
I2256	[(IP39 × I1445) I2066] IP39	Iowa State Univ.
I5125	(IP39 × Tendermost) IP39	Iowa State Univ.
IL11	Narrow Grain Evergreen	Univ. Illinois
IL14	Narrow Grain Evergreen	Univ. Illinois
IL18b	Country Gentleman	Univ. Illinois
IL27a	Country Gentleman	Univ. Illinois
P39	Golden Bantam	Purdue Univ.
P51	Golden Bantam	Purdue Univ.
VCO108§	Sungold	Ontario‡
V576	Golden Bantam	Ontario
V642	Sugar and Gold	Ontario
V663	(NJ143Y × V2) Iochief	Ontario
V679	{[(Me100 × V1) × open] V643 } Gold Cup	Ontario
V7726	North Star × Gold Cup	Ontario
VI5177¶	I45 × I5116	Ontario
VI5492¶	Me2RT × I5125	Ontario

†: Connecticut Agricultural Experiment Station.

‡: Horticultural Research Institute of Ontario, Vineland Station, and Horticultural Experiment Station, Simcoe, Ontario, Canada.

§: Originally from Ottawa Research Station, Canada.

¶: Originally from Iowa State University.

Historically important is Redgreen, a white single-cross that was the first hybrid sweet corn to be grown extensively (Smith 1955). Other famous hybrids derived from this program were Spancross, Marcross, Carmelcross, and Lincoln (Table 2).

Table 2. Some historically important public hybrids of sweet corn based on the gene *su1* (adapted from Singleton and Jones, 1939; Singleton, 1948; Smith, 1955; Kerr, 1979b; Kerr, 1979c; Kerr, 1981a; Kerr, 1981b).

Hybrid	Pedigree	Originator
	Extra early	
Spancross	C13 × C3	CAES†
Earlivee	V663 × V642	Ontario‡
Northernvee	(V642 × V641) V631	Ontario
	Early	
Marcross	C13 × C6	AES
	Early midseason	
Carmelcross	P39 × C13	AES
Goldenvee	VI5492 × VCO108	Ontario
	Midseason	
Lincoln	P39 × C23	CAES
Tablevee	V679 × V576	Ontario
	Late midseason	
Golden Cross Bantam	P39 × P51	Purdue Univ.
Iochief	I5125 × I453	Iowa State Univ.
	Late	
Pershing	C68 × Oh55	CAES
Evergreen 14x11	IL14 × IL11	Univ. Illinois
Country Gentleman 13	IL27a × IL18b	Univ. Illinois
Flavorvee	V7726 × VI5177	Ontario

†: Connecticut Agricultural Experiment Station.
‡: Horticultural Research Institute of Ontario, Vineland Station, and Horticultural Experiment Station, Simcoe, Ontario, Canada.

Another public program was instituted in 1920 at Purdue University by G.M. Smith in cooperation with the United States Department of Agriculture. The first hybrid released by this program was Golden Cross Bantam (Table 2), which was made available for seed production in 1933 (Smith 1955). The success of Golden Cross Bantam was immense and set the standard by which all the other hybrids were evaluated. Their parental inbreds (P39 and P51) were obtained by selfing different strains of Golden Bantam and became very popular, especially P39 that was an important foundation inbred for succeeding generations of inbreds (Gerdes and Tracy 1994).

Huelsen and Gillis began a program in Illinois in 1926 with special emphasis on tenderness (Smith 1955). This program used extensively Country Gentleman and several strains of Evergreen as its foundation and produced inbreds that were widely used by the Midwest processing industry and undoubtedly contributed to today's private pedigrees (Table 1). Some hybrids were also released (Table 2).

At Iowa State University Haber developed several important inbreds that are still in use today (Marshall and Tracy 2003), among them the parental inbreds of the famous hybrids

Iochief and Iobelle (Tables 1 and 2). These two hybrids are also important because they are the basis of the first supersweet hybrids, namely Illini Xtrasweet and Florida Sweet. Haber was a strong proponent of using Corn Belt Dent inbreds in the breeding of sweet corn for improving plant health and vigor (Marshall and Tracy 2003). We have used cold tolerant Spanish field maize populations and inbreds to adapt sweet corn to the European Atlantic coast (Malvar *et al.* 1997; Revilla *et al.* 1998).

Another public program in North America was organized later than those mentioned above. It was started in 1945 in Ontario, Canada, leaded by E.A. Kerr. As a result, several inbreds and hybrids were released (Tables 1 and 2). Earlivee (Kerr 1981a) and Flavorvee (Kerr 1981b) are single crosses that are still cultivated nearly 30 years after their release.

Through the 1960s, state agricultural experiment stations of land grant universities led the way in sweet corn breeding in the USA, although private companies made some significant contributions. During the past 30–35 years, cultivar development has shifted to the private sector, which has had an important impact on the development of new endosperm types and improved disease and insect resistance. One particular private-sector development was the hybrid Jubilee, released in 1960 by Rogers Seed Company of Nampa, Idaho. Jubilee became extremely important in the processing industry and remained the leading cultivar until the 1990s (Marshall and Tracy 2003).

The use of sweet corn germplam by breeders agrees with morphological and isozyme characterization of open-pollinated sweet corn cultivars. Thus, a group called Northeastern Sweet included most of the northeastern sweet corn varieties and this group is further subdivided in Yellow-kernel sweet corn where Golden Bantam is included and White-kernel sweet corn where Stowell's Evergreen is integrated (Revilla and Tracy 1995a, b). The Northeastern Sweet group was found to be close to the Northern Flint group of field corn. The sweet corn varieties from Mexico (Maiz Dulce, Dulcillo del Noroeste, etc) and from Peru (Chullpi complex) constitute other two distinctive groups related to Southern Dent and Corn Belt Dent groups of field corn. However, the sweet corn varieties of these two groups do not significantly contribute to modern commercial sweet corn. Finally, Country Gentleman does not appear to be close to any of the previous groups. The heterotic relationships between sweet corn open-pollinated cultivars agree with the previous classifications. Thus, a strong heterotic pattern was defined as 'Country Gentleman × Golden Bantam' while a weaker one was 'Stowell's Evergreen × Golden Bantam' (Revilla and Tracy 1997). However, such heterotic patterns are more theoretical than real because most inbreds have been released from Golden Bantam and inbreds related to Stowell's Evergreen have also important contributions from Golden Bantam.

1.2. Origin of the *Su1* Mutation

Maize geneticists have long debated the origin of the *su1* mutation. The point of controversy was whether one or two independent mutations have been selected and fixed (Tracy *et al.*, 2006).

Galinat (1971) and Mangelsdorf (1974), based on morphological evidence, argued that the *su1* allele was selected once in the Peruvian Andes and then introgressed into local maize throughout the hemisphere. In this hypothesis, Chullpi, a Peruvian race, was the original *su1* source. Mangelsdorf (1974) grouped the more than 300 races of maize found in America into a small number of lineages, using the concept of *lineage* as "descent in a line from a common progenitor". He described six such lineages, one of which is Chullpi, which, besides being the progenitor, according to him, of all sweet corns was also the progenitor of related starchy-seeded forms with globular ears. In some places this race is known as Chulpi or Chulpillo. Under Mangelsdorf's hypothesis, Chullpi was the progenitor of the race Maíz Dulce from Jalisco, México. Other sweet corn race, Dulcillo del Noroeste, was produced by the cross of Maíz Dulce and Reventador, a popcorn race (Wellhausen *et al.* 1952). From Dulcillo del Noroeste, the mutant *su1* was introgressed into northern races including Northern Flint, the progenitor of modern sweet corn (Revilla and Tracy 1995).

Other authors affirm that independent mutations were fixed twice, the first time thousands of years ago in the Andes, and again, more recently in what is now the northeastern USA (Erwin 1934, 1947, 1951). This theory is base on three main arguments: 1, lack of sweet corn remains in archaeological collections from eastern North America; 2, the observation in field maize of spontaneous mutations to the *su1* allele; and 3, the fact that there is no written record of sweet corn in the USA until the XIX century.

To help resolve this controversy, Tracy *et al.* (2006) sequenced the *Su1* gene from 57 accessions of *su1* maize comprising six geographic areas in the USA, Mexico, and Peru, and compared the results with normal *Su1* alleles identified previously. They concluded that five independent mutations have been selected in the history of sweet corn. Three of these five alleles were single base pair changes at highly conserved sites and a fourth was a transposon.

1.3. Other Mutants

Until the 1950s sweet corn was synonymous with maize homozygous for the allele *su1*. In 1953, J.R. Laughnan proposed the use of the mutant *shrunken2* (*sh2*) that brought about kernels with more concentration of sugars. Laughnan (1953) also noted that the use of this mutant "would have an advantage in extending the period during which the product could be held after picking since shrunken endosperms carry a genetic block which reduces greatly this conversion of sugars." In 1961, he released Illini Chief, the first supersweet hybrid that was the *sh2* version of the famous *su1* hybrid Iochief (Table 2) (Tracy 1997). Since seed of Illini Chief was difficult to produce, Illinois Foundation Seed Inc. produced a three-way cross [(R853 × R839) R825] that was named Illini Xtra Sweet (Table 2) with improved seed production capability (Tracy 1997). This hybrid is still sold today. A few years later the University of Florida released Florida Staysweet (Table 2), a *sh2* version of the *su1* hybrid Iobelle (Table 2) (Wolf 1978). Based on this developments and further breeding efforts, *sh2* hybrids are widely grown for the fresh market and especially for processing in North America, Europe and Japan (Marshall and Tracy 2003). As the *sh2* gene is not allelic to *su1*, *sh2* maize must be cultivated in isolation from *su1* sweet corn.

The almost complete lack of starch in kernels homozygous for *sh2* produces a defective grain that when mature germinates with more difficulty than standard *su1* kernels. This difficulty in germination is especially present under conditions far from the optimum. At the Misión Biológica de Galicia, located in Pontevedra (northwestern Spain), we sowed the inbreds R825, R839, R851, and R853 (the conversion to *sh2* of the famous *su1* inbreds I5125, P39, P51, and I453, respectively) in 1976. R839 and R851 did not germinate at all and the others germinated very badly. Dr. Laughnan sent us new stocks of R839 and R851 the next year, but the result was the same – these inbreds cannot germinate under the normal growing conditions usually found in May in Pontevedra which are characterized by cool temperatures and a wet soil. The germination of the other lines, i.e. R825 and R853, was decreasing year after year until they did not germinate at all. This is a problem that we had not experienced with the *su1* inbreds. Heterozygous supersweet material, that is *sh2* hybrids and populations, can, nevertheless, be grown. Modern *sh2* inbreds with improved agronomic performance are reasonably maintained in such conditions and new inbreds have actually been released from them.

Another significant development occurred in 1976, with the publication of an article by Gonzales *et al.* describing a new inbred, IL677a, that carried a new mutation (later called *sugary enhancer1*, *se1*) that had the combination of high contents of sugar (almost as high as *sh2* genotypes) and WSP, a desirable trait of *su1* varieties absent in *sh2* genotypes. This new mutation needs a *su1* homozygous background for expression. Later, the University of Illinois released new sugary enhancer inbreds (Rhodes *et al.* 1982; Juvik *et al.* 1983; Mikel *et al.* 1983; Juvik and D'Arcy 1988). Both preharvest and post harvest, *su1se1* loses sugar at a rate similar to *su1*, but *su1se1* starts with a higher sugar content than *su1*, so its quality is acceptable for a longer period of time (Ferguson *et al.* 1979; Marshall and Tracy 2003).

The three mutants *su1*, *sh2*, and *se1* are the major endosperm mutants used for producing sweet corn in temperate areas. However, other five mutants have being used: *amylose extender1* (*ae1*), *dull1* (*du1*), *waxy1* (*wx1*), *brittle1* (*bt1*), and *brittle2* (*bt2*). The hybrid 'Pennfresh ADX' (Table 2), proposed by Garwood and Creech (1979), carries the recessive endosperm combination ae1, du1, and wx1. They say that "this hybrid can be harvested over a longer period than traditional sweet corns which are homozygous for the sugary (*su1*) gene, and it retains its quality much longer following harvest." This hybrid is related to the popular standard hybrid 'Iochief' (Table 2). ADX hybrids have not been a commercial success (Marshall and Tracy 2003). The use of the mutants *bt1* and *bt2* has been used to develop super-sweet cultivars adapted to Hawaii (Brewbaker and Banafunzi 1975; Brewbaker 1977). Hybrids with these mutants are grown presently.

Table 3. Endosperm genes that have been used in commercial sweet corn cultivars (Neuffer *et al.*, 1997; Tracy, 2001).

Gene	Symbol	Phenotype
amylose extender1	*ae1*	Glassy, tarnished, high amylose content
brittle1	*bt1*	Mature kernels collapsed, angular oftentranslucent and brittle
brittle2	*bt2*	Similar to *sh2* (below)
dull1	*du1*	Glassy, tarnished
shrunken2	*sh2*	Inflated, transparent, sweet collapsed on drying becoming angular and brittle
sugary1	*su1*	Wrinkled and translucent
sugary enhancer1	*se1*	Observed only in *su1* lines: inflated, light colored
waxy1	*wx1*	Opaque, endosperm stains red with iodine

1.4. Summary of Mutant Characteristics

There are, then, eight mutants that affect endosperm carbohydrates and that have been used for producing sweet corn. Based on mature kernel phenotype, these mutants fall into two classes (Boyer and Shannon 1984). Class-1 mutants are *bt1*, *bt2*, and *sh2*. They are characterized for accumulating sugars at the expense of starch and phytoglycogen. Sucrose is greatly increased, and glucose and fructose are elevated to some extent. At 18–21 days after pollination, endosperm homozygous for a class-1 mutant has four to eight times the total sugar in field corn endosperm. Cultivars of these genotypes are usually called "supersweet" (Marshall and Tracy 2003). Class-2 mutants (*ae1*, *du1*, *su1*, and *wx1*) alter the types and amounts of polysaccharides produced. The *se1* gene does not fit into any of the classes. Table 3 summarizes the main characteristics of all those mutants.

1.5. Present Situation

A survey through internet of the catalogs of the main producers of sweet corn seed will show a bewildering array of types and colors. White hybrids are common and so bicolor varieties, i.e. cultivars with a mixture of white and yellow kernels, usually 80% yellow and 20% white.

Single-mutant hybrids *sh2* are the most common varieties used by the processing industry although other mutant combinations are also found. Cultivars homozygous for *su1* are still grown, especially for fresh consumption. Besides hybrids, one can find old open-pollinated varieties like Golden Bantam and Country Gentleman in some catalogs addressed to home gardeners and with a sense of nostalgia. For fresh consumption at great scale most varieties are single crosses with one (*su1* or *sh2*), two (*su1 se1*), or more mutants. Sugary enhancer sweet corn has become popular in local roadside markets (Marshall and Tracy 2003).

In the last years mixtures of different types of kernels on the same cob are frequent. One of the most common mixtures is what is called synergistic. Synergistic sweet corn is characterized by having *se1* and *sh2* kernels in a proportion of 75% of the former and 25% of the latter. Synergistic types have the tenderness of sugary enhanced types with enhanced sugar content. To cultivate synergistic sweet corn, isolation from *sh2* types is necessary to avoid the presence of starchy kernels in both types.

In most catalogs of the commercial companies one can find presently more types of sweet corn besides normal sugary (*su1*), sugary enhanced (*se1*), super-sweet (*sh2*), and synergistic (*se1+sh2*), for instance triple sweet, triple sweet plus, augmented super-sweet, xtra tender, etc. Probably, augmented super-sweet is the most frequent type. This kind of sweet corn adds the *su1* and *se1* alleles to a *sh2* parent.

2. Stand Establishment, Eating Quality and Characteristics Associated with Both Traits in Sweet Corn

Stand establishment (emergence and seedling vigour) and eating quality are fundamental in sweet corn production and in turn are related with the physical and chemical characteristics of the kernels.

2.1. Chemical and Physical Characteristics of Sweet Corn Kernels

The sugary genotypes have about twice the sugar content of field corn at 20 days after flowering (the eating stage) (Laughnan 1953; Creech 1965; Wann *et al.* 1971). In turn, the sugary enhancer and shrunken varieties have, approximately, twice or more of the sugar content of sugary varieties (Laughnan 1953; Soberalske and Andrew 1978; Douglass *et al.* 1993; Azanza *et al.* 1996). The level of sugars is influence by background effects of different sweet corn inbreds. Thus, the percentage of total sugars varied from 13% to 26% in seven sugary lines, the sucrose concentration varied from 191 to 312 mg/g in four sugary enhancer inbreds and from 221 to 363 mg/g in nine shrunken inbreds (Soberalske and Andrew 1978; Azanza *et al.* 1996). Regarding the carbohydrates, the sweet corn genotypes have lower starch than the wild types at the eating stage. For example, Creech (1965) reported 66% of starch in the wild endosperm, while in the *su1* and *sh2* endosperm the percentage of starch was 28 and 18%, respectively. Soberalske and Andrew (1980) also reported a slightly lower concentration of starch in the *sh2* endosperm than in the *su1*. In relation to background effects, considerable differences between *su1* lines were found for starch percentage, but not between *sh2* lines (Soberalske and Andrew 1980). The sugary varieties present high levels of water soluble polysaccharides (WSP) or phytoglycogen, 30 to 40% at 18 days after pollination (Andrew *et al.* 1944 ; Creech 1965). Appreciable amounts of WSP occurs only in genotypes with the *su1* allele, thus *sh2* genotypes have a negligible content of WSP (Soberalske and Andrew 1980). Sugary enhancer varieties have also significant amounts of WSP because they have the *su1* alleles in addition to the *se1* (Ferguson *et al.* 1978; Azanza *et al.* 1996). In addition to the main effect of *su1* on WSP concentration, other genes of minor effects seem to have an influence on the trait. Thus, the percentage of WSP varied from 13% to 29% in seven sugary inbreds (Soberalske and Andrew 1980) and the concentration of WSP varied from 83 mg/g to 201 mg/g in four sugary enhancer inbreds (Azanza *et al.* 1996). As conclusion, the *su1*, *se1* and *sh2* mutations have a major effect on the levels of sugars and polysaccharides in the kernel, but several genes of minor effect have also a relevant joint effect on those traits. The presence of those minor genes implies that selection for carbohydrate or sugar composition is possible within each mutant. Congruent with this hypothesis, an indirect response for starch concentration was obtained in a recurrent selection program for emergence and kernel weight carried out in a *sh2* population (Juvik *et al.* 1993). Also, Jones (1919) and Tracy and Chang (2007) were able to effectively select for endosperm appearance within sugary populations. Other example of effectiveness of selection for endosperm composition are the sugary hybrids selected for increased sugar, as for example, Earlivee or Silver Queen, which have higher sugar content than other standard sweet corn such as Iochief (Marshall 1987). In addition to main components, some chemical compounds were also found in sweet corn kernel in small concentrations, for example, dimethyl sulphide (DMS) (Wong *et al.* 1994). Azanza *et al.* (1996) did not found difference between the DMS concentration of genotypes with the *su1*, *se1* or *sh2* mutation.

The kernels of *su1*, *se1*, and *sh2* mutants differ not only in chemical but also in physical characteristics. Thus, from eating stage (20 days after pollination) to maturity, *su1* seeds are heavier than *sh2*, the difference increasing with the maturity of the seeds (Andrew 1982; Styer and Cantliffe 1983a). The genetic background and interaction of the genetic background with the mutants play a significant role in the expression of kernel weight (Andrew 1982; Styer and Cantliffe 1983a). The pericarp tenderness can be defined as the resistance of pericarp to fragmentation by chewing (Huelson 1954). The pericarp tenderness is negatively related with pericarp thickness (Bailey and Bailey 1938; Ito and Brewbaker 1981). At eating stage sugary hybrids have thicker pericarp than shrunken (Zan and Brewbaker 1999), but at maturity, seeds of *sh2* have thicker or equal pericarp than *su1* when compared in an isogenic background (Tracy and Schmidt 1987; Zan and Brewbaker 1999). However, at eating stage, the pericarp toughness, measured as the force required to puncture the pericarp, was significantly greater in the *sh2* inbreds than in the *su1* and the *se1* inbreds in the experiment of Azanza *et al.* (1996). This indicates that, at least in the differentiation of the pericarp tenderness of these mutants, the thickness of the pericarp is not an important factor. In addition to the effect to the main mutants, the genetic background is also important in the expression of pericarp thickness (Tracy and Schmidt 1987). Shrunken genotypes have been also reported to have a greater number of large airspaces between the pericarp and aleurone layer and lower numbers of tightly packed starch granules than sugary genotypes (Styer and Cantliffe 1983b).

2.2. Germination and Seedling Vigour of Sweet Corn Genotypes

The stand establishment of sweet corn hybrids could be particularly compromised in areas of cool and humid springs as are typical in Atlantic Europe (north of Spain, west of France, the Netherlands, South of United Kingdom, etc), as well as in Canada, some areas of the US Corn Belt, etc. To study emergence and seedling vigour in cold conditions different approach can be used, mainly, evaluations in early spring, although not always cold conditions are obtained, and evaluations in cold chamber with an environment that simulates, but it is not the true field environment. Different experiments have shown significant, but not strong correlations (r = 0.60, approximately), between emergence in the cold chamber and emergence in the field (Waters and Blanchette 1983; Menkir and Larter 1987; Ordás *et al.* 2006). Waters and Blanchette (1983) have analyzed different cold test and found that combining the seedling-growth cold test (total seedling dry weight) with an electrolyte leakage test gives the highest correlation with field emergence. Due to the complexity of factors influencing seed vigour and the wide variation in field conditions, a combination of vigour test methods may be preferable to a single test method (Styer and Cantliffe 1983a). Ordás *et al.* (2006) have found that the best genotypes under cold chamber conditions had an acceptable performance under field conditions, but the opposite was not true. Therefore, they propose to choose first the genotypes that show the best performance in the cold chamber and then, among these, select the best genotypes in the field.

When studying the genetics of cold tolerance at early stages of plant development of sweet corn cultivars relevant questions are: what is the main effect of the mutant (*su1*, *se1* or *sh2*)?, what is the effect of the genetic background where the mutant is located?, and, are there mutant × background interactions? In the case of sugary genotypes, the evaluation of isogenic lines with both the wild and *su1* allele have shown that the sugary endosperm have lower emergence and/or seedling vigor than the wild type (Nass and Crane 1970; Zan and Brewbaker 1999). However, Rowe and Garwood (1978), also analysing isogenic lines, have found large differences between genetic backgrounds and the existence of mutant × background interaction. They also found that in a concrete genetic background the sugary endosperm was even superior to the normal endosperm. Similarly, Hotchkiss *et al.* (1997), that screened for cold tolerance 35 open-pollinated sweet corn cultivars, field corn synthetics, and high altitude germplasm, have found large genetic variation between cultivars. In addition, some sugary cultivars from USA were found to have a cold tolerance comparable to the tolerant checks. Although those varieties are closer to cultivated sweet corn than the ancient tropical populations commonly used as sources of cold tolerance, they are still far from the strict commercial standards for sweet corn. This limits its usefulness for applied breeding programs at the short or medium term. Unfortunately, among the main historical varieties, Country Gentleman and Stowell's Evergreen were sensitive to cold temperatures, both *per se* and in hybrid combinations, and Golden Bantam was similar to them. Using seed from ears segregating for *su1*, Haskell and Singleton (1949) have found no significant difference in the ability of wild and *su1* kernels to withstand cold treatment. For this and the previously commented experiments it can be concluded that the genetic background and the mutant × background interaction play important roles in cold tolerance of sugary varieties and that the choice of the variety may overcome the deleterious effect of the mutant.

Once it is known that genetic variation exists for cold tolerance and sources of tolerance are identified, the next step for designing an efficient breeding program it is to determine the inheritance of the trait. This was acomplished by evaluating a diallel among six open-pollinated sweet corn cultivars in a cold chamber. The variation detected was primarily due to general combining ability, and secondarily to specific combining ability and reciprocal effects. Under this scenario, it could be expected that selection for hardiness would be effective for sweet corn varieties. This was the case for several sugary varieties selected for hardiness in early spring in Britain (Mather and Haskell 1949) and several sugary inbreds developed in North Spain (Ordás *et al.* 1994).

The analysis of isogenic lines and hybrids has shown that the *sh2* mutant had lower emergence and seedling vigor than *su1* (Young *et al.* 1997; Zan and Brewbaker 1999; Ordás *et al.* 2009). This conclusion was corroborated by other experiments (Hannah and Cantliffe 1977; Styer *et al.* 1980; Styer and Cantliffe 1983a; Douglass *et al.* 1993; Hassell *et al.* 2003). Furthermore, as said above, we found the *sh2* inbreds difficult to maintain in our Atlantic conditions (North Spain) which has delayed their incorporation to our breeding program. However, similarly to the experiments with *su1*, the experiments with *sh2* have shown significant differences between genetics backgrounds for cold tolerance which suggests the possibility of improvement of cold tolerance of shrunken varieties by selection (Andrew 1982). Accordingly with this expectation, a significant progress from mass selection for field emergence was obtained in a *sh2* population (Bell *et al.* 1983). In our sweet corn breeding program, given the problems that we had to maintain the *sh2* lines, we started a selection program with an open pollinated variety and simply by multiplying the populations repeatedly each year in our conditions we observed that the population improved their germination and seedling vigour. However, all commercial *sh2* hybrids are probably descended from Laughnan's original stock (Marshall and Tracy 2003) and, therefore, the narrow genetic base of the *sh2* germplasm limits its breeding potential.

Adapted standard sweet corn could be useful for improving the adaptation of *sh2* sweet corn if the genes that confer adaptation when combined with *su1* are also favourable when combined with *sh2*. To check this we evaluated two diallel sets of *su1* and *sh2* near-isogenic inbred lines at early field planting and at cold chamber (Ordás *et al.* 2009). For most of the traits related to adaptation, there were significant mutant × genotype interaction and the estimates of GCA of each version of the same inbred were different, probably due to epistasis. This indicates that cold tolerance sugary genotypes cannot be useful for improving *sh2* sweet corn and the conversion of *su1* genotypes will give unpredictable results. Regarding the *se1* allele, although less experiments have been carried out with this mutant, it is clear that has lower germination and seedling vigour than *su1* and similar or only slightly better stand than *sh2* (Douglass *et al.* 1993; Young *et al.* 1997; Hassell *et al.* 2003; Ordás *et al.* 2009). The variability between *se1* genotypes indicates that cold tolerance could possible be improved to equal that observed in some of the best *su1* varieties (Douglass *et al.* 1993). However, the *se1* germplasm have a narrow genetic base because all *se1* inbreds probably derive from IL677a, the inbred where the *se1* mutant was identified (Gonzales *et al.* 1974). Therefore, for the improvement of the adaptation to cold conditions of the *se1* varieties the use of another source of germplasm could be desirable. Sugary inbreds were evaluated as sources of favourable alleles for the improvement of *se1* hybrids at the field and at the cold chamber (Ordás *et al.* 2004; Ordás *et al.* 2006). These reports conclude that the emergence and seedling vigour of the *se1* hybrids can be improved by the *su1* inbreds, particularly those obtained in situ in the Atlantic environment (Ordás *et al.* 1994). The sugary × sugary enhancer hybrids, produced by crossing a sugary line with sugary enhancer line, constitute an interesting alternative to sugary and sugary enhancer hybrids for cool areas because they have better plant and ear types and better eating quality than the sugary hybrids and better emergence, early vigour and shorter flowering time than sugary enhancer hybrids (Ordás *et al.* 2008).

In addtion to genetic improvement, some research has been done to improve the germination and early vigor of seedling by using seed treatments, for example with *Trichoderma harzianum* Rifai, which is a biocontrol fungus that reduces root rot and increase root growth. This enhancement can restore some stress-induced growth reduction and may directly reverse oxidative injury (Bjorkman et al 1998). Other alternative treatment is matriconditioning that improved the performance of sweet corn in early field plantings at suboptimal temperatures, particularly when including pesticides and/or gibberellin during conditioning. As said before, electrolyte leakage was greater in *sh2* corn compared to the *su1* corn, and matriconditioning reduced the leakage for both genes (Khan et al 1995). In order to improve *sh2* seed quality, early harvest is considered a way for the avoidance of seedborne fungi. Maximum kernel dry weight is reached between 43 and 66 days after silking, depending on the growth cycle and weather conditions. Along this period, moisture decreases linearly. Untreated seed harvested several days after the grain had reached maximum weight produced an 80% stand in the field the following year. Early harvest caused high germination and low vigor and late harvest increased seedling vigor (Wilson and Trawatha 1991).

2.3. Eating Quality of Sweet Corn Mutants

The primary components of fresh sweet corn eating quality associated with consumer preference are flavour, texture and aroma (Flora and Wiley 1974; Azanza *et al.* 1994). The relative importance in determining overall liking is flavour 45%, texture 30% and aroma 25% (Azanza *et al.* 1996a). Sweetness is an important component of flavour (Culpepper and Magoon 1927; Flora and Wiley 1974; Azanza *et al.* 1996a). Different aromas have been found in sweet corn: rotten egg, fruity, corn-like, grassy, etc (Flora and Wiley 1974; Azanza *et al.* 1996c). The corn-like aroma is the pleasant aroma that consumers would expect in sweet corn, while other aromas could be responsible of a low overall liking. However, although some aromas could be repulsive at high concentration, the combination of several aromas at low concentration could produce a pleasant overall aroma (Flora and Wiley 1974). In some organoleptic tests, *sh2* and *se1* hybrids were preferred over *su1* hybrids (Showalter and Miller 1962; Zan and Brewbaker 1999), while in others some sugary hybrids obtained similar scores to those of *sh2* and *se1* hybrids (Evensen and Boyer 1986). Shrunken and sugary enhancer varieties are sweeter than sugary varieties, otherwise sugary enhancer and sugary varieties are tender and creamy than shrunken and, in addition, they have the typically sweet corn flavour which in *sh2* varieties is less intense. One advantage of shrunken varieties is that maintain a high eating quality for a longer period than sugary or sugary enhancer (Tracy 2001).

2.4. Relationship between Chemical and Physical Properties of the Kernels and Germination

Both mature and immature whole kernel sugar content have been shown to negatively correlate with emergence (from r = -0.50 to r = -0.70, approximately) (Douglass *et al.* 1993; Azanza *et al.* 1996b; Young *et al.* 1997). On the contrary, starch content is positively correlated with germination (r = 0.70) (Azanza *et al.* 1996b). However, because sugar and starch content are interrelated is difficult to evaluate the extent to which each individual factor relates to poor germination (Young *et al.* 1997). Different hypothesis have been proposed to explain the relationship between sugar or starch and emergence. The higher levels of sugars to starch could create a higher osmotic potential, giving the potential for more water uptake during the initial phase of imbibition. Rapid movement of water into seed would result in accelerated hydration, increasing the potential for damage or leakage (Styer and Cantliffe 1983b). Reduced kernel starch could represent a limited carbon supply for the developing of the seedling and impact vigour. The mobilization of carbohydrate reserves could be also reduced due to a delay in α-amylase transcription, as a consequence of low starch and high initial sucrose levels, which could affect germination and seedling vigour of sweet corn cultivars, especially *sh2* genotypes (Young *et al.* 1997). The breakdown of storage carbohydrates into energy components, such as adenosine triphosphate (ATP) to aid in the synthesis of protein and nucleid acids should be limiting in sweet corn mutants, particularly in *sh2*, since the endosperm reserve is reduced (Bell *et al.* 1983). However, studies have shown that there is little difference in ATP levels between *sh2* and normal germinating kernels (Styer *et al.* 1980). The germination has been positively related to kernel size or weight (Andrew 1982). The seeds of small size and weight may allow more water to be absorbed per unit area increasing the risk of damage or leakage. In the case of the *sh2* seeds the greater number of large airspaces between the pericarp and aleurone layer can allow the pericarp to be broken easily as well as facilitate water movement (Styer and Cantliffe 1983b). The leakage of electrolytes during imbition could be other factor that negatively effect emergence. The leakage of the seed is usually estimated by measuring the electrical conductivity of the leachate (SLEC). In the case of sweet corn, negative correlations between SLEC and emergence have been reported (r = 0.60, approximately) (Waters and Blanchette 1983). The SLEC of *sh2* genotypes is higher than the SLEC of *su1* genotypes (Styer and Cantliffe 1983b; Schmidt and Tracy 1988). SLEC has been associated to kernel weight (r = 0.82). The difference between inbreds and the inbred × endosperm interaction have been significant for SLEC (Schmidt and Tracy 1988). Solute leakage during germination reduces metabolic energy available for embryo growth and may provide substrates for the growth of pathogenic microorganism (Juvik *et al.* 1993). In the mass selection program of a *sh2* population for field emergence and seed weight, a significant correlation was obtained between both traits (Bell *et al.* 1983). However, the relationship between emergence and kernel weight is confounded because of the effect of selection which changed simultaneously both traits. SLEC, carbohydrate and starch content also changed as an indirect response of selection, although it is not possible to estimate the individual relationship between those traits and emergence because of the high intercorrelation of the variables and the correlations with kernel weight which was also directly selected (Juvik *et al.* 1993).

2.5. Relationship between Chemical and Physical Properties of the Kernels and Eating Quality

Little work has been done on the relation between carbohydrate composition of the endosperm and sensory quality. In the evaluation of six hybrids, including sugary and sugary enhancer hybrids and a sugary hybrid with the *sh2* allele segregating in the kernels, the sugar concentrations were not correlated with sensory scores on the day of harvest (Evensen and Boyer 1986). However, in a evaluation of sugary, sugary enhancer and sugary × sugary enhancer hybrids in three environments, the measurement of the soluble solids content which consist largely of sugars and the table quality test were consistent in two of the three environments (Ordás *et al.* 2008). Azanza *et al.* (1996c), evaluating 103 $F_{2:3}$ families developed from a cross between a sugary and a sugary enhancer inbred, have carried out the more complete experiment to known the relationship between composition and sensory quality. They have found that the concentration of sucrose was significantly correlated with sweetness ($r = 0.80$), although the correlation between sucrose and overall liking was only moderate ($r = 0.40$). The correlation between other sensory attributes or chemical components and overall liking was not high either. This is an indication of the complexity of overall liking which is the sum of many factors of minor effect.

The textural eating quality of sweet corn depends mainly of three factors: pericarp tenderness (Bailey and Bailey 1938; Kramer *et al.* 1949; Huelson 1954), level of water soluble polysaccharides (WSP) or phytoglycogen (Culpepper and Magoon 1927), and moisture content (Wann *et al.* 1971). In the evaluation of 103 $F_{2:3}$ sugary and sugary enhancer families, the kernel tenderness, estimated as the force required to crush the kernels, was correlated with the perception of tenderness ($r=0.87$) by selected panellists and the tenderness was related with overall liking ($r=0.48$) (Azanza *et al.* 1996c). WSP gives the endosperm of sugary varieties the smooth texture and creaminess, characteristic of traditional sweet corn varieties (Tracy 2001). However, Azanza *et al.* (1996c) did not found significant correlation between WSP and overall liking: either there was little genetic variation between families for WSP or in the presence of an adequate WSP level other factors increased in importance in determining consumer quality. Some of the chemical compounds present in sweet corn in small concentration produce aromas that influence eating quality (Flora and Wiley, 1974). Thus, DMS was described by panellists as having pleasant and "corn like" aroma (Flora and Wiley, 1974). However, Azanza *et al.* (1996c) did not found significant correlation between

DMS and overall liking. They identified the V179 compound as associate to grassy aroma, although the correlation was not high (r = 0.36).

3. Sweet Corn Breeding for Resistance to Insect Attack

3.1. Introduction

Losses due to pest (insect and diseases) and costs of controlling them, tend to be much higher in sweet corn than in field corn for several reasons: tolerance of consumers to damage on the ears is much lower, the planting season is spread over a longer period of time, and sweet corn has a high proportion of Northern Flint in its background, a race that has lower pest resistance than Corn Belt Dents (Tracy 2001). High levels of resistance to common diseases have been achieved in sweet corn by conventional breeding; however, genetic resistance to insect attack has been difficult to incorporate to elite material (Marshall and Tracy 2003). We precisely will focus on resistance to insect attack in sweet corn in the next paragraphs (for a review of resistance to diseases in sweet corn see Marshall and Tracy 2003).

Many insect can cause economic damage to sweet corn, but the European corn borer (*Ostrinia nubilalis*, ECB) and corn earworm (*Helicoverpa zea*) are the most widespread and serious insect pest of sweet corn (Marshall and Tracy 2003). Infestations by insects, mainly ECB and corn earworm, typically cause damage on 5 to 25% of the ears in early season planting, and often greater than 50% ear damage in late season plantings, if not controlled (Dively 2007). In the Mediterranean area, a major insect pest of corn is the Mediterranean corn borer (MCB, *Sesamia nonagrioides*) which, given its voracity, could also represent a serious problem for sweet corn cultivation. In the evaluation of six standard sweet corn hybrids in six environments under natural MCB infestation, 70% of the ears had damage in the kernels and the mean percentage of damaged kernels per ear was 20% (Velasco *et al.* 2002). ECB and MCB are usually two although the number of generations is higher in southern, warmer latitudes, and lower in northern, cooler latitudes (Velasco *et al.* 1999a, Iowa State University 2006). The first generation larvae attack the plant at the whorl stage, while the second generation at silking or later.

Larvae of the corn borers feed on leaf, sheath-collar, stalk, silk, husks, cobs, shanks and kernels of the maize plant, depending on the stage of plant development at oviposition. Direct ear feeding and contamination which cause a reduction of ear quality is the most critical problem for sweet corn generated by corn borers and other insects (Carlson and Andrew 1976; Joyce and Davis 1995; Gardner *et al.* 2000). Ear damage tolerance levels vary, depending on the processor or fresh market outlet. Fresh market outlets generally require less ear damage than processors, although most processors prefer little or no damage and, in some cases, provide a bonus for insect-free corn (Iowa State University 2006).

3.2. Insecticides and Transgenic Technology

Because the worm-contaminated sweet corn is unacceptable to customers, farmers use multiple insecticide applications to minimize corn borer damage: 2 to 4 sprays in the case of processing corn or 4 to 7 sprays in the case of fresh market (Dively 2007). Of surveyed states in the United States 89%, approximately, of the fresh-market sweet corn and 74% of the processing corn acreage is sprayed with insecticides (Gardner *et al.* 2000). Pyrethroids broad-spectrum insecticides are most commonly used, but substantial amounts of other insecticides are used as well (Dively 2007). Genetically modified maize expressing the Cry1Ab or Cry1-F toxins from *Bacillus thuringiensis* (Bt) Berliner has been successfully used for suppression of ECB populations (Pereira *et al.* 2008). Also, when MCB larvae were fed with leaves of two Bt hybrids mortality reached the 100% of the individuals before 15 days (Butrón *et al.* 2005). Analysis of more than a decade of global resistance monitoring data shows that resistance to Bt corn was not detected in populations of ECB and MCB (Gassmann *et al.* 2009). However, high levels of resistance to Cry1Ab and Cry1-F Bt toxins have been selected for under-laboratory conditions in a number of ECB colonies (Siqueira *et al.* 2004; Pereira *et al.* 2008). The success of Bt crops exceeded expectations of many, but does not preclude problems in the future as pests would evolve resistance to such crops (Tabashnik *et al.* 2003). The high-dose/refuge (HDR) strategy consists in plant a non-Bt refuge area which allows the random gene exchange between insect subpopulations exposed and not exposed to an insecticide. HDR strategy has been recommended as one of the best resistance management strategy for delaying the development of resistance to Bt crops (Tyutyunov *et al.* 2008). Mandatory regulations were imposed in 2000 for Bt maize in US; the current size of the mandatory refuges for Bt maize are between 20% and 50% of the area under Bt crops (Bourguet *et al.* 2005). In the case of sweet corn, adoption of Bt varieties has been limited due to concerns over consumer acceptance of genetically modified organism (GMO) foods, although consumer attitudes are changing in US and most consumers would purchase Bt sweet corn if it reduces their exposure to pesticides (Dively 2007). Thus, adoption of Bt sweet corn has increased in recent years totalling about 5% of the US acreage (Dively 2007). However in several countries of Europe, as for example France, there is still a strong rejection to GMO foods. In Bt sweet corn varieties, the control of ECB and MCB is virtually 100%, but corn earworm densities are not eliminated (Edelson and Mackey 2006). According to Dively (2007), it is not uncommon to find 20-30% of the Bt ears infested or showing minor tip damage which affects the quality of the product, particularly for fresh markets outlets. Therefore, there is the possibility that transgenic sweet corn may need supplemental control of the corn earworm, particularly under high pest pressure (Farrar *et al.* 2009).

3.3. Natural Genetic Resistance

3.3.1. Evaluation Techniques

Natural genetic resistance could be an alternative or complement other control tactics as pesticides or transgenic technology. In general, important prerequisites to any host-plant resistance study are the development of techniques to maintain adequate insect populations that can inflict damage on the test plants, and the evaluation of resulting effects on the insect or the plants (Wiseman *et al.* 1974). Evaluation of genotypes of maize for resistance is often difficult because of inadequate or uneven natural infestation (Linduska and Harrison 1977), while the artificial infestation with eggs masses or larvae insures uniformity of time and level of treatment (Andrew and Carlson 1976). The development of laboratory techniques for rearing lepidopterous insects and the production of large quantities of relatively inexpensive eggs masses and larvae have stimulated research on the use of artificial infestation to simulate oviposition and illustrate differences among genotypes for relative levels of resistance (Grier and Davis 1983). Artificial infestation has been used for evaluating resistance to ECB (Andrew and Carlson 1976), MCB (Velasco *et al.* 1999a) and corn earworm (Wiseman *et al.* 1974) in sweet corn. Multiple aspects have to be considered in the evaluation of genotypes for resistance: the generation of the insect, the stage of development of the insect in the moment of the infestation, the technique used for the infestation, the number of insects per plant, the positions in the plant where the insects are placed, the stage of development of the plant in the moment of the infestation, the moment when the damage is measured, and the way in which the damage is measured.

The ear damage caused by the first generation of ECB is much lower than the ear damage caused by the second generation (Chiang and Hodson 1963; Carlson and Andrew 1976) and there is not relationship between the ear damage of first and second generation (Andrew and Carlson 1976). Furthermore, Carlson and Andrew (1976) have not found genetic differences between genotypes for ear resistance to ECB first generation attack. For those reasons, many of the experiments have been focused on second generation attack. However, it is worth to comment that Gardner *et al.* (2000), contrarily to the previous experiments, have found similarities for corn infested at the silk and the whorl stages.

Regarding the stage of the insect, masses of hatching or black-headed eggs, (Wiseman *et al.* 1974; Carlson and Andrew 1976; Grier and Davis 1983; Jarvis 1988; Velasco *et al.* 1999a; Malvar *et al.* 2002, etc) or neonate larvae (Shehata *et al.* 1975; Linduska and Harrison 1977; Lamb *et al.* 1994; Joyce and Davis 1995; Warnock and Davis 1998, etc) are commonly used. The larvae are usually mixed with air-dried corn cob grits and a device called Davis inoculator or "bazooka" is used to infest the corn plants (Wiseman *et al.* 1980). In screening for ECB resistance, two or four masses of, approximately, 40 eggs are usually used (Carlson and Andrew 1976; Andrew and Mosely 1980). Grier and Davis (1983) have shown that 9 egg masses maximize the ability to differentiate among genotypes, although the cost of labor and materials for growing, infesting, and evaluating the experimental plants would greatly increase, making such number of eggs masses impracticable. In screening for MCB and corn earworm, usually one mass egg of approximately 40 is enough, given the voracious nature of the larvae (Wiseman *et al.* 1974; Velasco *et al.* 1999a). When ECB larvae are used instead of eggs for artificial infestation, usually 40 or 50 larvae per plant are introduced (Warnock and Davis 1998; Gardner *et al.* 2000). Joyce and Davis (1995) have found that line-by-infestation level interaction was not significant in an experiment with different level of infestation, from 50 to 200 larvae per plant. This confirms that 40 or 50 is an adequate larvae number for screening for ECB resistance. However, in the case of corn earworm, because of its voracity, three larvae are enough for infesting and Wiseman *et al.* (1974) have shown that the damage produce by three larvae is equivalent to that produced by a 30 eggs mass.

ECB and corn earworm eggs masses or larvae are usually placed under the husk near the tip of the developing shoot or in the silks in order to minimize attrition due to weather, to eliminate the need of several infestation, and because in sweet corn the major concern is ear damage (Wiseman *et al.* 1974; Carlson and Andrew 1976). However, the ear damage caused by natural infestation could be overestimated because in nature not all plants are infested and several of the eggs depositions are placed not on the silks or husk, but on the leaves (Iowa State University 2006). Furthermore, when infesting directly in the ear, mechanisms of resistance that could exist outside the ears are missing. Velasco *et al.* (1999a) have compared

MCB artificial infestation made among the silk, maximizing the ear damage, with the infestation made between shank and stem, as naturally occurs in the case of MCB. They have found, for ear damage, inbred × infestation method interaction and, hence, identifying the most resistant inbred depends on the infestation method.

The infestation to simulate second generation attack is usually made at silking and can be made simultaneously for all genotypes when most of the genotypes have silked or separately for each genotype at the moment of silking of each genotype. In the first case, if the genotypes differed in maturity, the plant genotypes are infested in different developmental stages which could affect the damage produced by the larvae. In the second case, the larvae in different genotypes are exposed to different external environmental conditions during their growth which also could affect the damage produced by the larvae. Anyway, in several studies of ECB resistance later genotypes show less ear damage than earlier genotypes (Carlson and Andrew 1976; Andrew and Carlson 1976; Grier and Davis 1980; Andrew and Mosely 1980; Gardner *et al.* 2000). However, Velasco *et al.* (2004a) have shown that infestation date had not a clear effect on ear resistance to MCB attack and that the two earliest varieties of the experiment were the most resistant. The moment where the damage is measured varies between experiments: about 30 days after infestation (Carlson and Andrew 1976), when 278 heat units had accumulated since infestation (Grier and Davis 1980), when the ear reach the eating stage (Malvar *et al.* 2002), etc. This last option seems to have more sense because the damage produced after the eating stage have not commercial relevance.

The evaluation of ear damage also varied between experiments. Some author used an ear damage index which is the average of a tip rating and an ear tunnel count below the tip (Carlson and Andrew 1976). In other experiments the number of damaged kernels (Grier and Davis 1983) or the percentage of ears with less than 1% of damage (Joyce and Davis 1995) were used as a measure of the damage. With a similar purpose different scales were also used, for example, a scale that combine ear damage and ear damage location (Warnock *et al.* 1998) or a scale that only takes account the ear damage independently of the location (Malvar *et al.* 2002). The scales that include damage location have been used to meet processing industry standards in which damage in the tip is less important than damage in the side of the ear. Warnock *et al.* (1998) questioned whether the use of such scales, that give high ratings to ear with little total feeding damage but with damage well located below the ear tip, might overlook potentially useful germplasm. These authors compared a scale that combines ear feeding damage and damage location with an alternative scale based solely on ear feeding damage and found that both scales consistently separated sweet corn genotypes into resistant, marginally resistant, and susceptible classes. However, they conclude that plant breeders should carefully evaluate the efficacy of individual visual scales before incorporating them into selection programs. Velasco *et al.* (1999a) compared different measured of damage, including number of tunnels, length of tunnels, percentage of ears with damaged kernels, and a visual scale based on the percentage of damaged kernels. The visual scale showed the highest correlation with most traits; the authors conclude that the visual scale seems to be the best indicator of resistance and should be used in preliminary test. Other traits that are important in field corn do not have relevance in sweet corn. For example, stalk tunnelling could produce ear stress and, as consequence, a reduction in kernel or ear size, but this effect has not been reported in sweet corn (Grier and Davis 1980). Other possible consequence of stalk tunnelling is stalk breakage, but in sweet corn ears are harvested before the stalks start to dry out and for that reason the stalk breakage is not important (Chiang and Hodson 1963).

3.3.2. Breeding for Natural Resistance to Insect Attack

As in any breeding program, the first step in sweet corn breeding for corn borer resistance will be to know the genetic variation for this characteristic and to identify sources of resistance. In an experiment in which 16 sweet corn hybrids were evaluated under ECB artificial infestation in two years, highly significant variation was found between the hybrids: some of them showed relative ear susceptibility, while others presented relative ear resistance. The difference between hybrids was appreciable: 5.60 vs. 0.87 tunnels in the ears for the comparison of the more divergent hybrids. Furthermore, there was not significant genotype × environment interaction, indicating that the results were consistent over environments. Gardner *et al.* (2000), evaluating commercial processing hybrids, obtained similar results: significant genetic variation and consistency between environments. For ear damage estimated on a 1-9 scale the differences between hybrids were substantial, for example 2.1 vs. 5.4 between the most resistant and the most susceptible hybrid. Based on these results, the authors conclude that in the short term, sweet corn genotypes with identified resistance to ECB can be incorporated in commercial corn production to minimize corn borer damage. Sweet corn hybrids less prone to corn borer damage, although not totally resistant, will require less insecticide for corn borer control (Carlson and Andrew 1976). Gardner *et al.* (2000) also proposed to incorporate the resistance genotypes in a long-term breeding program in which the degree of resistance could be improved to higher levels. For MCB ear resistance, significant genetic variation was also found between 20 sweet corn populations (Velasco *et al.* 1999b). However, the difference between populations for MCB ear damage estimated on a 1-9 scale (2.1 vs. 3.6 between the more divergent populations) was lower than the difference between hybrids detected for ECB ear damage. There were also significant genetic differences for ear damage between the 21 sweet corn hybrids from a diallel of seven inbreds, but the difference between hybrids was not large (1.5 vs. 2.2 between the more divergent hybrids) (Velasco *et al.* 2002). In this experiment, similarly to the experiments of hybrid resistance to ECB attack, there was not significant genotype × environment interaction indicating consistency between environments.

Once it is known that genetic variation exists for ear damage and sources of resistance are identified, the next step for improving the tolerance to insect attack it is to determine the inheritance of the trait. Different genetic designs can be used to study the inheritance of a trait: diallel, mean generation analysis, etc., and several of them have been used to study the inheritance of ear resistance to insect attack in sweet corn. Grier and Davis (1980) reported broad sense heritabilities of 57 to 71%, depending on the infestation method, for ECB ear resistance when 63 F_3 random families derived from a commercial sweet corn hybrid were evaluated. In the analysis of two diallels of 10 sweet corn inbreds, it was found that the general combining ability (GCA) was significant in both diallels for ECB ear resistance, while the specific combining ability (SCA) was only significant in one diallel and its value was much lower than that of GCA (Andrew and Mosely 1980). Based on those results, the authors suggested that additive effects were responsible for ear resistance and that improvements in resistance should be attainable. However, Grier and Davis (1980) and Lamb *et al.* (1994) indicated that nonadditive effects could influence the inheritance of ear resistance as F_1 heterosis was detected in their germplasm. Furthermore, in a mean generation analysis using two different crosses, Warnock *et al.* (1998) have found that epistatic and dominance effect also condition ear resistance and that heritability for ear resistance is low (17%). The authors concluded that multiple genes interact to affect ECB ear resistance and that the inheritance of the trait is complex. For MCB, the diallel analysis of seven inbreds has shown that GCA for ear resistance explained much of the genetic variation between the hybrids (Velasco *et al.* 2002). A mean generation analysis (Velasco *et al.* 2004b) confirmed the importance of additive effects for MCB ear resistance, while the importance of dominance effect depended of the particular cross. Similar results have been found for ear resistance to corn earworm attack in sweet corn: inheritance of resistance is complex, but more additive variance was evident among sweet corn crosses (Widstrom *et al.* 1967). As conclusion, additive effects seem to play a significant role in ear resistance to ECB, MCB or corn earworm attack which encourages the use of recurrent selection for improving the trait. However, recurrent selection experiments in the literature for improving ear resistance to insect attack in sweet corn are limited. Wann and Hills (1975) carried out tandem mass selection in a sweet corn composite for earworm resistance and agronomic characters and found a slight but steady increase of ear resistance through four generations of selection. In the experiment the effective selection pressure for earworm resistance was 25% of a segment of the population and the authors suggested that the selection pressure for earworm resistance be maintained at around 10%.

Revilla et al. (2005) compared ECB y MCB susceptibility in seven pairs of sugary and shrunken near-isogenic inbreds. They found that ear damage was not significantly different between sugary and shrunken strains and, therefore, they concluded that theoretical and practical results of sugary breeding for corn borer resistance could be capitalized for shrunken corn breeding.

4. Using Field Corn for Sweet Corn Breeding: Challenges, Benefits And Troubles

Sweet corn has been cultivated in all continents and widespread worldwide, suggesting a large potential adaptation. Such an amazing adaptation for a defective mutant has been possible by two main procedures: spontaneous mutation in adapted field corn varieties, and production of new sweet corn varieties from crosses between sweet corn genotypes and local field corn varieties. Either way, field corn (with the wild allele *Sugary1*, *Su1*) has been the main source of favorable alleles for adaptation and agronomic performance of sweet corn (with the mutant allele *sugary1*, *su1*).

Before genetics was involved in maize breeding, spontaneous mutations were already selected by farmers and used to release improved varieties that were good enough for the colonization of most of America and some other regions of the planet by this crop. Indeed, the recessive mutant *su1* frequently appears in field corn backgrounds and quite often is successfully fixed by mass selection. With the advent of genetics, plant breeders developed genetic and experimental designs, along with advanced statistical methods that speeded the previous procedure. However, even though the newest methodologies are so advanced, their success still depends on the viability of the old *su1* mutant for specific genetic backgrounds, because *su1* can be lethal or deleterious when introduced in some genotypes of field corn. Anyway, sweet corn breeders often search alleles for cold tolerance in the germplasm collection of field corn. Actually, field corn has been successfully used to improve sweet corn in America and abroad. Tracy found that Latin American populations (1990a) and field corn inbreds (1990b) could increase agronomic performance and other traits related to yield. In Europe, Cartea *et al.* (1996a, b) and Malvar *et al.* (1997a, b) found that US and Spanish field corn populations and inbreds could be useful sources of favorable alleles to improve yield and agronomic performance and Revilla *et al.* (1998a) found sources of cold tolerance for improving sweet corn among Spanish field corn.

As previously commented, another major limitation of sweet corn performance is the higher susceptibility to pests compared to field corn. Many studies have shown the potential of flint and dent temperate corn germplasm for improving sweet corn resistance to European corn borer (Joyce and Davis 1995), resistance to corn earworm (Guo *et al.* 2001, 2004), and stalk and root quality (Treat and Tracy 1994). Similarly to what happened for cold tolerance, as a general rule field corn hybrids are more resistant to ECB than sweet corn hybrids (Jarvis 1988). However, the most resistant sweet corn inbreds often do not differ significantly from the most resistant field corn inbreds (Velasco *et al.* 1999a). Nevertheless, the field corn background is normally involved on corn borer tolerance as well; in fact, sweet corn genotypes seem to be more resistant when field corn is in their pedigree (Velasco *et al.* 1999b).

The use of field corn for improving sweet corn can have unspecific objectives and follow intuitive procedures or could be directed to specific targets and make use of a systematic plan, as adaptation or pest resistance.

4.1. Favorable Alleles for Adaptation

There are several methods for choosing the best base germplasm for breeding programs, such as general combining ability (GCA) (Sprague and Tatum 1942) and predicted three-way cross (PTC) (Sprague and Eberhart 1977) that can identify the best potential donors and receivers of favorable alleles. Therefore, the choice of potential donors of favorable alleles for improving an otherwise elite variety can rely on the average performance of parents in hybrid combinations, the heterotic pattern of potential parents, and breeders' intuition. Complementation of potential parents is relatively easy to visualize for some traits but almost impossible for quantitative characters with complex inheritance. Heterotic grouping is also difficult when the genotypes involved do not already belong to an established pattern or when adaptation is deficient or the conditions are stressful because heterosis is a function of the performance of parents and offspring, and the reduced value of poorly adapted or stressed parents biases the estimation of heterosis.

Therefore, the methods based on performance of phenotypes per se or in crosses constrained to optimum conditions provide a information generally biases. To solve these difficulties Dudley (1984, 1987) developed a statistical procedure for estimating the relative number of loci ($\mu G'$) for which a potential donor line is different from each parent of an elite single cross that should be improved. Other methods have been developed based on Dudley's approach: UBND (Gerloff and Smith 1988a,b) is an estimate of a minimum upper bound on μG, NI (Bernardo 1990) estimates the relative number of favorable alleles that can be gained from the inbred minus the relative number of favorable alleles that can be lost from parents of the hybrid during selection, and PNGg and PNGceg (Metz 1994) estimate the relative number of loci where favorable alleles can be gained as a proportion of the relative number of loci where favorable alleles can be either gained or lost.

The identification of favorable alleles in field corn for improving agronomic performance of sweet corn is a special problem because the concept of elite hybrid qualitatively differs when applied to sweet corn than when applied to field corn. In fact, a sweet corn hybrid cannot be considered elite when compared to a similar field corn genotype carrying the wild type allele *Su1* based on agronomic performance, yield or disease or pest resistance. Actually, most field corn inbreds would appear as potential donors of favorable alleles for improving most sweet corn hybrids.

Furthermore, the choice of field corn inbreds as potential donors for improving the agronomic performance of sweet corn assumes some risks, particularly the viability of the allele *su1* cannot be guaranteed when combined to same genes. More, inbred performance may change in a sweet corn background compared with a field corn background. And, finally, quality is very important in sweet corn and depends on specific combinations of genes, which involves the risk of losing quality when field corn genotypes are used to obtain new sweet corn varieties.

Nonetheless, Malvar *et al.* (1997a) identified American and Spanish field corn inbreds as useful sources of favorable alleles to improve yield and agronomic performance of sweet corn. Similarly, Malvar *et al.* (1997b) identified US and European field corn inbreds as potential donors for improving sweet corn hybrids for vigor and adaptation to European conditions. Both reports showed that some European flint inbreds were potential sources of early vigor and adaptation for improving sweet corn hybrids, but some American dent inbreds were better for improving adult plant appearance.

Cartea *et al.* (1996a b) identified field corn populations as potential sources of favorable alleles for improving agronomic performance and adaptation of sweet corn. The improvement of cold tolerance by using favorable alleles from sweet corn was also intended by Revilla *et al.* (1998). Those experiments showed that the differences among field and sweet corn genotypes were large and most field corn populations could be sources of favorable alleles for sweet corn breeding, although some populations had outstanding performance.

The obvious question is how reliable are these estimations of favorable alleles of field corn for improving sweet corn. High theoretical correlations have been reported between most of the estimators, and different estimators are more appropriate for each study for field corn hybrids or for sweet corn hybrids (Cartea *et al.* 1996a, b; Malvar *et al.* 1997a, b). Several authors (Zanoni and Dudley 1989b; Hogan and Dudley 1991; Pfarr and Lamkey 1992a, b) have published that methods for identifying donors for improving elite hybrids reliably predicted the performance of the best and worst potential donors of favorable alleles. Malvar *et al.* (2001, 2004) also found a good agreement between predictions and realized results of donors for improving yield and adaptation. The verifications of predictions of the estimator of favorable alleles have shown so far that the estimator $\mu G'$ correctly predicted the best donor of favorable alleles for improving yield and agronomic performance of sweet corn hybrids. Realized response, however, did not coincide with the predictions, particularly for the ranking of donors. The prediction of the best recipient reasonably fitted the realized results, and actually some inbreds are being produced from those experiments, although the quality is always a major concern.

Our experience with the methods of identification of favorable alleles is that the predictions are reasonably consistent and reliable. However, the outcome of the subsequent breeding programs has shown that the new *su1* varieties released from field × sweet corn crosses are unpredictable, being *su1* frequently genetically unstable, and the quality of the field corn varieties do not match the expectations.

4.2. Capitalizing Field Corn Heterotic Patterns

At difference of field corn, heterotic patterns among sweet corn inbreds are not, at least knowingly, used in sweet corn breeding. However, based on the experience with field corn, theoretically we can expect that the establishment of heterotic patterns would improve agronomic performance and adaptation of sweet corn in new regions. Hypothetically, field corn heterotic patterns might be transferable into sweet corn. The main field corn heterotic pattern used in breeding programs is Reid × Lancaster (Reif *et al.* 2005). There are also alternative heterotic patterns used worldwide; for example, European flint × U.S. dent, that is frequently used in Europe. Similarly, other heterotic patterns have been defined across European germplasm, particularly a general heterotic pattern northern × southern Europe (Revilla *et al.* 2002, 2006). The heterotic pattern northern × southern Spain is part of the northern × southern Europe heterotic pattern (Ordás 1991).

Davis *et al.* (1988) converted the field corn heterotic pattern Reid Yellow Dent × Lancaster Sure Crop into a similar, though weaker sweet corn heterotic pattern (NE-HY-13A × NE-HY-13B). On the other hand, experimental data showed that the best crosses to establish enhanced sweet corn heterotic patterns involving Spanish maize would be Golden Bantam × northern Spain and Stowell's Evergreen × southern Spain (Revilla *et al.* 2000). Based on those results, our group obtained two synthetics that combine a sweet and a field heterotic pattern: EPS31, made by crossing Golden Bantam with a composite from northern Spain, and EPS32, made by crossing Stowell's Evergreen to a composite from southern Spain. We evaluated the main four sweet corn heterotic patterns reported among *su1* cultivars (Golden Bantam × Country Gentleman, Golden Bantam × Stowell's Evergreen, NE-HY-13A × NE-HY-13B, and EPS31 × EPS32) together with their seven parents under field conditions and also in a growth chamber under cold conditions. Golden Bantam × Stowell's Evergreen was the most cold tolerant in the cold chamber and field emergence and early vigor were highest for EPS31 x EPS32. We realized that the objective of capitalizing on the field corn heterotic patterns by conversions to sweet corn failed because the heterosis available in field corn heterotic patterns was not transmitted to the new sweet corn heterotic patterns.

Another potential value of field corn heterotic patterns consists on their potential resistance to pests. As previously stated, besides adaptation, the main problems for sweet corn in Europe are corn borers. Butrón *et al.* (1999) found that hybrids with the highest yield under infestation did not have the highest yield under insect-free conditions. Some field corn synthetics, particularly the Mediterranean background, had more favorable variety effects for resistance than the sweet corn cultivars, Country Gentleman being the poorest. Among the sweet × field corn crosses, Corn Belt Dent × Golden Bantam, Corn Belt Dent × Stowell's Evergreen, and Atlantic germplasm × Country Gentleman had the highest heterotic effects for yield under corn borer infestation. The crosses Atlantic field corn × Golden Bantam and Mediterranean field corn × Stowell's Evergreen had the lowest heterotic effects. The crosses Atlantic field corn × Golden Bantam and Mediterranean field corn × Stowell's Evergreen appear the best for improving yield under infested conditions of the sweet corn heterotic pattern Stowells Evergreen × Golden Bantam.

The practical results in terms of releasing sweet corn with larger agronomic performance adapted to cold conditions or resistant to corn borers depends on *su1* viability and stability. The sweet corn varieties obtained from these genetic studies suffer from the same drawbacks than any other new *sugary1* cultivar, and they would require preliminary cycles of intrapopulation recurrent selection for agronomic performance and flavor before initiating an interpopulation recurrent selection program to enhance heterosis (Revilla *et al.* 2000). One of the main handicaps for converting field corn heterotic patterns into sweet corn is that field × sweet corn crosses could produce off-flavors due to gene interactions, dosage effects, disruption of linkage groups, etc. Consequently, commercial sweet corn breeding does not rely on heterotic patterns for developing commercial hybrids and we should not expect a great success of any new heterotic pattern we can release.

4.3. Instability of Sweet Corn Mutants

Sweet corn inbred lines are less stable than their isogenic versions carrying the wild type allele. This phenomenon of variability among versions of the same inbred line are common to other inbreds, although is more apparent for *su1* inbreds because are more susceptible to unfavorable conditions. Variability among strains of maize inbreds has been observed since the beginning of the inbred–hybrid system (Jones 1945) and has been confirmed with the advent of molecular markers (Tracy *et al.* 2000). When inbred lines are conserved in diverse environments, we can often find degenerative single-gene changes that can result even in heterosis in crosses to their respective wild types. Based on previous works and their own experiments, Tracy *et al.* (2000) concluded that most observed changes involve several loci are produced at different rates depending on the genotype, but he did not find clear causes of such changes. Fleming *et al.* (1964) considered that the divergence they observed in conservation of maize inbred lines in different locations was due to residual heterozygosity or mutation. Instability of genotypes can be due to selection, errors, and outcrosses (Kameswara Rao and Jackson 1996; Parzies *et al.* 2000). Revilla *et al.* (2004) reported variability during conservation of maize inbred lines and suggested natural selection for viability and vigor within inbred lines during storage. The problem of new variability generated during inbred conservation can be increased by the stress due to lack of adaptation. Such is the case of sweet corn, which is poorly adapted to cold humid areas. In a study of the changes produced during conservation of sweet corn inbred lines from USA in north-western Spain, we found differences between strains in the DNA that were not related to agronomic performance. Residual heterozygosity or outcrossing cannot explain these results. Lack of adaptation could have caused stress-induced mutagenesis. Natural selection could have eliminated unfavorable selective mutations, but neutral mutations can be found at the molecular level and favorable mutations could have been selected at the agronomic level.

Looking at the stability of the gene itself, *su1* spontaneously appears in field corn at a relatively high frequency compared to other mutants. From history and experience we know that *su1* can be easily fixed by self-pollination and new sweet corn genotypes can be established. In fact, spontaneous mutation is one of the explanations of the wide adaptation of sweet corn and the extraordinary resemblance between sweet corn cultivars grown in a region and the autochthonous field corn populations of that area. However, breeders also have the opposite experience, i.e. *su1* can be hardly maintained in some genetic backgrounds. Even though this is a quite common phenomenon, the regulation of the stability or lethality of *su1* is not yet understood.

When sweet corn breeders use field corn to improve sweet corn agronomic performance, frequently they have to deal with the reduced viability of *su1* plants within some field corn genetic backgrounds (Tracy1990, 2000; Butrón *et al.* 2008). Certainly, *su1* is lethal or semi-lethal for some specific sweet x field corn crosses (Revilla *et al.* 2000a, 2006; Tracy 2000) and the frequency of *su1* plants is reduced during selection programs (Martins and Da Silva 1998). One interesting point is that the defective allele *su1* is even more unstable under stress and produces genetic abnormalities, which happen more often some years, probably when the weather is farthest away from the optimum conditions. On the other hand, the donor of *su1* has also a crucial role in the survival rate or fitness of the mutant when introduced into a new wild genotype, which will be equivalent to the susceptibility to stress. The causes of differences in the survival rates of *su1* when introduced into different field corn genotypes are poorly known, and the dependence on the climatic conditions is even more ignored. Research on these issues is under way following the approach of a genetically regulated response to adverse conditions.

Presumably, the physiological cause of the reduced viability of *su1* is related to the condition of defective mutant. The deleterious effects of *su1* are mainly related to the amount of carbohydrates available in the kernel (Tracy and Chang 2007); therefore, different field corn recipients carrying *su1* could accumulate variable amounts of carbohydrates, which should be critical for plant development during the growth cycle, and particularly during the heterotrophic period, when plant growth depends on the nutrients of the grain. But the whole cycle is affected by this relationship between the production of nutrients by the green tissue and the accumulation in the kernel. The plant growth is strongly affected by the source-sink relationship. The sink effect of grain filling strongly affects vegetative nutrition in maize, and the nutrient demand is considerably reduced in *su1* and even more in shrunken2 (*sh2*) (Treat and Tracy 1994). These authors demonstrated that *sh2* improved stalk performance as a consequence of a reduction on grain yield. Such effects were smaller for *su1* and concluded that the significant reduction in reproductive sink caused by the *sh2* endosperm was not associated with decreased biomass. The effects and benefits of nutrient distribution strongly depend on genotype (Kamau *et al.* 2008b) which also affects the sink side (Kamau *et al.* 2008a).

The viability of *su1* plants has been investigated mainly at germination; for example, Revilla *et al.* (2000a, 2006) concluded that the main cause of reduction in *su1* frequency was the poor germination of *su1* kernels compared with *Su1* and that *su1* plants were more viable when Corn Belt Dent germplasm was crossed with the sweet corn population Stowell's Evergreen than when Spanish germplasm was crossed with the sweet corn populations. Differences between sweet and field corn are perceptible at the adult plant stage, particularly for ear traits. Actually, the *su1* plants have poorer performance than the *Su1* plants, particularly at earlier stages of development, and early vigor seems to be the most limiting factor for *su1* plant development, after germination. The differences in performance between *su1* and *Su1* plants decreased during plant growth and were not significant for most adult traits. Specific combinations of donors and recipients of the *su1* allele had different effects on performance of *su1* plants. Therefore, the introduction of sweet corn mutants into field corn genetic backgrounds depends on the genotypic effects on mutant fitness. The previous results suggest that the genetic regulation of *su1* and *sh2* viability depends on a complex mechanism involving additive, dominance, and epistatic genetic effects, along with environmental effects and the corresponding genotype x environment interactions. However, this research is still being carried out.

Some authors have reported partial reversion of sweet corn to starchy phenotypes under unfavorable conditions (Erwin 1934). Sweet corn breeders have frequently faced the lost of quality due to the appearance of starchy phenotypes when the temperature was low during sweet corn germination. Several notes concerning mutations and partial reversions at the *Sugary1* locus due to transposons have been published in the Maize Genetics Cooperation Newsletter, but no complete reversion has been reported so far.

4.4. Incompatible Genes

So far we have seen that *su1* can be easily introduced into any maize genetic background through backcrossing but is not viable in all maize varieties, being lethal in homozygosis for same genotypes. Besides, certain varieties converted to *su1su1* genotype become unstable and display an astonishing sort of defective kernels, partial reversions, and genetic failures, suggesting that the converted genotypes are undergoing a high mutation rate. Those observations suggest that there must be some genetic regulation of mutation and mutant viability. Furthermore, there could be specific genes that regulate mutation, reversion and compatibility among mutants and genotypes.

Until now, we have observed that viability of *su1* depends on the recipient genetic background (Revilla *et al.* 2000, 2005, 2006). The previous work demonstrates that differences of viability depend on the donor (*su1*), the receptor (*Su1*) and their interaction. In extreme situations, we have observed partial reversion of genotypes homozygous for *su1* that produce a phenotype resembling the wild type *Su1*. Moreover, genotypes that had been fixed for *su1* sometimes display an unusual variability without an obvious genetic explanation. Such awesome variation causes complications for practical breeding when the objective is to use field corn for improving adaptation or resistance to stress or to pests of sweet corn (as previously described) because the converted variety that was considered stable becomes unstable after several years and produces undesirable phenotypes that causes the rejection of the released variety. Such phenomenon is the responsible of the unpredictability of conversions of field corn into sweet corn.

The causes of the diverse viability and stability of the *su1* mutant depending on the donor and the receptor and their interaction suggest that there should be a genetic mechanism involved in the regulation of the viability of *su1*. Alternatively, there could be alleles that are incompatible with that mutant. From the theoretical perspective, this problem involves an exciting challenge because it might imply that the variability that appears in a genotype might be genetically regulated. Such a hypothesis will have implications in other important scientific issues such as degenerative diseases, ageing or even evolution.

Although the existence of genes involved on the regulation of the viability of *su1* or genes incompatible with *su1* is still a hypothesis, there are some experimental results that support this proposition. Revilla *et al.* (2006) reported differences in the viability of *su1* among diverse field corn inbreds. Butron *et al.* (2008) reported the results from a program of conversions of field corn populations into sweet corn by backcrossing; the program was discontinued due to the increasing problems found in the converted populations as the breeding program was carried out. But the most striking evidence is provided by the continuously appearance of partially reverted grains that produce the field corn genotypes converted to sweet corn under stressful conditions.

5. Molecular Genetics and Breeding

5.1. Location of the *Sugary1*, *Shrunken2* and *Sugary Enhancer1* Genes

The locus *Su1* is estimated to be between position 36,823,500 and position 39,067,700 on Chromosome 4 based on the map: IBM2 2008 Neighbors 4, while the locus *Sh2* is between 207,118,100 and position 207,123,000 on Chromosome 3 (MaizeGDB). At difference of sugary or shrunken phenotypes which has been clearly associated to one concrete gene and allele (*su1* and *sh2*), the study of genetics of sugary enhancer has resulted more difficult than originally anticipated (Schultz and Juvik 2004). Ferguson *et al.* (1978) found that the sugary enhancer phenotype segregated in a F_2 testcross generation in a ratio 3:1, indicating that a single recessive gene was responsible for the trait. A RFLP mapping population, consisting in 214 $F_{2:3}$ families derived from IL731A (a sugary enhancer inbred) × W678 (a sugary inbred), was used to localize the sugary enhancer gene. Three different analyses of the same population have shown that the sugar content at eating stage was associated to the distal end of the long arm of chromosome 2 (12.1 cM distal to probe umc36a) (Tadmor *et al.* 1995; Azanza *et al.* 1996c; Juvik *et al.* 2003). The percentage of phenotypic variation explained by this region ranged from 15 to 39% in the three experiments. Tadmor *et al.* (1995) scored the sugary enhancer phenotype based on lighter kernel colour and a more finely wrinkled pericarp texture. Based on those characteristics the sugary enhancer was also located 12.1 distal to umc36a. Katzir *et al.* (1999) develop two pairs of near isogenic lines (NIL) to map the sugary enhancer gene. The sugary inbreds IL678a and IL451b were repeatedly crossed to IL677a, the inbred where the sugary enhancer mutant was discovered. After each backcross selfed kernels were tested for the presence of sugary enhancer characteristics and, therefore, NILs were obtained that are expected to differ in the presence of the sugary enhancer allele. According to the evaluation of the NILs, the sugary enhancer phenotype did not show an association with the markers in chromosome 2, but showed an association with markers at chromosome 3 and 6. Those markers were associated to sugar content in previous experiments, particularly the marker umc50 at chromosome 3 (Tadmor *et al.* 1995; Azanza *et al.* 1996c; Juvik *et al.* 2003). However, the proportion of phenotypic variance explained by those markers is much lower than the proportion of phenotypic variance explained by the marker umc36a in chromosome 2. Furthermore, the markers at chromosome 3 and 6, at difference of the umc36a marker, were not associated to visual sugary enhancer score in a previous experiment (Tadmor *et al.* 1995). A population of 308 $F_{2:3}$ families derived from the near isogenic parents IL451b-sugary × IL451b-sugary enhancer was evaluated to fine map the sugary enhancer gene (Schultz and Juvik 2004). However, polymorphic DNA markers from any of the regions previously shown to be associated with sugars (chromosome 2, 3 and 6) were not found. La Bonte and Juvik (1991) using B-A translocations and sugar analyses mapped the sugary enhancer gene on the long arm of chromosome 4. That region was also associated with sugar content in the experiment of Tadmor *et al.* (1995), but the proportion of phenotypic variance explained by the marker at chromosome 4 (5%) was much lower than that explained by the marker at chromosome 2 (25%). Katzir *et al.* (1999), to explain the complexity of molecular tagging of sugary enhancer, hypothesize that more than one genotype can result in the sugary enhancer phenotype. That is, those authors propose, in contradiction to the previous hypothesis of a single gene mutation, that more than one locus, at which elevated sugar content are associated, explain the sugary enhancer phenotype.

5.2. Location and Effects of Quantitative Trait Loci (QTL) on Kernel Characteristics, Stand Establishment and Eating Quality

The QTL analyses corroborate the quantitative nature of traits related to kernel characteristics, stand establishment and eating quality and confirm the importance of the genetic background on those traits. Azanza *et al.* (1996c), analyzing the mapping population derived from IL731A (a sugary enhancer inbred) × W678 (a sugary inbred), detected a total of 74 unlinked chromosomal regions that influence different characteristics associated with sweet corn eating quality. The average number of QTL influencing any given trait was five, with numbers ranging from two to nine. The average percentage of phenotypic variance explained by all markers was, approximately 20%, although, depending on the trait, varied from 7.9 to 51.6%. However, given the relatively low number of families (214) and markers (93) used, a bias is expected in both the number of QTLs and the percentage of phenotypic variance explained. The number of QTL is expected to be underestimated, while the percentage of explained variance is expected to be overestimated. It has been suggested that, in QTL experiments, the remaining variation, that is, the variation not explained by the markers, comes from environmental variance, QTL with effects too small to be detected or epistatic variance (Azanza *et al.* 1996c; Holland 2007). Regarding QTL associations with kernel chemical and physical characteristics, Azanza *et al.* (1996c) found that QTL influencing starch concentration were independent of those affecting the concentration of sucrose or phytoglycogen, while all the QTL associated with a decrease in phytoglycogen concentration increased sucrose concentration. Within the chromosomal regions related to sensory attributes, out of the four different QTL affecting panel overall liking, two were also associated to other sensory attributes. Regarding the relationship between kernel characteristics (chemical or physical) and sensory attributes determined by selected panellists, several chromosomal regions were associated simultaneously to both. For example, marker umc50 at chromosome 3 is related to the kernel chemical composition (starch, sucrose and phytoglycogen), sensory attributes (sweetness, starchiness, etc) and overall liking. Also, the region around marker umc36A, possible candidate for the sugary enhancer gene, was associated to both, sucrose concentration and sweetness perception by the panellist, but, surprisingly, the region was not associated to overall liking. Juvik *et al.* (2003) also analyzed the IL731A × W678 mapping population for kernel chemical characteristics, but in addition to those characteristics Juvik *et al.* (2003) analysed characteristics related to seedling stand establishment instead of characteristics related to eating quality. Furthermore, Juvik *et al.* (2003) also analysed a *sh2* mapping population. In the sugary × sugary enhancer population the sugary enhancer morphological marker, scored based on kernel colour and pericarp texture of mature-dry kernels, was associated with emergence. The allele that was detrimental for emergence was favourable for sugar content: the difference between homozygous classes was a 29% reduction in emergence. The authors conclude that the improved eating quality of sugary enhancer comes with the cost of reduced emergence. Other two regions, at chromosome 8 and 7, were found to carry significant QTL for emergence in all locations, but again the allele that improves the emergence, decreases the sugar content. However, for the marker at chromosome 8, the allele that increases the emergence by 22%, only decreases the sugar content by 8% and, therefore, the authors believe that the QTL associate to the marker could be valuable for improving emergence. In the *sh2* population, two regions located on chromosome 2 and 8 were associated to emergence in all environments, but not to fresh harvest kernel sweetness or sensory panel overall liking. A QTL for emergence was located in chromosome 1 in some of the environments. The QTL on chromosome 8 exerted similar effects in the sugary enhancer and in the shrunken population confirming the interest of the QTL for marker assisted selection (MAS). The low values of the phenotypic variance explained by the QTL, for example 17.7 for seedling emergence in the *sh2* population (Juvik *et al.* 2003) the theoretical expectation of the efficacy of MAS or marker assisted backcrossing (MAB) is low. Because methods for obtaining unbiased estimates, like cross validation, were not used the true values of the proportion of phenotypic variance are probably underestimated and, therefore, the expectation of the efficacy of MAS or MAB is even lower.

5.3. MAS and MAB Experiments

MAS and phenotypic selection (PS) were applied to three $F_{2:3}$ populations with the *sugary1*, *sugary enhancer1* or *shrunken 2* endosperm mutations (Yousef and Juvik 2001). One cycle of selection was applied for both single and multiple traits including seedling emergence, kernel sucrose concentration, kernel terderness, and hedonic rating (taste panel preference). A total of 52 paired comparisons were made between MAS and PS composite populations. In spite of the low expectation of the efficacy of MAS, in 38% of the paired comparisons MAS resulted in significantly higher gain than PS across populations, while PS was significantly greater in only 4% of the cases. With MAS, simultaneous improvement in emergence and sucrose was observed in the sugary enhancer population and a population was obtained which provide germplasm that is of particular use in breeding programs to develop sweet corn with improved emergence and eating quality. In addition to the previous experiment, a recombinant inbred line derived from the *sh2* population that was used to detect QTL for seedling emergence (Juvik *et al.* 2003) was used as a donor parent to backcross the three QTL detected on chromosome 1, 2 and 8 into three elite commercial sweet corn inbreds (Yousef and Juvik 2002). Again the results obtained with the assistance of the markers were very promising: the introgression of the QTLs increased the emergence between 28.2 and 40.8%, depending on the particular marker. Furthermore, the marker-QTL associated effects were reproducible across generations and populations indicating that QTL identified in one population can exert similar effects in different genetic backgrounds. Therefore, it can be concluded that incorporating DNA markers to traditional sweet corn breeding programs could expedite selection progress and be cost-effective (Yousef and Juvik 2001).

5.4. Association Studies

Although, the QTL studies previously commented give interesting information about regions involved in different traits, their resolution is rather low, on the order of 20 cM or more. This distance can correspond to millions of bases in maize and therefore, real evaluation of individual genes has not occurred. One practical consequence of the low resolution of the QTL studies is that the efficacy of MAS is reduced due to recombination. Other problem of the QTL studies carried out with bi-parental populations is that only two alleles are compared. By contrast, association mapping can provide a high-resolution alternative for the evaluation of candidate genes and has the potential to evaluate a wide range of alleles by Wilson *et al.* (2004), who phenotyped 102 maize inbreds that represent most of the diversity available to breeding programs around the world. The authors sequenced six maize candidate genes involved in starch biosynthesis, including *su1* and *sh2*, in 102 inbreds to test the association between alleles and starch concentration and starch composition quality. The study identified, in the *Sh2* gene, a polymorphism consisting in a 1-bp deletion in intron eight at site 3674, *Sh2*-3674(del1), as significant associated to kernel composition, but failed to detect associations in the *Su1* gene. The association studies, due to the cost and time of DNA sequencing, were limited to specific candidate genes. It is expected that in the near future, because of the next-generation sequencing technologies, wide genome association studies will be conducted. Thus, genes related to emergence or eating quality different from sugary or shrunken will be identified, and favourable alleles will be easily and effectively incorporated into breeding programs.

6. Conclusion

Sweet corn breeding has successfully used different approach. Sweet corn breeders can be proud for obtaining the first commercial single corn hybrids. Sweet corn breeding has also productively used the variation generated by new mutations as is the case of *sh2*, a mutant that has produced, and nowadays still generates, huge economical benefits. Also, the introgression of exotic germplasm has been successful, the most notorious example being the sugary enhancer varieties obtained by introgression from a Bolivian variety. An important aspect of sweet corn breeding is the negative relationship between stand establishment and eating quality because the chemical or physical properties of the kernel that favor the former trait are detrimental for the latter. However, it has been shown that to select for both agronomic performance and eating quality is possible. In addition to mutants with large effects, genetic background has an important role on sweet corn characteristics, with a significant mutant by genotype interaction that affects sweet corn performance and even mutant viability; these effects hamper the use of field corn for sweet corn breeding. Losses due to pests and diseases are higher in sweet corn than in field corn because consumer tolerance to the presence of damage in the ears is much lower. Genetically modified maize has been successfully used to avoid insect damage, although natural genetic resistance could be an alternative or complement other control means as pesticides or transgenic technology. Based on the successfully used of MAS or MAB in some experiments, some authors conclude that incorporating DNA markers to traditional sweet corn breeding programs could expedite selection progress and be cost-effective. In the near future, the next-generation sequencing technologies will facilitate the detection of genes related to emergence or eating quality and favorable alleles will be more effectively incorporated into breeding programs.

REFERENCES

Andrew, RH. Factors influencing early seedling vigor of shrunken-2 maize. *Crop Sci,* 1982 22, 263-266.

Andrew, RH; Brink, RA; Neal, NP. Some effects of the waxy and sugary genes on endosperm development in maize. J Agric Res, 1944 69, 355-371.

Andrew, RH; Carlson, JR. Evaluation of sweet corn inbreds for resistance fro European corn borer. *J Amer Sco Hort Sci,* 1976 101, 97-99.

Andrew, RH; Mosely, PR. Second generation European corn borer damage relationships in sweet corn. *Crop Sci,* 1980 20, 559-562.

Azanza, F; Bar-Zur, A; Juvik, JA. Variation in sweet corn kernel characteristics associated with stand establishment and eating quality. *Euphytica*, 1996b 87, 7-18.

Azanza, F; Juvik, JA; Klein, BP. Relationships between sensory quality attributes and kernel chemical-composition of fresh-frozen sweet corn. *J Food Qual*, 1994 17, 159-172.

Azanza, F; Klein, BP; Juvik, JA. Sensory characterization of sweet corn lines differing in physical and chemical composition. *J Food Sci,* 1996a 61, 253-257.

Azanza, F; Tadmor, Y; Klein, BP; Rocheford, TR; Juvik, JA. Quantitative trait loci influencing chemical and sensory characteristics of eating quality in sweet corn. *Genome*, 1996c 39, 40-50.

Bailey, DM; Bailey, RM. The relation of the pericarp to tenderness in sweet corn. Proc Amer *Soc Hortic Sci,* 1938 36, 555-559.

Bell, RD; Darrah, LL; Zuber, MS. Progress from mass selection for field emergence and seed weight in a *sh2* population of maize. *Crop Sci*, 1983 23, 461-464.

Bernardo, R. An alternative statistic for identifying lines useful for improving parents of an elite single cross. *Theor Appl Genet*, 1990 80, 105-109.

Bjorkman, T; Blanchard, LM; Harman, GE. Growth enhancement of shrunken-2 (*sh2*) sweet corn by Trichoderma harzianum 1295-22: Effect of environmental stress. *J Amer Soc Hortic Sci,* 1998 123, 35-40.

Bourguet, D; Desquilbet, M; Lemarie, S. Regulating insect resistance management: *The case of non-Bt refuges in the US Environ. Manage*, 2005 76, 210-220.

Boyer, C; Shannon, JC. The use of endosperm genes for sweet corn improvement. *Plant Breed. Rev.*, 1983 1, 139-161.

Brewbaker, JL. 'Hawaiian Super-sweet #9' corn. *HortScience*, 1977 12, 355-356.

Brewbaker, JL; Banafunzi, N. 'Hawaiian #6' corn. *HortScience*, 1975 10, 427-428.

Creech, RG. Carbohydrate synthesis in maize. *Adv. Agron.*, 1968 20, 275-322.

Butrón, A; Álvarez, A; Revilla, P; Malvar, RA; Rodríguez, VM; Ruiz de Galarreta, JI; Ordás, A. Agronomic performance of sweetcorn populations derived from crosses between sweetcorn and field corn. *Spanish J Agric Res*, 2008 6, 378-384.

Butrón, A; Malvar, RA; Velasco, P; Vales, MI; Ordás, A. Combining abilities for maize stem antibiosis, yield loss and yield under infestation and no infestation with pink stem borer. *Crop Sci*, 1999 39, 691-696.

Butrón, A; Ordás, B; Revilla, P; Sandoya, G; Ordás, A; Malvar, RA. Is leaf or sheath antibiosis involved in the resistance of maize composite EPS12 to *Sesamia nonagrioides*?. *Can Entomol*, 2005 137, 350-355.

Carlson, JR; Andrew, RH. Parameters for determination of damage to sweet corn genotypes from first and second generation European corn borer. *Crop Sci,* 1976 16, 40-42.

Cartea ME, RA Malvar, P Revilla, A Ordás. 1996a. Identification of field corn populations to improve sweet corn for Atlantic European conditions. *Crop Science* 36:1506-1512.

Cartea ME, RA Malvar, P Revilla, A Ordás. 1996b. Improvement of early vigor and adaptation of sweet corn for the European Atlantic coast with open-pollinated field corn populations. *Maydica* 41:119-125.

Chiang, HC; Hodson, AC. European corn borer damage to sweet corn as affected by the date of planting. *J Econ Entomol,* 1963 56, 243-248.

Creech, RG. Genetic control of carbohydrate synthesis in maize endosperm. *Genetics*, 1965 52, 1175-1186.

Creech, RG. Carbohydrate synthesis in maize. *Adv. Agron.*, 1968 20, 275-322.

Culpepper, CW; Magoon, CA. A study of the factors determining quality in sweet corn. *J Agr Res*, 1927 34, 413-433.

Davis, DW; Brewbaker, JL; Kaukis, K. Registration of 'NE-HY-13A' and 'NE-HY-13B' complementary cultivars of maize germplasm. *Crop Sci*, 1988 28, 381.

Diveli, GP. Does it pay to grow Bt sweet corn. 2007 January. Available from: *http://www.rogersadvantage.com/pdf/GreatLakesBtCornInfo.pdf.*

Douglass, SK; Juvik, JA; Splittstoesser, WE. Sweet corn seedling emergence and variation in kernel carbohydrate reserves. *Seed Sci. Technol,* 1993 21, 433-445.

Dudley, JW. A method of identifying lines for use in improving parents of a single cross. *Crop Sci,* 1984 24, 355-357.

Dudley, JW. Modification of methods for identifying inbred lines for improving parents of elite single crosses. *Crop Sci,* 1987 27, 944-947.

Erwin, A. A rare specimen of *Zea mays* var. *Saccharata*. *Science*, 1934 79, 589.

Erwin, a: Sweet corn in the pre-Columbian period in the upper Missouri region. *J. Am. Soc. Agron.*, 1947 39, 833-834.

Erwin, AT. Sweet corn –mutant or historic species? *Econ. Bot.*, 1951 5:302-306.

Edelson, JV; Mackey, C. Controlling corn earworm on sweet corn. *Arthropod Management* Test, 2006 31, Report E20.

Evensen, KB; Boyer, CD. Carbohydrate composition and sensory quality of fresh and stored sweet corn. *J Amer Soc Hort Sci,* 1986 111, 734-738.

Farrar, RR; Shepard, BM; Shapiro, M; Hassell, RL; Schaffer, ML; Smith, CM. Supplemental control fo lepidopterousj pests on Bt transgenic sweet corn with biologically-based spray treatments. *J Insect Sci,* 2009 9, 1-10.

Ferguson, JE; Dickinson, DB; Rhodes, AW. Analysis of endosperm sugars in a sweet corn inbred (Illinois 677a) which contains the sugary enhancer gene and a comparison of *se* with other corn genotypes. *Plant Physiology*, 1979 63, 416-420.

Ferguson, JE; Rhodes AM; Dickinson, DB. The genetics of sugary enhancer (*se*), an independent modifier of sweet corn (*su*). *J Hered,* 1978 69, 377-380.

Fleming AA; Kozelnicky, GM; Browne, EB. Variation between stocks within long-time inbred lines of maize (Zea mays L.). *Crop Sci,* 1964 4, 291-295.

Flora, LF; Wiley, RC. Sweet corn aroma, chemical components and relative importance in the overall flavor response. *J Food Sci,* 1974 39, 770-773.

Galinat, WC. The evolution of sweet corn. *Mass. Agric. Stn. Bull. 591*, 1971.

Gardner, J; Hoffmann MP. Resistance to European corn borer in processing sweet corn. *HortScience,* 2000 35, 871-874.

Garwood DL; Creech, RG. 'Pennfresh ADX' hybrid sweet corn. *HortScience*, 1979 14, 645.

Gassmann, AJ; Carriere, Y; Tabashnik, BE. Fitness cost of insect resistance to Bacillus thuringiensis. *Annual Review of Entomology*, 2009 54, 147-163.

Gerdes, JT; Behr, CF; Coors, JG; Tracy, WF. *Compilation of North American maize breeding germplasm.* Madison, WI: Crop Science Society of America; 1993.

Gerdes, JT; Tracy, WF. Diversity of historically important sweet corn inbreds as determined by RFLPs, morphology, isozymes, and pedigrees. *Crop Sci.*, 1994 34:26-33.

Gerloff, JE; Smith, OS. Choice of method for identifying germplasm with superior alleles 1. Theoretical results. *Theor Appl Genet,* 1988a 76, 209-216.

Gerloff, JE; Smith, OS. Choice of method for identifying germplasm with superior alleles 2. Computer simulation results. *Theor Appl Genet*, 1988b 76, 217-227.

Gonzales, JW; Rhodes, AM; Dickinson, DB. A new inbred with high sugar content in sweet corn. *HortScience*, 1974 9, 79-80.

Grier, SL; Davis, DW. Infestation procedures and heritability of characters used to estimate ear damage caused by second-brood European corn borer (*Ostrinia nubilalis*, Hubner) on corn. *J Amer Soc Hort Sci,* 1980 105, 3-8.

Grier, SL; Davis, DW. Optimizing European corn borer (Lepidoptera: Pyralidae) infestation of maize ears for kernel damage evaluations. *J Econ Entomol*, 1983 76, 429-431.

Guo, BZ; Zhang, ZJ; Li, RG; Widstrom, NW; Snook, ME; Lynch, RE; Plaisted, D. Restriction fragment length polymorphism markers associated with silk maysin, antibiosis to corn earworm (Lepidoptera: Noctuidae) larvae, in a dent and sweet corn cross. *J Econ Entomol,* 2001 94, 564-571.

Guo, BZ; Zhang, ZJ; Butrón, A; Widstrom, NW; Snook, ME; Lynch, RE; Plaisted, D. Lost *P1* allele in *sh2* sweet corn: quantitative effects of *P1* and *a1* genes on concentration of maysin, apimaysin, methoxymaysin, and chlorogenic acid in maize silk. *J Econ Entomol,* 2004 97, 2117-2126.

Hannah, LC; Cantliffe, DJ. Percentage stand and sugars in four Florida sweet corns. *Proc Fla State Hort Soc*, 1977 90, 412-413.

Haskell, G; Singleton, RW. Use of controlled low temperature in evaluating the cold hardiness of inbred and hybrid maize. *Agron J*, 1949 41; 34-40.

Hassell, RL; Dufault, RJ, Phillips, TL. Low-temperature germination response of su, se, and sh(2) sweet corn cultivars. *HortTechnology*, 2003 13, 136-141.

Hogan, RM; Dudley, JW. Evaluation of a method for identifying sources of favorable alleles to improve an elite single cross. *Crop Sci*, 1991 31, 700-704.

Holland, JB. Genetic architecture of complex traits in plants. *Current Opinion in Plant Biology*, 2007 10, 156-161.

Hotchkiss, JR; Revilla, P; Tracy, WF. Variation of cold tolerance among open-pollinated sweet corn cultivars. *HortScience,* 1997 32, 719-723.

Huelson, WA. *Sweet corn.* New York, USA: Interscience Publ Inc; 1954.

Iowa State University. European corn borer. 2006. Available from: *http://www.ent.iastate.edu/pest/cornborer/.*

Ito, GM; Brewbaker, JL. Genetic advance through mass selection fro tenderness in sweet corn. *J Amer Soc Hort Sci,* 1981 106, 469-499.

Jarvis, JL. Damage by European corn borer (Lepidoptera: Pyralidae) to popcorn kernels compared with damage to kernels of dent corn and sweet corn. *J Econ Entomol,* 1988 81, 1214-1216.

Jones, DF. Selection of pseudo-starchy endosperm in maize. *Genetics*, 1919 4, 364-393.

Jones, DF. Heterosis resulting from degenerative changes. *Genetics*, 1945 30, 527-542.

Juvik, JA; D'Arcy, CJ. *Sugary (su)* and *sugary enhancer (se)* sweet corn inbreds with resistance to maize dwarf mosaic virus. *HortScience*, 1988 23, 412-413.

Juvik, JA; Jangulo, MC; Headrick, JM; Pataky, JK; Tracy, WF. Kernel changes in a *shrunken2* maize population associated with selection for increased field emergence. *J Amer Soc Hort Sci*, 1993 118, 135-140.

Juvik, JA; Mikel, MZ; Carey, EE; Rhodes AM. Release of six Illinois sweet corn inbreds with the sugary enhancer (*se*) gene. *HortScience*, 1983 18, 965-966.

Juvik, JA; Yousef, GG; Han, T; Tadmor, Y; Azanza, F; Tracy, WF; Barzur, A; Rocheford, TR. QTL influencing kernel chemical composition and seedling stand establishment in sweet corn with the *shrunken2* and *sugary enhancer1* endosperm mutations. *J Amer Soc Hort Sci,* 2003 128, 864-875.

Joyce MS; Davis, DW. Transmittability of ear resistance to European corn borer in sweet corn testcrosses and resistance stability. *J Amer Soc Hort Sci,* 1995 120, 107-111.

Kamau, DM; Spiertz, JHJ; Oenema, O. Carbon and nutrient stocks of tea plantations differing in age, genotype and plant population density. Plant and Soil, 2008a 307, 29-39.

Kamau, DM; Spiertz, JHJ; Oenem, O; Owuor, PO. Productivity and nitrogen use of tea plantations in relation to age and genotype. *Field Crops Research*, 2008b 108, 60-70.

Kameswara Rao, N; Jackson, MT. Seed longevity of rice cultivars and strategies for their conservation in genebanks. *Ann Bot,* 1996 77, 251-260.

Katzir, N; Tzuri, G; Meir, A; Chemke, T; Goldman, T; Juvik, J; Bar-Zur, A; Tadmor, Y. RAPD analysis of two pairs of *su1se/su1Se* near-isogenic lines for the identification of chromosomal regions affecting the *sugary enhancer* phenotype. *Maydica*, 1999 44, 149-153.

Kerr, EA. Sweet corn inbreds released in 1967 by the Horticultural Research Institute of Ontario. *Hortic. Res. Inst. Ont. Rep*. 1967 1968, 40-45.

Kerr, EA. High quality sweet corn inbreds released in 1971 and 1972 by the Horticultural Research Institute of Ontario. *Can. J. Plant Sci*., 1973 53, 177-179.

Kerr, EA. More high quality yellow sweet corn inbreds released from Simcoe, Ontario. *Can. J. Plant Sci*., 1979a 59, 887-890.

Kerr, EA. Goldenvee sweet corn. *Can. J. Plant Sci*., 1979b 59, 891-892.

Kerr, EA. Tablevee sweet corn. *Can. J. Plant Sci*., 1979c 59, 893-894.

Kerr, EA. Northernvee and Earlyvee sweet corns. *Can. J. Plant Sci*., 1981a 61, 471-472.

Kerr, EA. Flavorvee sweet corn. *Can. J. Plant Sci*., 1981b 61, 487-488.

Khan, AA; Ilyas, S; Ptasznik, W. Integrating low water potential seed hydration with other treatments to improve cold tolerance. *Annals of Botany,* 1995 75, 13-19.

Kramer, A; Guyer, RB; Ide, LE. Factors affecting the objective and organoleptic evaluation of quality in sweet corn. *Pro Am Soc Hort Sci,* 1949 54, 342-356.

La Bonte, DR; Juvik, JA. Characterization of *sugary1* (*su1*) sugary enhancer (*se*) kernels in segregating sweet corn populations. *J Am Soc Hort Sci,* 1990 155, 153-157.

Lamb, EM; Davis, DW; Andow, DA. Mid-parent heterosis and combining ability of European corn borer resistance in maize. *Euphytica*, 1994 72, 65-72.

Laughnan, JR. The effect of the sh2 factor on carbohydrate reserves in the mature endosperm of maize. *Genetics*, 1953 38, 485-499.

Linduska, JJ; Harrison, FP. Corn earworm: artificial infestation as a means of evaluating resistance in sweet corn. *J Econ Entomol,* 1977 70, 565-567.

MaizeGDB. Available from: *http://www.maizegdb.org/.*

Malvar RA; Cartea, ME; Revilla, P; Ordás, A. Identification of field corn inbreds adaptaed to Europe to improve agronomic performance of sweet corn hybrids. *Crop Sci,* 1997a 37, 1134-1141.

Malvar, RA; Cartea, ME; Revilla, P; Soengas, P; Ordás, A. Verification of predictions from estimators of favorable alleles to improve yield of sweet corn hybrids. *Maydica*, 2004 49, 49-55.

Malvar, RA; Revilla, P; Cartea, ME; Butrón, A; Ordás, A. Checking performance of field corn inbreds as donors of favorable alleles to improve early vigor and adaptation of sweet corn hybrids to European conditions. *Maydica*, 2001 46, 187-193.

Malvar, RA; Revilla, P; Cartea, ME; Ordás, A. Field corn inbreds to improve sweet corn hybrids for early vigor and adaptation to Europen conditions. *Maydica*, 1997b 42, 247-255.

Malvar, RA; Revilla, P; Velasco, P; Cartea, ME; Ordás, A. Insect damage to sweet corn hybrids in the south Atlantic European coast. *J Amer Soc Hort Sci*, 2002 127, 693-696.

Mangelsdorf, PC. *Corn: its origin, evolution, and improvement.* Cambridge, MA: The Belknap Press of Harvard University Press; 1974.

Marshall, SW. Sweet corn. In: Watson SA, Ramstad PE. *Corn: Chemistry and Technology.* St. Paul, Minnesota, USA: American Association of Cereal Chemists, Inc; 1987; 431-445.

Marshall, SW; Tracy, WF. Sweet corn. In: White PJ, Johnson LA. *Corn: Chemistry and Technology, 2nd edition*. St. Paul, Minnesota, USA: American Association of Cereal Chemists, Inc; 2003; 537-569.

Martins, MEQP; Da Silva, WJ. Genetic and genotypic frequencies of endosperm mutants in maize populations under natural selection. *J Hered,* 1998 89, 516-524.

Mather, K; Haskell, G. Breeding cold hardy sweet corn in Britain. J Agr Sci, 1949 39, 56-63.

Metz, G. Probability of net gain of favorable alleles for improving an elite single cross. *Crop Sci,* 1994 34, 668-672.

Mikel, MA; D'Arcy, CJ; Rhodes, AM; Carey, EE; Juvik, JA. Sugary (*su*) sweet corn germplasm with resistance to the maize dwarf mosaic virus. *HortScience*, 1983 18, 964-965.

Nass, HG; Crane, PL. Effect of endosperm mutants on germination and early seedling growth rate in maize (*Zea mays* L.). *Crop Sci,* 1970 10, 139-140.

Neuffer, MG; Coe, EH; Wessler, SR. *Mutants of maize.* New York: Cold Spring Harbor Laboratory Press; 1997.

Ordás, A. Heterosis in crosses between American and Spanish populations of maize. *Crop Sci,* 1991 31, 931-935.

Ordás, A; Revilla, P; Malvar, RA; Cartea, ME. Development of sweet corn hybrids adapted to environmental conditions of the northwest of Spain. *Maydica*, 1994 39, 171-175.

Ordás, B; Malvar, RA; Soengas, P; Ordás, A; Revilla, P. *Sugary1* inbreds to improve *sugary enhander1* hybrids of sweet corn for adaptation to cold areas with short growing seasons. *Maydica*, 2004 49, 279-288.

Ordás, B; Padilla, G; Malvar, RA; Ordás, A; Rodríguez, VM; Revilla, P. Cold tolerance improvement of *sugary enhancer1* hybrids of sweet corn. *Maydica*, 2006 51, 567-574.

Ordás, B; Revilla, P; Ordás, A; Malvar, RA. Hybrids sugary × sugary enhancer of sweet corn: a valuable option for cool environments. *Scientia Hort,* 2008 118, 111-114.

Ordás, B; Revilla, P; Soengas, P; Ordás, A; Malvar, RA. *Sugary enhancer1* inbreds to improve the quality of *sugary1* hybrids of sweet corn. *J Amer Soc Hort Sci,* 2005 130, 68-74.

Ordás, B; Rodriguez, VM; Romay, MC; Malvar, RA; Ordás, A; Revilla, P. Adaptation of super-sweet corn to early planting: Mutant by genotype interaction. *J Agric Sci,* 2009 (accepted).

Pereira, EJG; Lang, BA; Storer, NP; Siegfried, BD. Selection for Cry1F resistance in the European corn borer and cross-resistance to other Cry toxins. *Entomologia Experimentalis et Applicata*, 2008 126, 115-121.

Pfarr, DG; Lamkey, KR. Evaluation of theory for identifying populations for genetic improvement of maize hybrids. *Crop Sci,* 1992a 32, 663-669.

Pfarr, DG; Lamkey, KR. Comparison of methods for identifying populations for genetic improvement of maize hybrids. *Crop Sci,* 1992b 32, 670-677.

Parzies, HK; Spoor, W; Ennos, RA. Genetic diversity of barley landrace accessions (Hordeum vulgare ssp. vulgare) conserved for different lengths of time in ex situ gene banks. *Heredity*, 2000 84, 476-486.

Reif, JC; Hallauer, AR; Melchinger, AE. Heterosis and heterotic patterns in maize. *Maydica*, 2005 50, 215-223.

Revilla, P; Abuín, MC; Malvar, RA; Soengas, P; Ordás, B; Ordás, A. Genetic variation between Spanish and American versions of sweet corn inbred lines. *Plant Breed,* 2005 124, 268-271.

Revilla, P; Boyat, A; Álvarez, Á; Gouesnard, B; Soengas, P; Ordás, A; Malvar, RA. Heterotic patterns among French and Spanish maize populations. *Maydica*, 2006a 51, 525-535.

Revilla, P; Malvar, RA; Abuín, MC; Ordás, B; Soengas, P; Ordás, A. Genetic background effect on germination of *su1* maize and viability of the *su1* allele. *Maydica*, 2000a 45, 109-111.

Revilla, P; Malvar, RA; Cartea, ME; Ordás, A. Identifying open-pollinated populations of field corn as sources of cold tolerance for improving sweet corn. *Euphytica*, 1998 101, 239-247.

Revilla, P; Malvar, RA; Cartea, ME; Soengas, P; Ordás, A. Heterotic relationships among European maize inbreds. *Euphytica*, 2002 126, 259-264.

Revilla, P; Malvar, RA; Rodríguez, VM; Butrón, A; Ordás, B; Ordás, A. Variation of *sugary1* and *shrunken2* frequency in different maize genetic backgrounds. *Plant Breed,* 2006b 125, 478-481.

Revilla, P; Rodríguez, VM; Malvar, RA; Butrón, A; Ordás, A. Comparison among sweet corn heterotic patterns. *J Amer Soc Hort Sci,* 2006c 131, 388-392.

Revilla, P; Tracy, WF. Isozyme variation and phylogenetic relationships among open-pollinated sweet corn cultivars. *Crop Sci,* 1995a 35, 219-227.

Revilla, P; Tracy, WF. Morphological characterization and classification of open-pollinated sweet corn cultivars. *J Amer Soc Hort Sci*, 1995b 120, 112-118.

Revilla, P; Tracy, WF. Heterotic patterns among open-pollinated sweet corn cultivars. *J Amer Soc Hort Sci*, 1997 122, 319-324.

Revilla, P; Velasco, P; Malvar, RA; Ordás, A. Variability among maize inbred lines for seed longevity. *Gen Res Crop Evol,* 2004 53, 771-777.

Revilla, P; Velasco, P; Vales, MI; Malvar, RA; Ordás, A. Cultivar heterosis between sweet and Spanish field corn. *J Amer Soc Hort Sci*, 2000b 125, 684-688.

Rhodes, AM; Carey, EE; Dickinson, DB. Illinois sweet corn inbreds with the *su* genotype. *HortScience*, 1982 17, 411-412.

Rowe, DE; Garwood, DL. Effects of four maize endosperm mutants on kernel vigor. *Crop Sci,* 1978 18, 709-712.

Schmidt, DH; Tracy, WF. Endosperm type, inbred background, and leakage of seed electrolytes during imbition in sweet corn. *J Amer Sco Hort Sci,* 1988 113, 269-272.

Schultz, JA; Juvik, JA. Current models for starch synthesis and the *sugary enhancer1* (*se1*) mutation in *Zea mays*. *Plant Physiology and Biochemistry,* 2004 42, 457-464.

Shehata, MA; Hargraves, JC; Davis, DW. A new technique for determining resistance of sweet corn to second-brood European corn borer. *HortScience*, 1975 10, 277-278.

Showalter, RK; Miller, LW. Consumer preference for high-sugar sweet corn varieties. *Pro Fla State Hort Soc,* 1962 75, 278-280.

Siqueira, HAA, Moellenbeck, D; Spencer, T; Siegfried, BD. Cross-resistance of Cry1Ab-selected *Ostrinia nubilalis* (Lepidoptera: Crambidae) to *Bacillus thuringiensis* delta-endotoxins. *J Econ Entomol*, 2004 97, 1049-1057.

Singleton, WR; Jones, DF. Sweet corn inbreds. Conn. Agric. Exp. Stn. Circ. 1936 112, 50-58.

Singleton, WR; Jones, DF. Early sweet corn hybrids Spancross, Marcross and Carmelcross. *Conn. Agric. Exp. Sta. Circ.* 1939 138, 1-11.

Singleton, RW. Noyes Darling, first maize breeder. *J. Hered.*, 1944 35, 265-267.

Singleton, RW. Hybrid sweet corn. Conn. Agric. *Exp. Stn. Bull.* 518 1948.

Smith, GM. In: Sprague GF, editor. *Corn and corn inprovement,* 1ª ed. New York: Academic Press, Inc.; 1955; 441-463.

Soberalske, RM; Andrew, RH. Gene effects on kernel moisture an sugars of near-isogenic lines of sweet corn. *Crop Sci,* 1978 18, 743-746.

Soberalske, RM; Andrew, RH. Gene effects on water soluble polysaccharides and starch of near-isogenic lines of sweet corn. *Crop Sci,* 1980 20, 201-204.

Styer, RC; Cantliffe, DJ. Relationship between environment during seed development and seed vigor of two endosperm mutants of corn. *J Amer Soc Hort Sci,* 1983a 108, 717-720.

Styer, RC; Cantliffe, DJ. Changes in seed structure and composition during development and their effects on leakage in two endosperm mutants of sweet corn. *J Amer Soc Hort*, 1983b 108, 721-728.

Styer, RC; Cantliffe, DJ; Hannah, LC. Differential seed and seedling vigor in *shrunken-2* compared to three other genotypes of corn at various stages of development. *J Amer Sco Hort Sci,* 1980 105, 329-332.

Tabashnik, BE; Carriere, Y; Dennehy, TJ; Morin, S; Sisterton, MS; Roush, RT; Shelton, AM; Zhao, J. Insect resistance to transgenic Bt crops: lessons from the laboratory and field. *J Econ Entomol,* 2003 96, 1031-1038.

Tadmor, Y; Azanza, F, Han, T; Rocheford, TR; Juvik, JA. RFLP mapping of the *sugary enhancer1* gene in maize. *Theor Appl Genet*, 1995 91, 489-494.

Tracy, WF. Sweet corn. In: Hallauer AR. *Specialty corns*, 2nd edition. Boca Raton, Florida, USA: CRC; 2001; 155-198.

Tracy, WF. Potential contributions of five exotic maize populations to sweet corn improvement. *Crop Sci*, 1990a 30, 918-923.

Tracy, WF. Potential of field corn germplasm for the improvement of sweet corn. *Crop Sci,* 1990b 30, 1041-1045.

Tracy, WF. History, breeding, and genetics of supersweet corn. *Plant Breed. Rev*., 1997 14, 189-236.

Tracy, WF. Vegetable uses of maize (corn) in pre-Columbian America. *HortScience*, 1999 34, 812-813.

Tracy, WF. Sweet corn. In: Hallauer AR, editor. *Specialty corns, 2nd edition.* New York: CRC Press; 2001; 155-193.

Tracy, WF; Chang YM. Effects of divergent selection for endosperm appearance in a *sugary1* maize population. *Maydica*, 2007 52, 71-79.

Tracy, WF; Schmidt, DH. The effect of endosperm type on pericarp thickness in near-isogenic lines of sweet corn. *Crop Sci,* 1987 27, 692-694.

Tracy, WF; Talbert, LE; Gerdes, JT. Molecular variation and F_1 performance among strains of the sweet corn inbred P39. *Crop Sci,* 2000 40, 1763-1768.

Tracy, WF; Whitt, SR; Buckler, ES. Recurrent mutation and genome evolution: Example of *sugary1* and the origin of sweet maize. *Plant Genome*, 2006 1:S49-S54.

Treat, CL, Tracy, WF. Endosperm type effects on biomass production and on stalk and root quality in sweet corn. *Crop Sci,* 1994 34, 396-399.

Tyutyunov, Y; Zhadanovskaya, E; Bourguet, D; Arditi, R. Landscape refuges delay resistance of the European corn borer to Bt-maize: A demo-genetic dynamic model. *Theor Pop Biol,* 2008 74, 138-146.

Velasco, P; Malvar, RA; Butron, A; Revilla, P; Ordás, A. Ear feeding resistance of sweet corn inbreds to pink stem borer. *J Amer Soc Hort Sci*, 1999a 124, 268-272.

Velasco, P., R.A. Malvar, P. Revilla, A. Butrón, and A. Ordás. Ear resistance of sweet corn populations to *Sesamia nonagrioides* (Lepidoptera: Noctuidae) and *Ostrinia nubilalis* (Lepidoptera: Pyralidae). *J. Econ. Entomol*, 1999b 92, 732-739.

Velasco, P; Revilla, P; Butrón, A; Ordás, B; Ordás, A; Malvar, RA. Ear damage of sweet corn inbreds and their hybrids under multiple corn borer infestation. *Crop Sci*, 2002 42, 724-729.

Velasco, P; Revilla, P; Cartea, ME; Ordás, A; Malvar, RA. Resistance of early maturing sweet corn varieties to damage caused by *Sesamia nonagrioides* (Lepidoptera: Noctuidae). *J Econ Entomol,* 2004a 97, 1432-1437.

Velasco, P; Soengas, P; Revilla, P; Ordás, A; Malvar, RA. Mean generation analysis of the damage caused by *Sesamia nonagrioides* (Lepidoptera: Noctuidae) and *Ostrinia nubilalis* (Lepidoptera: Crambidae) in sweet corn ears. *J Econ Entomol,* 2004b 97, 120-126.

Wann, EV; Brown, GB; Hills, WA. Genetic modifications of sweet corn quality. *J Amer Soc Hort Sci*, 1971 96, 441-444.

Wann, EV; Hills, WA. Tandem mass selection in a sweet corn composite for earworm resistance and agronomic characters. *HortScience*, 1975 10, 168-170.

Warnock, DF; Davis, DW. Comparison of two visual scales for estimating European corn borer ear damage in maize. *HortScience*, 1998 33, 1048-1049.

Waters, L; Blanchette, BL. Prediction of sweet corn field emergence by conductivity and cold test. *J Amer Sco Hort Sci*, 1983 180, 778-781.

Wellhausen, EJ; Roberts, LM; Hernandez, XE. *Races of maize in Mexico*. Cambridge, MA: Harvard University Press; 1952.

Wilson, DO; Trawatha, SE. Physiological maturity and vigor in production of florida staysweet shrunken-2 sweet corn seed. *Crop Sci,* 1991 31, 1640-1647.

Wilson, LM; Whitt, SR; Ibañez, AM; Rocheford, TR; Goodman, MM; Buckler, ES. Dissection of maize kernel composition and starch production by candidate gene association. *Plant Cell*, 2004 16, 2719-2733.

Widstrom, NW; Davis, JB. Analysis of two diallel sets of sweet corn inbreds for corn earworm injury. *Crop Sci,* 1967 7, 50-52.

Wiseman, BR; Davis, FM; Campbell, JE. Mechanical infestation device used in fall armyworm plant resistance programs. *Florida Entomologist*, 1980 63, 425-432.

Wiseman, BR; Widstrom, NW; McMillian, WW. Methods of application and numbers of eggs of the corn earworm required to infest ears of corn artificially. *J Econ Entomol,* 1974 67, 74-75.

Wolf, EA. Florida Staysweet. Fla. Agric. *Exp. Stn. Circ*. S-259; 1978.

Wong, AD; Juvik, JA; Breeden, DC; Swiader, JM. Shrunken2 sweet corn yield and the chemical components of quality. *J Amer Soc Hort Sci,* 1994 119, 747-755.

Young, TE; Juvik, JA; DeMason, DA. Changes in carbohydrate composition and α-amylase expression during germination and seedling growth of starch-deficient endosperm mutants of maize. *Plant Sci,* 1997 129, 175-189.

Yousef, GG; Juvik, JA. Comparison of phenotypic and marker-assisted selection for quantitative traits in sweet corn. *Crop Sci*, 2001 41, 645-655.

Yousef, GG; Juvik, JA. Enhancement of seedling emergence in sweet corn by marker-assisted backcrossing of beneficial QTL. *Crop Sci*, 2002 42, 96-104.

Zan, GH; Brewbaker, JL. Seed quality of isogenic endosperm mutants in sweet corn. *Maydica*, 1999 44, 271-277.

Zanoni, U., and J.W. Dudley. 1989. Comparison of different methods of identifying inbreds useful for improving elite maize hybrids. *Crop Sci*. 29:577-582.

In: Horticulture in the 21st Century
Editors: Abigail N. Sampson, pp. 81-120

ISBN 978-1-61668-582-9

Chapter 3

GRAFTING: AN ENVIRONMENTALLY FRIENDLY TECHNIQUE TO OVERCOME SOIL-BORNE DISEASES AND IMPROVE THE OUT OF SEASON PRODUCTION OF WATERMELON, CUCUMBER AND MELON

Fotios A. Bletsos*[1] *and Harold C. Passam

National Agricultural Research Foundation (NAGREF), Agricultural Research Centre of Northern Greece, P.O.Box 60458, GR-57001 Thermi-Thessaloniki, Greece
Laboratory of Vegetable Production, Agricultural University of Athens, Iera Odos 75, Votanikos, 118 55 Athens, Greece

ABSTRACT

Due to the limited availability of suitable land and the high consumer demand for out of season produce, growers tend to cultivate Cucurbit crops (watermelon, cucumber and melon) repeatedly on the same soil and often under unfavourable environmental conditions (e.g. low temperatures, insufficient light, the presence of soil-borne pathogens etc.), resulting in reduced yield and quality. However, many of these constraints may be largely overcome by grafting plants at the seedling stage on to compatible rootstocks. Moreover, this beneficial practice has the additional advantage of being 'friendly' towards the environment. Suitable rootstocks are those that exhibit resistance to soil-borne pathogens, tolerance to low temperatures and improve the growth vigour of the scion without adversely affecting fruit quality. Hence, by grafting, growers can cultivate sensitive cultivars with desirable yield and quality characteristics out of season and in pathogen-infested soil. Grafting is carried out in modern, commercial nurseries either by hand, or by semi- or fully-automated robots. To produce uniform, robust, grafted plants at an economically acceptable level, nurseries must employ advanced management technology to ensure seedling uniformity, precision in grafting and subsequent care of the grafted seedlings. The cost of grafting is more than compensated for by the resulting higher and earlier yield than that of non-grafted plants. Additionally, because it is an

---

[1] fbletsos@nagref.gr.

'environmentally friendly' technique, grafting is particularly applicable to the production of organic produce, for which today there is increasing consumer demand.

INTRODUCTION

Because the area of land available for vegetable cultivation is limited, crops are often repeatedly grown on the same soil resulting in a rapid increase in soil pathogen populations (O'Brien 1983) and an increase in the incidence of soil-borne diseases such as *Verticillium* and *Fusarium*, which are characterised by yellow or copper-coloured lesions on the leaves, discoloration and occlusion of the vascular bundles and consequent loss of yield and quality (Lee 1994; Oda 1995; Bletsos and Thanassoulopoulos 1999; Bletsos 2005a). Until 2005 these diseases were normally confronted by the sterilization of soil with methyl bromide.

Due to the susceptibility of most commercial vegetable cultivars to soil pathogens, the banning of methyl bromide in developed countries from 2005 and in developing countries from 2015 (Montreal Protocol 1997) will make the repeated cultivation of vegetables in the same soil increasingly unprofitable due to soil infestation. Soil pathogens are difficult to confront and the creation of resistant cultivars is a laborious and lengthy procedure. On the other hand, the grafting of commercial cultivars (e.g. Cucurbits) on to resistant rootstocks offers a practical and 'environmentally friendly' means of cultivating desirable, but pathogen sensitive, cultivars in pathogen infested soils (Lee 1994; Oda 1995; Bletsos 2005c).

HISTORY

Grafting is a technique in which the cut stems of two different plants are joined via rejeuvented cambium leading to their physical union and subsequent development as a single plant (Janick 1986).

Grafting was applied to tree crops even in ancient times. Both Aristotle (384-322 B.C.) and Theophrastus (371-287 B.C.) refer to grafting and knowledge of this process was also possessed by the Chinese from 1560 B.C. In Roman times fruit trees were grafted on a commercial scale and St. Paul refers to olive grafting in an Epistle to the Romans (Hartmann *et al.* 2002).

The Chinese grafted squash on to squash (self-grafting) from the 5th century A.D. (Anon 530-545) to produce large pumpkins, as did the Koreans from the 17th century. The technique they employed was to bring the shoots of two neighbouring plants into contact so as to effect a union after which only one shoot was permitted to develop. Afterwards they re-grafted two grafted plants and removed one of the two shoots, thus leaving the plant to develop further on four root systems. Only one or two pumpkins were allowed to develop per plant so that in the absence of competition they would become very large and could later be used as containers for storing rice (Hong 1710; PSNCK 1982).

The first example of heterografting (i.e. the grafting of two different species) relates to the grafting of watermelon (*Citrullus lanatus* [Thunb.] Matsum. and Nakai) on squash (*Cucurbita moschata* Duch.) in the early 1920s (Ashita 1927). The technique of watermelon grafting was quickly passed on to Japanese and then Korean growers. By the beginning of the

1930s Japanese and Korean farmers were grafting watermelons on bottle gourds (*Lagenaria siceraria* [Molina] Standl.) on a commercial scale so as to overcome the reductions in yield caused by soil pathogens (Lee 1994). Cucumber (*Cucumis sativus* L.) was first grafted towards the end of the 1920s, but commercial production of grafted cucumbers began only later, during the 1960s.

Korean farmers started grafting vegetables on a small commercial scale from the early 1950s. Initially they grafted seedlings at an advanced stage of development and the success rate was less than 50%. Moreover, an average worker grafted only about 150 plants per day (Ashita 1930, 1934). Later, grafting was performed at an early stage of development and moist straw or paper was applied to keep the graft in place. Using this method a worker could graft 800-1200 seedlings per day. Nowadays, various types of plastic clips are commercially available and can be applied to the site of grafting so as to hold the rootstock and scion in place. This assists the worker and enables him/her to graft more than 1500 seedlings a day, whereas a modern robot can graft over 10,000 seedlings a day. Consequently, from the 1990s 93% of watermelons, 72% cucumbers and 30% melons are cultivated on grafted plants (Oda 1995; Lee 1994).

In Europe, vegetable grafting started commercially from the early 1990s, while in the USA it has more recently been adopted for the confrontation of soil-borne pathogens and nematodes in organic farming. In the developing countries of Central America, (e.g. Mexico and Guatemala) projects are being financed by the UN to promote the adoption of vegetable grafting. In Greece, grafting was introduced for watermelon during the 1970s and is now commercially used for 90% of all watermelons, 30% cucumbers, 20% melons.

GRAFTING TECHNOLOGY

Rootstock and Scion Selection

Rootstocks for vegetable grafting should exhibit good compatibility, resistance to low temperature, soil pathogens (*Fusarium, Verticillium*) and nematodes, promote scion growth, increase production and not reduce produce quality. Rootstocks may belong to the same or a related genus, or they may be F_1 hybrids produced from related species. Many rootstocks bearing some or all of the above characteristics are available commercially. Growers select the rootstock that is most suitable for the season they are interested in, the cultivation method (greenhouse or open field), the soil and climatic conditions of the area and the cultivar(s) to be grown (Singh *et al.* 2004).

Grafting Methods

Growers and nursery producers specializing in grafted Cucurbits employ grafting techniques that are recognized throughout the world. Briefly, these are as follows (Figure 1):

a) Tongue approach grafting
b) Hole insertion grafting

c) Splice grafting

a)

b)

– Figure continued

c)

Figure 1. The principal grafting techniques used for the grafting of Cucurbits: (a) Tongue approach grafting; (b) Hole insertion grafting; (c) Splice grafting

Tongue Approach Grafting

Tongue approach grafting constitutes the oldest and arguably the simplest of techniques for the grafting of Cucurbits (Hong 1710). It is favoured by growers who are relatively inexperienced in grafting and/or who may not possess a greenhouse capable of precisely controlling the micro-climate, as well as by commercial enterprises. Although it requires more room within the greenhouse and more labour compared with other methods, no special tools or equipment are required. Tongue approach grafting has a high success rate because the two seedlings each retain their root system until the union of the scion and rootstock. At the time of grafting seedlings all have the same height and stem diameter, and this is achieved by regulation of the sowing dates. The seedlings are grafted at an early stage, preferably at the hypocotyls of both plants.

The true leaves and the apical meristem of the rootstock are carefully excised prior to grafting in order to reduce unnecessary loss of nutrients during unwanted shoot growth of the rootstock and to accelerate its union with the scion. Frequently, one of the cotyledons of the rootstock is excised together with the leaves, thus assuring complete removal of the aerial parts of the rootstock and reducing shading of the grafted seedlings within the confined space of the greenhouse.

Grafting is performed by incising the rootstock in a downward direction (towards the root) and the scion in an upward direction (towards the shoot apex) in each case at an angle of 30-40° with reference to the vertical axis, thus forming a 'V' with a depth that will assist the union of as many vascular bundles of the two seedlings as possible. The cut surfaces of the scion and rootstock are held in place by a special plastic clip and the grafted seedlings are planted at a spacing of 5.5 x 5.5 cm in trays of compost, sprayed with water and placed under shade for 1-2 days (Figure 1a). Subsequently they are transferred to an unshaded area of the greenhouse and held under normal growing conditions. About 10-12 days after grafting the hypocotyl of the scion is cut just below the site of grafting and after a further 3-5 days the percent success of the grafting procedure is assessed. The root and lower part of the hypocotyl of the scion quickly degenerates (dries out) and the plastic clip is removed a few days later, either before or after transplantation. Using this technique an experienced worker can graft about 1500 seedlings a day compared with a robot's output of about 600 plants per hour.

Advantages and Disadvantages

Although tongue approach grafting is a simple technique with a high percent success rate, today it is not widely employed by commercial nurseries specialized in the production of grafted seedlings mainly because:

a) It is highly labour intensive, requiring large labour inputs not only at the stage of grafting but also for the subsequent excision of the hypocotyl of the scion and the removal of the clips
b) Compared with other methods, it requires more room within the greenhouse for the development of the grafted plants
c) There is a danger of the scion rooting if the grafted seedlings are transplanted too deep and the graft comes into contact with the soil, rendering the plant susceptible to soil pathogen attack (Lee 1994).

Hole Insertion Grafting

This technique is adopted according to the plant species that is to be grafted, the preferences and experience of the grafter when performed by hand, and the type and function of robot if done mechanically.

The seedlings that are to be used as rootstock and scion must be uniform and robust at the site of grafting. At the time of grafting, the true leaf and apical meristem of the rootstock are excised with the aid of a sharp scalpel. Using a small wooden or plastic rod, or a small mechanical drill, a hole is opened in the rootstock hypocotyl between the cotyledons at a 20-30° angle away from the vertical axis. The scion is cut at a slant or tapered and placed within this hole taking care to avoid insertion into the pith cavity since this permits the subsequent development of adventitious roots from the scion down through the pith and into the soil thereby nullifying the graft (Figure 1b).

At the time of grafting the watermelon seedlings are smaller than those of the squash or gourds that are used as rootstocks. The watermelon seeds are sown 7-8 days after the gourd or 3-4 days after the squash, and grafting takes place 7-8 days later while the cotyledons of the watermelon are still unopened and firm. This facilitates the insertion of the scion into the hole.

Advantages and Disadvantages

Hole insertion grafting requires greater expertise in comparison with tongue approach grafting and precise climate control within the greenhouse during the post-grafting stage. Large producers prefer the hole insertion technique because it does not require additional labour to remove the clips or cut the hypocotyl of the scion. The grafted seedlings are initially more vigorous than those produced by the tongue approach technique because hole insertion grafting permits a greater number of vascular bundles of the scion and rootstock to unite (Oda 1994).

Splice Grafting

Most experienced growers and commercial producers of grafted seedlings employ splice grafting for the production of grafted Cucurbit seedlings. The splice grafting technique (also known as 'one cotyledon grafting' or 'slant grafting') is relatively simple and can be performed by hand, machines or robots (Lee and Oda 2003). The plants are ready to graft when there is at least one true leaf on both the scion and the rootstock. The apical meristem and one cotyledon are removed from the rootstock by cutting the hypocotyl at a 45° angle so as to leave a good surface area for the graft. The scion is cut at an opposite 45° angle about 2 cm below the cotyledons to facilitate clamping. The two halves of the grafted plant are then held in place by a plastic clip (Lee *et al.* 1998) (Figure 1c). After grafting, the plants are placed under shade and high humidity at about 25^oC for three days.

Advantages and Disadvantages

The principal advantage of splice grafting is that because most or all the vascular bundles of the rootstock fuse with those of the scion, the resulting grafted seedlings are strong and healthy. It is less labour intensive than the tongue approach technique and requires few tools. As with the other techniques, good climate control is required after grafting.

Each producer selects the grafting technique which is most suitable on the basis of experience and the facilities that are at his/her disposal, while preferences may change from year to year as new techniques evolve and new machines or robots are developed. Thus, although during the early years of grafting all watermelons were grafted by the tongue approach technique, more recently splice grafting has been adopted in combination with the removal of the rootstock from the substrate at the time of grafting and transplantation of the grafted plant to a special rooting substrate.

Tools, Clips and Grafting Aids

Specially designed knives and drills are used for hand-grafting seedlings by growers and nursery operators throughout the world. In addition, to maintain the cut surfaces in contact after grafting a number of specially designed, simple clamps and clips are available (Oda 1999), for example:

a) Simple plastic clips which usually incorporate a metal spring or similar system to assist the opening and closing of the clip. They are used mainly for grafting by the tongue approach method (Figure 2A, B, C, D).
b) Plastic clips for use by robots (Figure 2E).

Although all these types of clip may differ slightly in size and shape depending on the manufacturer, they are invariably cheap, simple to apply and may be used more than once.

Additionally, after grafting the seedlings are supported by special wooden or plastic rods which keep the plant upright and promote the union of the scion and rootstock (Figure 2F).

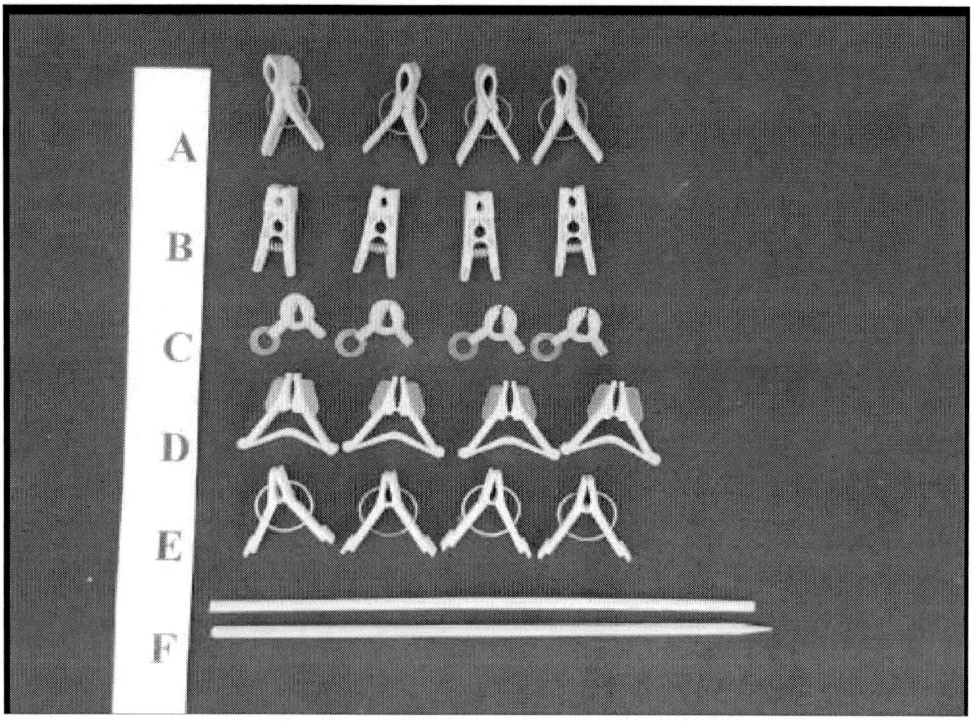

Figure 2. Various types of clip used for the grafting of Cucurbits: A, B, C, D) Clips used for the tongue approach method of grafting E) Clips sed by grafting robots, F) Plastic and wooden (cane) rods used to support of the grafted seedlings.

Acclimatization

With the term acclimatization, we mean the healing of the cut surfaces of the rootstock and scion and the hardening of the grafted seedlings to promote their survival on transplantation to the field or greenhouse (Oda 1999; Oda *et al.* 2001). The survival rate of grafted seedlings relates to the grafting technique employed and the conditions under which graft healing and subsequent acclimatization (hardening) take place.

The method of acclimatization depends on the means at the grower's disposal. Denna (1962), who did not have any special acclimatization facilities, enclosed the grafted seedlings in plastic bags so as to maintain a high relative humidity (RH) until the union of the rootstock and scion was achieved. On a small scale, Bletsos (unpublished data) acclimatized grafted seedlings in a self-designed acclimatization unit constructed of inverted plastic trays covered with transparent plastic. The RH within the unit was maintained at up to 100% by spraying with water. Similarly, growers frequently construct simple, low, plastic-covered tunnels in which the RH is maintained close to 100% by sprinklers spaced at 1.5 m along the length of the tunnel, while at the same time overhead shading nets reduce incoming radiation, lower the temperature and light intensity, and so reduce the incidence of wilting and promote healing (Nobuoka *et al.* 1996).

In modern commercial nurseries, the grafted seedlings are typically planted in trays containing 50-70 seedlings each and placed in specially constructed tunnels within which the temperature and RH are maintained at 24°C and 100% respectively. There they remain without additional watering until union of the rootstock and scion is complete. This takes about 6-10 days, depending on the type of graft. Partial shading may be employed during the day to prevent an excessive increase in temperature. The healing of the union between the rootstock and scion of the grafted seedlings is achieved by good control of air movement, temperature and RH within the tunnel, which allows the plants to develop without wilting (Nobuoka *et al.* 1997). To assist acclimatization both the plastic and the net for shading are gradually opened towards the end of the healing period. Subsequently, the grafted seedlings are transferred to the greenhouse where the temperature is held at 20/18°C (day/night). Approximately 25-37 days from sowing, or 14-21 days from grafting, the grafted plants are ready for transplanting to the field or greenhouse.

Estimation of Grafting Success

Grafting is deemed successful once the complete union of the vascular systems of the rootstock and scion has been achieved, thus allowing the unhindered transfer of water and nutrients from the rootstock to the scion and the transfer of photosynthates and growth substances from the scion to the rootstock. In small commercial nurseries with no specialized equipment, the success rate is estimated by visual observation by the grower and nursery workers. However, in modern, well-equipped commercial nurseries, the success of grafting is determined by methods such as:

a) *Thermic cameras*. The grafted seedling and a thermic camera are placed in a growth chamber at 25°C and 75% RH with overhead lighting. The light intensity is changed and the fluctuation in leaf temperature is detected by the thermic camera and recorded on a computer screen. In successfully grafted seedlings water flows freely from the roots to the leaves and the leaf surface temperature is 2-3°C lower than that of unsuccessfully grafted seedlings where inadequate water transfer to the scion reduces transpiration. In cases of partly successful grafting, intermediate leaf temperatures are recorded due to the lower rate of water transfer in comparison with the succesfully grafted seedlings (Torii *et al.* 1992).
b) *Measurement of electric current transfer* from the scion to the rootstock through the surface of the scion-rootstock union. Current transfer relates to the histological changes that occur during healing at the site of union (Lu *et al.* 1996).

To summarize, it is clear that the percent survival of grafted seedlings depends on the care that they receive both prior to and after grafting takes place. For example, prior to grafting the seedlings (both scion and rootstock) are exposed to direct sunlight for 2-3 days and watered normally so that the hypocotyl does not become too tall and/or thin. Care is taken that both the scion and rootstock have the same height and thickness (Oda *et al.* 1994). During grafting we try to maximize the area of the cut surfaces that are to be brought into contact and these surfaces are retained in contact by the application of an appropriate clip (Oda 1995). The subsequent acclimatization is carried out under carefully controlled conditions (temperature, RH and air movement) followed by gradual hardening of the plants prior to transfer to the cultivation area.

With the evolution of grafting technology changes are occurring in the priorities of plant breeders. Instead of creating cultivars with resistance to soil-borne pathogens and low temperature stress (a lengthy and expensive procedure), breeders are now increasingly devoting efforts to produce resistant rootstocks for grafting. The improvement of such rootstocks is particularly effective because they are largely derived from wild species with large genetic variation.

Micrografting

With the term micrografting we mean the grafting of meristematic tissues produced *in vitro* in the laboratory under aseptic conditions. With this method, the shoots of one plant are grafted on to the shoots or roots of another plant and the two plant parts unite to create a new plant that combines the characteristics of both plants. This method was first employed by Morel and Martin (1952) for the production of healthy, virus-free dahlias. On a commercial scale, it is used for fruit-tree crops (mainly citrus) and the production of disease- (especially virus-) free plants (Obeidy and Smith 1991). It is also used for the production of virus-free plants from virus-infected plants because the apical meristems of the virus-infected shoots that are used for tissue culture are typically free of virus and their compatibility with the rootstock is apparent from an early stage, i.e. within a few days (Obeidy and Smith 1991).

The micrografting of vegetables, which has only recently begun on a commercial scale, aims at reducing the high cost of grafting using the methods described above. This is due not only to the high cost of labour but also to the high cost of hybrid seed. For micrografting only a few seeds of the rootstock and the desired variety or hybrid (scion) are required. The seeds are sterilized by immersing first in ethanol for 1 min. and then in a commercial hypochlorite solution (2% chlorine) for 15-20 min., followed by washing with distilled, sterile water and drying on sterile, absorbent paper. Next they are germinated in Petri dishes containing MS nutrient medium (Murashige and Skoog 1962).

When the seedlings are about 2-3 weeks old the shoot, cotyledons etc. are cut into small pieces (explants) and placed in MS nutrient medium so as to differentiate and root. This process of micropropagation is repeated a number of times with the result that several thousand transplants can be produced within a short period of time (Sarowar *et al.* 2003). The micropropagated plants (rootstock and scion) are genetically and phenotypically identical and true to type (Bhojwani and Razdan 1983). After a further three weeks they are grafted in the laboratory by one of the classical techniques for vegetable grafting described above.

The micrografting success rate was relatively high (80–90%) for the commercial cucumber hybrid Stella F_1 grafted on to the squash rootstock TZ 148 F_1 (Sapountzakis and Tsaftaris, 2002), three commercial tomato F_1 hybrids (Garnet 622, Jumbo and Marvel) grafted on the rootstock RT 79 (Grigoriadis *et al.* 2005) and four traditional Greek melon varieties (Kokkini Banana, Thraki, Peplos and Amynteou) grafted on the squash rootstock TZ 148 F_1 and the Greek squash cultivar Lefki kolokytha (Pontiki 2005).

Despite the benefits of micropropagation indicated above, it is nevertheless an expensive and labour intensive procedure, and for this reason attempts have recently been made to mechanise and automate the process. Micropropagation in liquid culture in bio-reactors (Akita and Ohta 1998; Ziv *et al.* 1998; Konstas and Kintzios 2003), instead of on a solid substrate (MS), permits better control of the propagation conditions, and the cutting, selection and transfer of explants to the medium can be carried out by robot (Aitken-Christie *et al.* 1995; Gambley *et al.* 1997).

The progressive introduction of seedlings to the vegetable propagation trade, displacing the traditional dependence of growers on seed, combined with the demand of large commercial nurseries for certified plants so as to produce and sell a uniform, guaranteed product, will doubtless help to increase the adoption of micrografting for the production of grafted vegetable seedlings.

Simple Grafting Machines and Robots

Grafting is a procedure that requires a high labour input. In order to meet the market demand commercial producers of grafted seedlings graft millions of seedlings each year within a short space of time. Although the filling of the growing boxes with substrate, seed sowing, irrigation, fertilizer application and the control of the growth environment of the seedlings may be done automatically, the process of grafting itself is still carried out manually to a high percent (Kobayashi 1991). Because hand-grafting is both time-consuming and expensive, simple machines and automated robots that can graft at a rate of 600-1200 plants an hour have recently been introduced.

Since the principal vegetable species that are grafted are the Cucurbits, the first machines to graft vegetables were designed for this group of plants. The first grafting robot was invented in Japan in 1987 by the company BRAIN (Cutting-off Cotyledon Grafting) for the grafting of Cucurbits by the tongue approach tecnique. It grafted one plant every 3 sec with a success rate of 95%. Two years later (1989) a second type of robot BRAIN was introduced, also for Cucurbit grafting. This robot consisted of two circular discs, each with 20 holes around the circumference in which seedlings were placed manually. The rootstock seedlings were hung from the right disc while those of the scion were hung from the left disc. The two discs revolved in opposite directions and transferred the seedlings to two smaller discs where the hypocotyls were cut by a sharp blade and the cut surfaces brought into contact and held in place by a plastic clip (Onoda *et al.* 1992; Kurata 1994; Suzuki *et al.* 1995; Oda *et al.* 1992; Kobayashi *et al.* 1996). Soon afterwards, the same company introduced a third, improved robot BRAIN. This robot automatically takes up discs with 48 circumferential holes in which the seedlings are held individually. The seedlings are transferred one by one to a second rotating disc where they are checked and arranged so that the cotyledons are pointing in the same direcion, after which grafting is performed (Kurata 1994).

In Korea, the company YuHo constructed a cheap and simple machine for Cucurbit grafting by the tongue approach technique. The machine, which costs about US$ 400, is widely used in Korea as well as in many other countries of Asia and Europe. It is operated by an experienced worker and grafts 600 seedlings an hour (Lee and Oda 2003).

Another type of robot grafts seedlings by a technique involving the horizontal cutting of the hypocotyls of the rootstock and scion. The cut surfaces are then brought into contact and held in place by the application of glue (alkyl 2-cyanoacrylate), which hardens immediately after application. However, this robot has a low success rate with the Cucurbits because the weight of the scion is insufficient to hold the cut surfaces in contact. The percent success increases when:

a) The scion is placed upon the rootstock in such a way that its cotyledons form a right angle with the cotyledons of the rootstock, because this brings more vascular bundles into contact
b) The hypocotyls of the rootstock and scion have the same diameter at the site of grafting
c) Fig-leaf (Malabar) gourd (*Cucurbita ficifolia*) is used as a rootstock and the seeds are treated with gibberellic acid (GA_3) prior to sowing so as to create an elongated hypocotyl (Oda *et al.* 1993, 1994, 2001; Oda 1995).

Today several types of simple vegetable grafting machines and fully-automated robots are available on the markets of Japan and Korea (Hwang *et al.* 1999; Kang 2000). The robots that are most widely used are those of the companies YuHo in Korea and ISEKI, Japan.

Recently a fully-automated robot has been constructed which selects the rootstock and scion seedlings separately and transfers them to the grafting machine. It grafts 750 plants an hour with 90% success. This robot is expected to be commercially available in the near future (Kubota and McClure 2008).

In Greece, commercial nurseries producing grafted vegetable plants are finding it increasingly difficult to locate experienced grafters to graft thousands of plants over a short time period. Therefore to meet the ever increasing demand for grafted vegetable seedlings as well as to reduce the production cost and sale price, these enterprises are more and more adopting the idea of using automated grafting robots. However, although the cost of purchase is reasonable, nurseries are still reluctant to buy robots due to a lack of post-sales back-up.

The Cost of Production of Grafted Seedlings

Modern commercial nurseries produce uniform, healthy plants irrespective of the grafting technique employed (e.g. by hand or robot). Growers prefer to buy grafted seedlings because, although the cost of purchase is high, they can avoid the specialist grafting work and the initial care of the grafted plants. In the USA, the cost of grafted seedlings is about the same as the cost of soil sterilization with methyl bromide, while the grafted plants produce 15-30% higher yield than self-rooted ones. Moreover, the cost of weed control for field crops is significantly less for crops grown from transplanted grafted plants than for those grown direct from seed (EPA 1997). According to Leonardi and Romano (2004), the cost of producing grafted seedlings is 3-5 times higher than that of self-rooted seedlings. In Japan, the cost of grafted seedlings is 4 times higher than the price of seed (equivalent to about 1 Euro per seedling) (Sakata *et al.* 2008) while in Spain the cost is 0.25-0.35 Euros without taking into account the price of the seed (Lopez-Galarza 2008, reported in Davis *et al.* 2008). In Greece grafted seedlings cost approximately three times more than self-rooted seedlings, i.e. 1.5 Euros per plant.

Transportation Conditions for Grafted Plants

When the grafted seedlings are destined for the domestic market, they are packed in cardboard boxes and transported to the grower by closed truck. The temperature inside the truck is maintained at 13-15°C to avoid stress, since the outside temperature at the time of transportation may be low. Under these conditions, seedlings can travel for four days without damage.

Grafting Physiology

Graft Compatibility

With certain exceptions, grafting compatibility mainly depends on the botanical relationship between the rootstock and the scion. Incompatibility and failure in grafting are two different things. With the term incompatibility, we mean that in cases where grafting is expected to be successful the rootstock and scion fail to join, resulting in the inability of the scion to grow and its early death. With the term grafting failure, we mean that the scion fails to grow due to the environmental conditions or a lack of experience by the grafter. Physiologically, incompatibility refers to a recognition failure of the cells of the scion and rootstock, a lack of response between the cut surfaces of the rootstock and scion, or to the effects of growth substances or toxins (Andrews and Marquez 1993).

Most Cucurbits can be successfully grafted on a number of rootstocks. If signs of grafting failure are noticed during grafting, this can be overcome by changing the grafting technique and/or by improving the environmental conditions to which the seedlings are exposed for a few days after grafting (Lee 1989, 1994).

Graft Union

The union of rootstock and scion in woody plants is achieved through the cambium cells, which multiply and increase in size. Callus is formed at the surface of the union by the interaction of the cambial cells of the rootstock and scion. This then differentiates to form xylem and phloem (Janick 1986).

In vegetables, the wounded cells at the cut surfaces of the rootstock and scion that come into contact during grafting die and form a dead tissue that is subsequently absorbed. Next, the adjacent living cells of the rootstock and scion extend into the dead zone to form callus, which assists their union. At the same time, elongation of the scion is supported by the interaction of the cells of the scion and rootstock. A new vascular cambium differentiates to produce xylem and phloem and unites the vascular elements of the rootstock and scion (Andrews and Marquez 1993).

In the Cucurbits, the main shoot (hypocotyl and epicotyl) is either hollow or contains pith and usually has six vascular bundles (Oda *et al.* 1993). During grafting, it is normally impossible to achieve the union of all the six vascular bundles of the rootstock with those of the scion. Therefore, research is largely aimed at finding the most suitable method to achieve the union of as many vascular bundles as possible, and within the shortest time. Horizontal grafting has a better success rate when the axis of the cotyledons of the rootstock and that of the scion are vertical (Oda *et al.* 1993).

Ion Translocation between the Scion and Rootstock

In grafted seedlings the ability to absorb and translocate ions, as well as to translocate photosynthates, hormones, alkaloids and viruses, is affected by the type of rootstock and scion. Vegetables are largely grafted on to wild species in order to gain the advantage of their well-developed root system. Thus grafted seedlings can usually take up water and inorganic salts more effectively than the corresponding self-rooted ones (Masuda and Gomi 1984). Nutrient uptake by plants grafted on to some rootstocks is especially higher in the case of the macro-nutrients P, N, Mg and Ca (Gruscenko and Drobkov 1952; Masuda and Gommi 1984; Ikeda *et al.* 1986; Kim and Lee 1989) and the micro-nutrients Fe and B (Brown *et al.* 1971; Zaiter *et al.* 1987; Gomi and Masuda 1981). Moreover, grafted seedlings have the ability to take up inorganic ions from the soil when the soil temperature is very low (Tachibana 1982; Masuda and Gomi 1984; Ahn *et al.* 1999).

The Cucurbits, like other plant species, synthesise growth regulatory substances (hormones) within the roots, especially cytokinins, which are translocated to the shoots where they exert a significant effect on plant growth. The amount of cytokinin produced within the root system relates to the extent of its development. Thus the well-developed root systems of the rootstocks used for Cucurbit grafting produce relatively large quantities of cytokinins for translocation to the scion and so increase shoot growth and yield (Kato and Lou 1989). The amount and synthesis of the cytokinins varies from species to species. For example, only zeatin and dihydro-zeatin have been identified in cucumber, whereas in squash large quantities of isopentenyl adenine and isopentenyl adenosine are present. The synthesis of cytokinins within the xylem sap is affected by the growth of the shoots of the scion (Lee *et al.* 1999b, 2001). Only in the xylem sap of *S. angulatus* are cytokinin concentrations too low to detect. This observation may therefore explain why both seedlings of *S. angulatus* and the scions grafted on to *S. angulatus* rootstocks are sickly in appearance and slower in growth compared with the same scions grafted on to other, more vigorous rootstocks. Using different combinations of grafting between *S. angulatus* and cucumber, it has been found that promoters of flowering are also translocated between rootstocks and scions (Takahashi *et al.* 1982; Suge 1984).

During vegetable grafting it is also important to take all the necessary precautionary measures to avoid contamination of the scion by possible viruses in the rootstock or *vice versa*, since several Cucurbit viruses may be transmitted by this process (Alexander 1963, 1971). Some of the serious viruses that are transmitted by seeds and are present in the seeds of the rootstock (e.g. races of tobacco mosaic virus) can kill the grafted seedlings if the rootstock seeds are not first subjected to dry heat to neutralize the virus (Kim and Lee 2000).

Growth Effects

The grafted fruit-bearing vegetable species are usually more vigorous and productive than the corresponding self-rooted plants. As a rule, cucumber fruit produced by cucumbers grafted on squash and cultivated in sandy soil have a higher dry matter content than the fruit of self-rooted cucumbers (Shimada and Moritani 1977).

Physiologocal Disorders

Most of the physiological disorders and other undesirable fruit characteristics of grafted plants are attributed to the effect of the rootstock on the scion (Chung 1995; Matsuda and Honda 1981). For example, the melon hybrid Prince F_1 grafted on to squash produces abnormal fruit (Chung 1995). Other common abnormalities that appear to a greater or lesser extent include a reduction in total soluble solids (^{o}Brix), the retention of green colour in the form of stripes on ripe melons, the appearance of yellow lesions that are not of pathogenic origin on the foliage, and the occurrence of yellow off-colours in the fruit flesh (Ryu *et al.* 1973; Lee 1989; Lee *et al.* 1998). Melons grafted on the rootstock Shintozwa (*C. maxima* x *C. moschata*) are more vigorous and productive than the self-rooted plants, but the fruit have a lower soluble solids content and are therefore less sweet. In addition the grafted plants produce a higher percentage of green-striped fruit than the self-rooted plants. When melons are grafted on to the rootstock Kongo F_1 (*C. moschata* x *C. moschata*) the fruit produced are larger and sweeter, but have an irregular shape. However, when grafted on the rootstock Shirokiknza (*C. moschata*) the melon scion becomes sickly in appearance, with yellow lesions on the foliage and yellow stripes or green lesions on the fruit (Matsuda and Honda 1981).

Stress Tolerance

Vegetables are also grafted because they can grow and produce satisfactorily under unfavourable climatic conditions, such as soil temperatures that are either lower or higher than normal, or excessively dry, wet or saline soil conditions. Therefore, Cucurbits that are to be cultivated under such adverse growth conditions are grafted on to tolerant rootstocks, which are selected according to the problem to be confronted (Matsubara 1989).

Cucumbers that are to be grown in regions or periods subject to either low temperatures (principally early crops) or high temperatures are grafted on *C. ficifolia* or Shintozwa (*C. maxima* x *C. moschata*) (Okimura *et al.* 1986). However, for cultivation on saline soils cucumbers are mainly grafted on to squash rootstocks (*Cucurbita* spp.) as well as on bottle gourd rootstocks (*Lagenaria siceraria*). The grafted cucumbers tolerate low temperatures because grafting increases the lipid content g fresh weight^{-1} and the unsaturated fatty acid level while lowering the ratio of esterol to total lipids (Horvath *et al.* 1983; Bulder *et al.* 1990; Bulder *et al.* 1991; Matsubara 1989).

When watermelons are to be grown during periods of water shortage (Matsubara 1989) or low soil temperatures, they are grafted on Shintozwa or squash (*C. moschata*) in preference to bottle gourd (*L. siceraria*) because growth is better in the first two cases. Watermelons grafted on other watermelons of the same species (*Citrullus lanatus*) or *Benincasa hispida* develop less than those grafted on the other rootstocks, but when the soil temperature is 20°C (optimum temperature 20-23°C) they grow better. Recent research has also shown that grafted watermelons tolerate low temperatures because the leaf antioxidant concentrations and antioxidative enzyme activities are higher than in self-rooted plants (Liu *et al.* 2003; Liu *et al.* 2004).

Flower Formation

When some vegetable species are grafted, there are changes in the number and sex of the flowers that form. For example, grafting was found to induce the flowering of sweet potato (*Ipomoea batatas*), but did not affect flowering in tomato (Coggins and Lesley 1968).

In the Cucurbits, it was observed that when gynomonoecious (female and hermaphrodite flowers) melons are grafted on andromonoecious (male and hermaphrodite flowers) melons (i.e. the same species) the flowers subsequently induced on the scion were male. However, when melon is grafted on to squash (i.e. another species) the scion again produces male flowers, but to a lower percentage (Mockaitis and Kivilaan 1964).

Watermelons grafted on bottle gourds form female flowers earlier than those grafted on other rootstocks provided that optimum temperatures are experienced during the early stage of vegetative growth (Kurata 1976b; Sakata *et al.* 2007); however if the temperature during the same stage is low flowering of the grafted plant is delayed (Yamasaki *et al.* 1994; Miguel 1997; Xu *et al.* 2005d).

In cucumber, the formation of male, female or hermaphrodite flowers is determined by both genetic and non-genetic factors (principally growth regulators). The role of growth regulators was indicated in experiments where cucumbers were grafted on to rootstocks of different genotypes of the same species. It was found that the sex expression of the flowers induced on the scion is controlled by factors translocated upwards from the rootstock (Friedlander *et al.* 1977). Similarly, Satoh (1996) observed a delay in the induction of female flowers in cucumbers grafted on hybrid squash due to factors produced by the roots of the squash.

Fruit Quality

Generally it has been found that the rootstocks affect the quality of the fruit formed on the scion.

Watermelons and melons grafted on rootstocks of the genus *Cucurbita* tend to produce larger fruit of an irregular shape and lower quality (reduced total soluble solids, °Brix) (Lopez-Galarza *et al.* 2004; Xu *et al.* 2005d; Qi *et al.* 2006; Xu *et al.*, 2006). However, when watermelons and melons are grafted on other, more suitable rootstocks the fruit produced are of the same quality as those produced by the corresponding self-rooted plants (Davis and Perkins-Veazie 2005; Xu *et al.* 2005b; Alan *et al.* 2007; Alexopoulos *et al.* 2007).

Cucumbers produced on cucumber scions grafted on to rootstocks of the genus *Cucurbita* retain the corolla of the flower at the fruit apex even at commercial ripeness (Bletsos 2005a). This characteristic is attributed to a lack of silicon in the leaves (Yamamoto *et al.* 1989). However, cucumbers that are grafted on other suitable rootstocks yield fruit of the same quality as those produced by the corresponding self-rooted plants (Hoyos 2001).

THE PRINCIPAL CUCURBIT SPECIES TO WHICH GRAFTING IS APPLIED

Aim of the Application of Grafting to Cucurbits

The grafting of Cucurbits is practised commercially in order to:

a) Reduce infection by soil-borne diseases (*F. oxysporum* Schitdl and *V. dahliae*) (Alabouvette *et al.* 1974; Paplomatas *et al.* 2002; Bletsos 2001; Bletsos *et al.* 2002; Pavlou *et al.* 2002; Tjamos *et al.* 2002; Hirai *et al.* 2002; Trionfelti-Nisini *et al.* 2002; Bletsos 2005c; Xu *et al.* 2005c) and nematodes (*Meloidogyne* sp.) (Giannakou and Karpouzas 2003; Miguel *et al.* 2005; Siguenza *et al.* 2005)
b) Increase plant vigour and production (Lee *et al.* 1998; Bletsos 2001; Bletsos *et al.* 2002; Giannakou and Karpouzas 2003; Bletsos 2005c; Miguel *et al.* 2005)
c) Increase plant tolerance to low and high temperatures (Tachibana 1982; Okimura *et al.* 1986; Ahn *et al.* 1999; Liu *et al.* 2003; Liu *et al.* 2004).

Watermelon (*Citrullus lanatus* [Thunb.] Matsum. and Nakai)

Rootstocks

Watermelon is grafted on to species of gourd (*Lagenaria* sp.), squash (winter-*C. maxima* and *C. moschata*, and summer-*C. pepo*) and their interspecific hybrids, watermelon (*Citrullus* sp.), *Benincasa hispida*, and *Cucumis metuliferus*.

Gourd Rootstocks (*Lagenaria* sp.)

Watermelons grafted on gourd (*Lagenaria* sp.) exhibit the following characteristics:

a) They are more vigorous (Ryu *et al.* 1973; Balaz 1982; He and Fu 1988; D' Amore *et al.* 1996; Olympios 2001; Paroussi *et al.* 2007)
b) They are more tolerant to low temperatures (Okimura *et al.* 1986)

c) They are not infected by *Fusarium* (*F. oxysporum* f.sp *niveum*), *Verticillium* (*V. dahliae*) or nematodes (*Meloidogyne* sp.) (Mijuskovic and Vucinic 1977; He and Fu 1988; Mondal *et al.* 1994; Liu *et al.* 1995; D' Amore *et al.* 1996)
d) The grafted plants have a well developed root system and are supplied with water and nutrient elements more efficiently in comparison with self-rooted plants
e) The fruit produced are of the same quality as fruit produced from self-rooted plants (Ryu *et al.* 1973; Balaz 1982; He and Fu 1988).

(a)

(b)

Figure 3. Bottle gourd (*Lagenaria siceraria*): (a) flower, (b) fruits

(a)

b)

Figure 4. *Benincasa hispida*: (a) flower, (b) fruit.

a)

b)

Figure 5. *Cucumis metuliferus*: (a) male and female flowers, (b) mature fruits .

Squash Rootstocks (*Cucurbita* sp.)

Watermelons grafted on to winter or summer squash species (*Cucurbita* sp.) and their hybrids are characterized by:

a) Greater growth vigour (Ryu *et al.* 1973; Mondal *et al.* 1994; Zheng 1995; D' Amore *et al.* 1996; Pulgar *et al.* 2000; Paroussi *et al.* 2007)
b) Tolerance to low temperatures (Okimura *et al.* 1986)
c) Resistance to infection by *F. oxysporum* f.sp. *niveum* and *V. dahliae* (Mondal *et al.* 1994; Kouvelos 2005)
d) Tolerance to high temperatures, in the case of rootstocks of the squash hybrids Shintozwa F_1, Chulgap F_1 etc.
e) A well-developed root system. This is particularly important for watermelons cultivated in sandy soils (Pulgar *et al.* 2000).

Watermelon Species (*Citrullus* sp.)

Watermelons grafted on watermelon species can be infected by *F. oxysporum* f.sp. *niveum* and do not tolerate low temperatures. However, they produce particularly sweet fruit.

B. hispida

Watermelons grafted on *B. hispida* are not infected by *F.oxysporum* f.sp. *niveum*, but they do not tolerate low temperatures (Mondal *et al.* 1994; Okimura *et al.* 1986).

Some rootstocks that are suitable for grafting watermelon, cucumber and melon are presented in Table 1.

Table 1. Commercially suitable rootstocks for grafting cultivated varieties and hybrids of watermelon, melon and cucumber

Commercial name	Scientific name (common name)	International representative	Characteristics*
Achille F_1	*Lagenaria* sp. (gourd)	Esasem, Italy	Excellent compatibility with watermelon. Resistant to Fon and tolerant to V (Va, Vd), N, and low temperatures.
Argentario F_1	»	Syngenta Seeds BV (SandG), Holland	Excellent compatibility with watermelon. Moderate plant vigour and well-developed root system. Can be grown in all soil types and promotes early fruit production.
Dias F_1	»	DAP, Holland	Excellent compatibility with watermelon. Plants are vigourous and tolerant to Fon 0, 1, 2.
Early Star F_1	»	Golden West, U.S.A.	Excellent compatibility with watermelon. It promotes earliness in the scion. Tolerant to V (Va, Vd), Fon 0, 1 and Fom 0, 1, 2.

Table 1 (Continued)

Commercial name	Scientific name (common name)	International representative	Characteristics*
Emphasis F_1	»	Syngenta Seeds BV (SandG), Holland	Excellent compatibility with watermelon. It exhibits good tolerance to low temperatures and moderate resistance to Fon 0, 1, 2 and N.
Friend F_1	»	Hybrids Hellas, Greece	Excellent compatibility with watermelon.
Macis F_1	»	Nunhems, Holland	Excellent compatibility only with watermelon. Resistant to Fon and tolerant to V (Va, Vd).
Mini Bottle F_1	»	Hybrids Hellas, Greece	Excellent compatibility only with watermelon. It promotes early fruit production by the scion.
Pelops F_1	»	Rijk Swan, Holland	Excellent compatibility only with watermelon. Resistant to Fon.
Silver F_1	»	Asgrow, U.S.A.	Excellent compatibility only with watermelon.
Sunny F_1	»	Wing Seed, Holland	Excellent compatibility only with watermelon. Tolerant to Fon 0, 1, 2.
Valet F_1	»	Takii, Holland	Excellent compatibility only with watermelon. High resistance to Fon.
W-135 F_1	»	Hybrids Hellas, Greece	Excellent compatibility only with watermelon. It promotes early fruit production by the scion.
All Star F_1	*Cucurbita* sp. (squash) (*C. maxima* x *C. moschata*)	Golden West, U.S.A.	Excellent compatibility with watermelon. Tolerant to Fon.
Atlantic F_1 (Prince)	»	Wing Seed, Holland	Excellent compatibility with watermelon, good compatibility with melon and cucumber. Tolerant to F (Fon, Fom, Foc), and N.
Carnivor F_1	»	Syngenta Seeds BV (SandG), Holland	Excellent compatibility with watermelon and melon. Resistant to F (Fon 0, 1, 2).
Ercole F_1 (Nun 6001)	»	Nunhems, Holland	Excellent compatibility with watermelon, melon and cucumber. Resistant to F (Fon, Fom, Foc), Vd and tolerant to N.
Ferro F_1	»	Rijk Swan, Holland	Excellent compatibility with watermelon, melon and cucumber. Vigorous plant with well-developed root system. Resistant to F (Fon, Fom, Foc).
G-200 F_1	»	Hybrids Hellas, Greece	Excellent compatibility with watermelon, melon and cucumber.
Gourd Black Seeded F_1	»	Esasem, Italy	Excellent compatibility only with cultivated cucumber hybrids. Vigorous plant with well-developed root system. It is recommended to sow the seeds of the rootstock 1-2 days before those of scion. Tolerant to F and low temperatures.
Kazako F_1	»	Syngenta Seeds BV (SandG), Holland	Excellent compatibility with watermelon. It is recommended for early and moderately early fruit production.

Kublai F_1	»	Syngenta Seeds BV (SandG), Holland	Excellent compatibility with watermelon and melon. It is recommended for moderately early fruit production in sandy soils.
Leon F_1	»	Golden West, U.S.A.	Excellent compatibility only with cultivated cucumber hybrids. Tolerant to F and resistant to Oidium. It promotes the growth vigour of the scion.
No 613 F_1	»	Hybrids Hellas, Greece	Excellent compatibility only with melon cultivars and hybrids.
Polifemos F_1	»	Esasem, Italy	Excellent compatibility with watermelon and melon. Resistant to Fon, and Fom 1, 2, 3, 4.
RS 841 F_1	»	Royal Sluis, Holland	Excellent compatibility with watermelon, melon and cucumber. Triploid hybrid resistant to F (Fon, Fom, Foc) and N.
Shintosa Camelforce F_1 (Nun-9075)	»	Nunhems, Holland	Excellent compatibility with watermelon, melon and cucumber. Resistant to F (Fon, Fom, Foc) and Vd, tolerant to N.
Squash N.3 F_1	»	Sakata, Japan	Excellent compatibility with watermelon, melon and cucumber. It results in vigorous growth of the scion and confers resistance to F (Fon, Fom, Foc) and soil moisture.
Strongtosa F_1	»	Syngenta Seeds BV (SandG), Holland	Excellent compatibility with watermelon and melon. Relatively resistant to F (Fon 0, 1, 2, Fom 0, 1, 2, Foc 0, 1, 2) and N, with tolerance to low and high temperatures.
Syntosa F_1	»	Fito, Spain	Excellent compatibility with watermelon, melon and cucumber. Tolerant to N, F (Fon 1, 2, Fom 1, 2, Foc 1, 2), Vd and low temperatures.
TZ-148 F_1	»	Clause, France	Excellent compatibility with watermelon, melon and cucumber. Resistant to F (Fon 1, 2, Fom 1, 2, Foc 1, 2), Vd, N and low temperatures.

* Resistance or tolerance to:
F: *Fusarium*
F.o.c. 0, 1: *Fusarium oxysporum* f.sp. *cucumerinum*, races 0, 1
F.o.m. 0, 1, 2: *Fusarium oxysporum* f.sp. *melonis*, races 0, 1, 2
F.o.n. 0, 1: *Fusarium oxysporum* f.sp. *niveum*, races 0, 1
V: *Verticillium*
Va: *Verticillium albo-atrum*
Vd: *Verticillium dahliae*
N: Nematodes (*Meloidogyne*)
Ma: *Meloidogyne arenaria*
Mi : *Meloidogyne incognita*
Mj: Meloidogyne javanica

Experimental Results

Watermelons, self-rooted or grafted on to commercial rootstocks of gourd (*L. clavata*, *L. siceraria* and *L. vulgaris*), were cultivated in the field. It was found that the grafted watermelons gave: (a) earlier fruit production, (b) 70-80% higher total yield and (c) larger fruits than the self-rooted plants (Ryu *et al.* 1973; Balaz 1982; He and Fu 1988; D' Amore *et al.* 1996; Olympios 2001; Paroussi *et al.* 2007).

In other trials, watermelons, self-rooted or grafted on to commercial rootstocks of gourd, were cultivated in soil infected by *F.o.*f.sp *niveum*, *V. dahliae* and nematodes. It was found that the grafted watermelons were either not infected or infected to a much lower degree than the self-rooted watermelons (Mijuskovic and Vucinic 1977; He and Fu 1988; Mondal *et al.* 1994; Liu *et al.* 1995; D' Amore *et al.* 1996). Moreover, Ryu *et al.* (1973), Mondal *et al.* (1994), Zheng (1995), D' Amore *et al.* (1996), Pulgar *et al.* (2000) and Paroussi *et al.* (2007) cultivated watermelons, self-rooted or grafted on commercial rootstocks of squash, and found that the grafted watermelons were more vigorous, gave earlier fruit production and 15-20% higher total yield in comparison with the self-rooted plants. These positive characteristics were attributed to the better-developed root system of the rootstock, which permitted better uptake of nutrients for transfer to the scion. The fruit of the grafted plants were larger than those of the self-rooted plants without otherwise affecting fruit quality.

According to Ryu *et al.* (1973), fruit produced from grafted watermelons are characterised by a lower percent total soluble solids (^{o}Brix), the occurrence of yellow areas or other abnormalities within the red flesh, and inferior flavour. These deficiencies were suggested to arise from reduced calcium absorption by the plants in relation to nitrogen. However, more recent studies have shown that watermelons grafted on to commercial rootstocks of gourd (*L. siceraria* and *L. vulgaris*) produce fruit with approximately the same total soluble solids content (^{o}Brix) as fruit produced from the self-rooted plants (Balaz 1982; He and Fu 1988).

In Greece, watermelon cv. Crimson Sweet was self-rooted or grafted on to commercial rootstocks of squash (Mamouth F_1) or gourd (Daco F_1, Max-2 F_1, Calago F_1 and Astra F_1), and cultivated during 2003-2004: (a) in soil which had been sterilized with methyl bromide and (b) in soil that had been infected by *F.o.* f.sp. *niveum* και *V. dahliae* (Paroussi *et al.* 2007). It was found that the grafted watermelons irrespective of the rootstock had fewer infections by soil-borne diseases and gave:

a) Larger fruit and higher yield (except watermelons grafted on Astra F_1) in comparison with the self-rooted plants cultivated in sterilized or infected soil. Watermelons grafted on 'Mamouth F_1' produced fruits that were 6.0% larger than the fruit produced by self-rooted plants, and 26.6% and 12.9% larger than those produced by self-rooted plants cultivated in soil infected by *F.o.* f.sp *niveum* and *V. dahliae*, respectively. The corresponding increase in total yield was 100%, 272% and 203%, respectively.
b) Fruit that were sweeter than those from the self-rooted plants (^{o}Brix)
c) Fruit with no abnormalities in the red flesh.

The results of this experiment show that the new compatible rootstock hybrids introduced by plant breeders have made a positive impact on scion growth and yield without adversely affecting fruit quality (Davis *et al.* 2008). Similarly, Alexopoulos *et al.* (2007) grafted watermelons on gourd (Long gourd, Early Max, Max-2 and F_{14}) and found that the grafted plants gave a higher yield and larger fruit than the self-rooted plants without affecting the total soluble solids content (^{o}Brix) of the fruit.

Cucumber (*Cucumis sativus* L.)

Rootstocks

Suitable rootstocks for grafting cucumber are: fig-leaf (Malabar) gourd (*Cucurbita ficifolia*), summer squash (*Cucurbita pepo*), winter squash (*Cucurbita maxima, Cucurbita moschata, Cucurbita mixta, Cucurbita pepo*) and their interspecific hybrids, bottle gourd (*Lagenaria siceraria*), wax or white gourd (*Benincasa hispida*), African horned cucumber (*Cucumis metuliferus*) and burcucumber (*Sicyos angulatus*).

Fig-Leaf (Malabar) Gourd (*Cucurbita ficifolia*)

Cucumbers grafted on *Cucurbita ficifolia* possess the following characteristics:

a) They are not infected by *Fusarium oxysporum* f.sp *cucumerinum* (*F.o.* f.sp. *cucumerinum)* (Owen 1956; Armstrong and Armstrong 1978; Jenkins and Wehner 1983; Bao *et al.* 1992; Ahn *et al.* 1999; Ko 1999, reported in Lee and Oda 2003; Ashraf 2003)
b) They are tolerant to low temperatures (Tachibana 1982, 1986, 1988; Ahn *et al.* 1999; Lee *et al.* 1999a)
c) They are tolerant to nematodes (Abou-Hadid and Medany 1992; Echebarria 2001)
d) They are more vigorous than the corresponding self-rooted plants (Masuda and Gomi 1984; Shimada and Moritani 1977; Tachibana 1986, 1989; Zijlstra *et al.* 1994; Lee *et al.* 1999a, 1999b; Itai and Birnbaum 1991).

Winter and summer squash and their interspecific hybrids

Cucumbers grafted on to squash (*Cucurbita* sp.) are characterised by:

a) Higher vigour than self-rooted plants of the same cultivar (see watermelon)
b) Resistance to infection by *Fusarium* and *Verticillium* (Weng *et al.* 1993; Pavlou *et al.* 2002; Paplomatas *et al.* 2002; Papadaki *et al.* 2005)
c) Tolerance to high temperatures, in the case of squash interspecific hybrids like Shintozwa F_1. Therefore, cucumbers are often grafted on to squash for summer cultivation
d) Higher quality fruit than those produced by self-rooted plants of the same cultivar (Yamamoto *et al.* 1989; Yamakawa 1982).

Gourd (*Lagenaria* sp.)

Cucumbers grafted on to gourd exhibit good growth in greenhouse crops but show incompatibility symptoms when cultivated in the open field (Oh 2000). Similarly, Papadaki *et al.* (2005) found that cucumbers (cv. Knossos) grafted on to gourd (*L. vulgaris.*) are more vigorous than the self-grafted plants when cultivated in soil infected by *F.o.* f.sp. *cucumerinum.*

Wax (White) Gourd (*Benincasa hispida*)

This rootstock is very sensitive to *Verticillium*. It is also not considered to be suitable for grafting watermelon in early spring due to a low emergence rate and its short, thin hypocotyl (Yetisir and Sari 2004).

African Horned Cucumber (*Cucumis metuliferus*)

Although this rootstock apparently shows resistance to nematode infection (Siguenza *et al.* 2005), more experimental work is required before it can be recommended to farmers.

Burcucumber (*Sicyos angulatus*)

Cucumbers grafted on to *S. angulatus* exhibit the following characteristics:

a) They show tolerance to low temperatures and are not infected by *Fusarium* or nematodes of the *Meloidogyne* species (Ko 1999; Papadaki *et al.* 2005)
b) The quality characteristics of the fruit are not reduced, as occurs when cucumbers are grafted on to vigorous squash rootstocks (e.g. 'Shintozwa F_1') (Lee 1992).

However, *S. angulatus* has not been used as a rootstock during the last few years because its seeds germinate slowly and the seedlings produced are weak and non-uniform at the grafting stage. This makes grafting difficult and inefficient and the scions of the grafted plants do not give higher yields (Lee *et al.* 1991; Lee *et al.* 1999a; Bletsos, unpublished data).

The rootstocks show significant differences in water and nutrient uptake from the soil or nutrient solution. Apart from differences in nutrient concentrations within the tissues, large variations in total cytokinin concentrations and cytokinin composition in the sap of the xylem are found depending on the rootstock and the scion (Tachibana 1988; Zijlstra *et al.* 1994; Lee *et al.* 1999c).

Seed companies market a large number of rootstocks, varieties and hybrids that are suitable for cucumber grafting (see Table 1).

Experimental Results

Cucumbers are infected by *F.o.* f.sp. *cucumerinum* at all stages of their growth. Plants infected early in the growth cycle do not produce fruit, while those infected at later stages produce small, abnormal fruit. For these reasons, *Fusarium* can cause a major loss of income during cucumber cultivation (Owen 1956; Armstrong and Armstrong 1978; Jenkins and Wehner 1983; Bao *et al.* 1992; Ahn *et al.* 1999). By contrast, cucumbers grafted on *Cucurbita ficifolia* are not so severely infected by *F.o.* f.sp. *cucumerinum* and give satisfactory production even when they are cultivated in infected soil (Ko 1999, reported in Lee and Oda 2003; Ashraf 2003).

Cucumbers cv. Knossos grafted on *S. angulatus* were infected by *F.o.*f.sp *cucumerinum,* but to a lesser degree than self-grafted plants of the same cultivar (Papadaki *et al.* 2005). During the years 1999-2001, cucumber hybrids Discover F_1 (De Ruiter), self-rooted and grafted on *C. ficifolia*, were cultivated in soil infected by *F.o.*f.sp. *cucumerinum*. It was found that the grafted cucumbers were not infected by the fungus and produced 4.4% higher marketable yield than the self-rooted plants (Ashraf 2003).

Cucurbita ficifolia self-rooted seedlings grow normally at a soil temperature of 15°C whereas the growth of self-rooted cucumber seedlings is severely restricted (Ahn *et al.* 1999). However, when cucumbers are grafted on to *C. ficifolia*, they exhibit better growth and give a higher yield than self-rooted cucumbers during low temperatures in winter cultivation (Lee *et al.* 1999a). Cucumbers grafted on *C. ficifolia* are taller and produce 15-32% higher yield than self-rooted cucumbers when grown in soil infected by nematodes. Fruit quality (taste, size, shape and °Brix) does not differ between grafted and non-grafted cucumbers (Abou-Hadid and Medany 1992; Echebarria 2001).

Cucurbita ficifolia has a well-developed root system with a great ability to take up water and nutrients from the soil or the nutrient solution of hydroponic crops (Masuda and Gomi 1984; Shimada and Moritani 1977) even at low temperatures (Tachibana 1986, 1989). Cucumbers grafted on *C. ficifolia* have luxuriant vegetative growth resulting in higher yields than in the corresponding self-rooted plants (Zijlstra *et al.* 1994; Lee *et al.* 1999a, 1999b). The well-developed root system of *C. ficifolia* not only supplies the scion with water and minerals but also plays a significant role in the hormonal control of vegetative plant growth via cytokinin, abscisic acid and gibberellin synthesis (Itai and Birnbaum 1991).

Cucumbers grafted on *Cucurbita* species and their interspecific hybrids are not infected by *F.o.*f.sp *cucumerinum* (Weng *et al.* 1993), *F.o.*f.sp. *radicis-cucumerinum* (Pavlou *et al.* 2002) or *V. dahliae* (Paplomatas *et al.* 2002), and they produce 29.7-89.7% higher yield than the corresponding self-rooted plants. The commercial interspecific hybrids of winter squash (Leon F_1, Mirage F_1, Mamouth F_1 and Vivere F_1) delay the development of *Fusarium* on the scion of cucumber cv. Knossos (Papadaki *et al.* 2005).

Cucumbers of the commercial hybrid DRL-3077 F_1 (De Ruiter) were either self-rooted or grafted on squash rootstocks Mamouth F_1 and Rubra F_1 and cultivated during 2002 (Bletsos 2005a). It was found that the grafted cucumbers:

a) Had shorter internodes, a lower percentage of fruits retaining the flower corolla, improved shape at the fruit apex, without affecting the length of the fruit neck
b) Did not produce higher early or total yield
c) Did not affect the fruit quality characteristics (vitamin C, nitrate concentration, pH).

During 2004-2005, cucumbers of the Greek cv. Knossos were either self-grafted or grafted on to commercial squash rootstocks (Leon F_1, Mirage F_1, Mamouth F_1, Vivere F_1), *S. angulatus* and gourd (*L. vulgaris*) and cultivated in soil that was sterilized with methyl bromide and in soil that was first sterilized with methyl bromide but then inoculated with *F.o.* f.sp. *cucumerinum* cultured in the laboratory. It was found that grafted cucumbers irrespective of the rootstock:

a) Produced the same marketable yield (early and total) as the self-grafted plants in sterilized soil, whereas in soil that was inoculated with *F.o.* f.sp *cucumerinum* they produced 7-48% higher marketable yield and 20-94% higher total yield than the self-grafted plants
b) Were less infected by *F.o* f.sp *cucumerinum* than the self-grafted plants in the inoculated soil and showed a 20-46% increase in plant height and a 15-34% increase in the size of the root system
c) Produced fruits of the same quality (^{o}Brix, pH, fruit length, neck length, colour, shape of the apex) as those produced by self-grafted plants. They also formed more female flowers than the self-grafted plants, especially when grafted on Mirage F_1 and Mamouth F_1. Cucumbers grafted on Mirage F_1, *S. angulatus* and *L. vulgaris* showed a reduction in fruit corolla retention which over the two years of the experiment amounted to 7.3, 18.6 and 24.3%, respectively (Papadaki 2006).

Cucumbers grafted on winter squash (*C. moschata*) produce fruit that do not retain residues of the corolla and are more glossy than those from self-rooted plants or cucumbers grafted on *C. ficifolia* (Yamamoto *et al.* 1989; Yamakawa 1982).

Melon (*Cucumis melo* L.)

Rootstocks

Melons are grafted on to the following rootstocks:

Squash species (winter and summer) and their interspecific hybrids

Melon grafted on squash species and their interspecific hybrids:

a) Are more vigorous than the corresponding self-rooted plants
b) Are not infected by *Fusarium*, *Verticillium* or nematodes (Buitellar 1974; Pardossi and Tognoni 1985; Corazza 1998; Ferrari 1998; Jebari 1994; Ruiz *et al.* 1997; Bletsos 2005b; Cohen *et al.* 2002)
c) Are not infected by *Monosporascus cannonballus* (Edelstein *et al.* 1999; Cohen *et al.* 2002)
d) Are tolerant to low temperatures
e) Produce fruits of the same quality as those produced by self-rooted plants (Leoni *et al.* 1990).

Fig-leaf (Malabar) Gourd (*Cucurbita ficifolia*)

Melons Grafted on *C. ficifolia*:

a) Are more vigorous and give higher yield than the self-rooted plants (see cucumber)
b) Are not infected by *F.o.* f.sp. *melonis* (see cucumber)
c) Show tolerance to low temperatures (see cucumber)
d) Show tolerance to nematodes (Ferrari 1998).

Burcucumber (*Sicyos angulatus*) and Wax (White) Gourd (*Benincasa hispida*)

Melon can be grafted on to *S. angulatus* and *B. hispida.* According to Louvet (1974), melons grafted on *B. hispida* are not infected by *Fusarium* and are tolerant to low temperatures.

Experimental Results

Melons that were grafted on to commercial melon rootstocks exhibited the following characteristics:

a) Better vegetative growth than the corresponding self-rooted plants
b) Earlier and 65-310% higher total yield than the self-rooted plants (Leoni *et al.* 1990).

Later, Jebari (1994) reported similar results from his experiments. Buitellar (1974) cultivated melons, self-rooted and grafted on commercial rootstocks, and found that the grafted melons produced a higher yield than the self-rooted plants and were resistant to *Fusarium* (races 1 and 2).

Pardossi and Tognoni (1985) cultivated self-rooted and grafted melons in soil that was infected by pathogens and found that the grafted melons were infected less than the self-rooted ones. Therefore, they recommended growers to graft melons on to resistant rootstocks in order to reduce yield losses due to plant infection by *Fusarium* and *Verticillium*. Corazza (1998) recommended grafting to reduce plant infection not only by *Fusarium* but also by nematodes, while Ferrari (1998) came to the same conclusion after finding that melons grafted on resistant rootstocks were less infected by *Fusarium*. Jebari (1994) cultivated the F_1 hybrid melon Ariana H2 grafted on to the resistant rootstock RS 841 (*C. maxima* x *C. moschata*) in soil that was infected by *Fusarium* and nematodes and found that the grafted melons gave earlier production and 2.5 times higher total marketable yield in comparison with the self-rooted plants.

Ruiz *et al.* (1997) cultivated melons grafted on to squash interspecific hybrids (Shintozwa F_1, RS 841 F_1 and Kamel F_1) and found that Na, N and K uptake was affected by the rootstock genotype while P absorption was affected by the scion and the scion-rootstock interaction. The yield of the grafted seedlings was also affected by the type of rootstock and the scion-rootstock interaction. Ferrari (1998) found that melons grafted on to *C. ficifolia* or *C. metuliferus* were more resistant to nematode infection than the self-rooted plants.

During 1997-1998, melons of the Greek cv. Thraki, Peplos, Amynteou and Kokkini Banana were cultivated as self-rooted plants or grafted on commercial squash rootstocks (TZ 148 F_1, Mamouth F_1) and the native squash cultivar Kalkambaki. It was found that the grafted melons did not give larger fruits or a higher marketable yield than the self-rooted plants. In addition, the fruit from the grafted melons was less tasty due to a lower sugar concentration (Traka *et al.* 1998).

During 2001-2002, melons of the commercial hybrid Galia F_1 grafted on to the commercial squash rootstocks Mamouth F_1 and Shintosa Camelforce F_1 (Nun 9075 RT F_1) were cultivated in soil infected by *F.o.*f.sp. *melonis*. It was found that the grafted melons:

a) Were less infected by *F.o.* f.sp. *melonis* and gave more vigorous growth, 3.6-4.2 times higher early yield and 2.1-3.8 times higher total marketable yield than the self-rooted plants
b) Produced larger and sweeter fruit than the self-rooted plants
c) Had a better developed root system than the self-rooted plants (Bletsos 2005b).

Cohen *et al.* (2002) cultivated grafted melons of the Israeli cv. Ofir, which is sensitive to *Fusarium*, in soil infected by *F.o.* f.sp *melonis*. They found that the melons grafted on to *Fusarium*-resistant melons rootstocks (Orca and Adir) were less infected by *Fusarium* resulting in a higher yield than melons grafted on squash rootstocks (Brava F_1 and TZ 148 F_1).

In 2005, melons of the Greek cv. Kokkini Banana, Zakynthou, Argous, Agiou Vasiliou, Thraki, Peplos, Amynteou and the hybrid Galia F_1 were cultivated as self-rooted plants or grafted on to the commercial squash rootstocks Mamouth F_1, Rubra F_1, Big Max F_1 and Vivere F_1. It was found that:

a) Melons of cv. Argous grafted on all rootstocks and cv. Peplos grafted on the rootstocks Mamouth F_1 and Big Max F_1 were more vigorous than the self-rooted plants
b) Melon cv. Kokkini Banana and Amynteou grafted on the rootstock Mamouth F_1, cv. Agiou Vasiliou grafted on Rubra F_1 and Big Max F_1, and cv. Peplos grafted on Big Max F_1 gave higher yields than the self-rooted plants
c) The cv. Thraki and Amynteou grafted on all the rootstocks produced sweeter (higher ^{o}Brix) and tastier fruits than the self-rooted plants.

The results of this experiment indicate therefore that when melons are grafted on to suitable rootstocks both yield and fruit quality can be improved (Bletsos 2005b). However, Leoni *et al.* (1990) found that melons grafted on squash rootstocks produce fruits with a bigger mesocarp and slightly lower sugar content.

The fungus *M. cannonballus* Pollack and Vecker develops at temperatures of 25-35°C (conditions under which *Fusarium* and *Verticillium* do not develop) and causes sudden wilting and ultimate loss of melon crops in warm regions of India, south Spain, SW regions of the USA, Arabia, Central America, Japan, Taiwan, and Tunisia. To date, however, this fungus has not been detected in Greece. Infection of melons by *M. cannonballus* can be successfully confronted by grafting on resistant squash or melon rootstocks (Edelstein *et al.* 1999; Cohen *et al.* 2002).

CONCLUSION

Grafted vegetables are cultivated principally in Asia and the European countries of the Mediterranean Basin under intensive cropping systems, and more recently in America too. The price of grafted vegetable seedlings is about three times that of self-rooted seedlings. Additionally, certain physiological abnormalities may be detected in grafted vegetables that do not occur in self-rooted plants of the same cultivars, although these may be avoided by selection of more suitable rootstocks.

Growers can purchase grafted seedlings from specialized nurseries for out of season cultivation (e.g. under low temperatures) and the demand for fertilizers and water may be less. Grafted seedlings are less susceptible to soil-borne pathogens, give earlier and higher yield and make the production of organic vegetables more competitive.

Care must be taken to select the most appropriate rootstocks so as to prevent the appearance of undesirable fruit traits, which may cause a loss of quality. The use of suitable robots for grafting and the improvement of the acclimatization conditions will certainly lead to a significant reduction in the price of grafted vegetable seedlings.

REFERENCES

Abou-Hadid, A.F. and Medany M.A. 1992. Cucumber grafting for avoiding some soilborne diseases in plastic houses. *Acta Hort.* 319, 413-418.

Ahn, S.J., Im, Y.J., Chung, G.C., Cho, B.H. and Suh S.R. 1999. Physiological responses of grafted-cucumber leaves and rootstock roots affected by low root temperature. *Sci. Hort.* 81, 397-408.

Aitken-Christie, J., Kozai, T. and Takayama S. 1995. Automation in plant tissue culture. General introduction and overview, pp 1-18. In: J. Aitken-Christie, T. Kozai, and M. A. L. Smith (eds.), Automation and environment control in plant tissue culture. Kluwer Acad. Publ., Dordrecht, The Netherlands.

Akita, M. and Ohta Y. 1998. A simple method for mass propagation of potato (*Solanum tuberosum* L.) using a bioreactor without forced aeration. *Plant Cell Rep.* 18, 284-287.

Alabouvette, C., Rouxel, F., Louvet, J., Bremeersch, P. and Mention M. 1974. Recherche d' un porte-greffe resistant au *Phomopsis* et au *Verticillium dahliae* pour la culture du melon et du concombre en serre. PHM 152, 19-24.

Alan, O., Zdemir, N. and Nen Y. 2007. Effect of grafting on watermelon plant growth, yield and quality. *J.Agron.* 6, 362-365.

Alexander, L.J. 1963. Transfer of a dominant type of resistance to the four known Ohio pathogenic strains of tobacco mosaic virus (TMV) from *Lycopersicum peruvianum* to *L. esculentum*. *Phytopathology* 53, 869 (Abstract).

Alexander, L.J. 1971. Host-pathogen dynamics of tobacco mosaic virus on tomato. *Phytopathology* 61, 611-617.

Alexopoulos, A.A., Kondylis, A. and Passam H.C. 2007. Fruit yield and quality of watermelon in relation to grafting. *J. Food Agr. Environ.* 5, 178-179.

Andrews, K.P. and Marquez C.S. 1993. Graft incompatibility. Hort. Rev. 15, 183-232.

Anon. 530-545. Methodology for Living and Farming (Jaeminyosool) Vol. 5. Gourd. A revised book published by Chung Hwa Book Co., Ltd. Taiwan.

Armstrong, G.M. and Armstrong G.K. 1978. Formae speciales and races of *Fusarium oxysporum* causing wilts of the Cucurbitaceae. *Phytopathology* 68, 19-28.

Ashita, E. (ed.). 1927. Grafting of watermelons. Korea (Chosum) Agricultural Newsletter 1, 9.

Ashita, E. (ed.). 1930. Grafting methods of watermelons. Korea (Chosum) Agricultural Newsletter 4, 50.

Ashita, E. (ed.). 1934. Benefits of using grafted watermelons. Korea (Chosum) Agricultural Newsletter 8, 7.

Ashraf, M.M.I. 2003. Effect of alternatives to the methyl bromide methods for the reduction of infection by soil-borne pathogens. PhD thesis, University of Thessaloniki, 191 pp.

Balaz, F. 1982. Possibilities of grafting certain watermelon cultivars on *Lagenaria vulgaris* to prevent Fusarium wilt. *Sauvemena Poljiprivieda* 30, 563-568.

Bao, J.R., Cohn, H.S. and Gu H.F. 1992. Preliminary study on the cucumber wilt disease. *Plant Protect.* 18, 24-25.

Bhojwani, S.S. and Razdan, M.K. 1983. Plant tissue culture. Elsevier. Amsterdam, pp 71-72.

Bletsos, F.A. 2001. Grafting of fruit-bearing vegetables for a sustainable and environment friendly agriculture. New Agric. Environment Friendly (Publications of Technology of Agriculturae) 2, 59-69.

Bletsos, F.A. 2005a. The effect of cucumber grafting on growth, yield and fruit characteristics. Proc. of 22nd Scientific Congress of the Panhellenic Society of the Horticultural Science, 19-21 October 2005, Patra, Greece, 12 (B), 459-462.

Bletsos, F.A. 2005b. The effect of melon grafting on growth, yield and fruit quality. Proc. of 22nd Scientific Congress of the Panhellenic Society of the Horticultural Science, 19-21 October 2005, Patra, Greece, 12 (B), 451-454.

Bletsos, F.A. 2005c. Use of grafting and calcium cyanamide as alternatives to methyl bromide soil fumigation and their effects on growth, yield, quality and Fusarium wilt control in melon. *J. Phytopathology* 153, 1-7.

Bletsos, F.A. and Thanassoulopoulos C.C. 1999. Influence of Verticillium and Fusarium wilts on melon (*Cucumis melo* L.) growth. Proc. Conference on Cucurbitaceous and Solanaceous Vegetable Diseases in the Mediterranean Area, EPPO/MPU, 11-14/10/99, Kerkyra, Greece. OEPP/EPPO Buletin 30, 191-194.

Bletsos, F.A., Moustafa, M.A. and Thanassoulopoulos, C.C. 2002. Replacement of methyl bromide application by alternative methods in covered vegetables. Proc. 2nd Balkan Symposium on Vegetables and Potatoes. 11-15 October 2000, Thessaloniki, Greece. Acta Hort. ISHS 579, 451-456.

Brown, J.C., Chaney, R.L. and Ambler, J.E. 1971. A new mutant inefficient in the transport of iron. *Physiol. Plant.* 25, 48-53.

Buitellar, K. 1974. Grafting melons. Groenten-en-Fruit 29, 29 (Abstract).

Bulder, H.A.M., Speek, E.J., Van Hasselt, P.R. and Kuiper P.J.C. 1991. Growth temperature and lipid composition of cucumber genotypes differing in adaptation to low energy conditioning. *J. Plant Physiol.* 138, 655-660.

Bulder, H.A.M., Van Hasselt, P.R., Kuiper, P.J.C., Speek, E.J. and Den Nijs A.P.M. 1990. The effect of low root temperature on growth and lipid composition of low temperature tolerant rootstock genotypes for cucumber. *J. Plant. Physiol.* 138, 661-666.

Chung, H.D. 1995. Studies on the occurrence, cause and prevention of fermentation in melon fruit, *Cucumis melo* L. var. makuwa Mak. cv. Gumssaragi-euncheon. *Res. Rep.* Kyungpuk Provincial RDA, Korea.

Coggins, C.W. and Lesley, J.W. 1968. Attempts to improve flower bud retention and development in tomatoes with grafts, nutrition and growth regulators. HortScience 3, 237-238

Cohen, R., Horev, C., Burger, Y., Shriber, S., Hershenhorn, J., Katan, J. and Edelstein M. 2002. Horticultural and pathological aspects of Fusarium wilt management using grafted melons. HortScience 37 (7), 1069-1073.

Corazza, L. 1998. Fusariosis of winter melon. *Colture-Protette* 27 (5), 39-41.

D' Amore, R., Morra, L. and Parisi B. 1996. Grafted watermelon: production results. *Colture-Protette* 25 (9), 29-31.

Davis, A.R. and Perkins-Veazie, P. 2005-2006. Rootstock effects on plant vigor and watermelon fruit quality. *Cucurbit Genet. Coop. Rpt.* 28-29, 39-42.

Davis, A.R., Perkins-Veazie, P., Sakata, Y., Lopez-Galarza, S., Maroti, J.V., Lee, S-G., Huh, Y-C., Sun, Z., Miguel, A., King, S.R., Cohen, R., and Lee J-M. 2008. Cucurbit grafting. *Critical Reviews in Plant Sciences.* 27, 50-74.

Denna, D.W. 1962. A simple grafting technique for Cucurbits. *Proc. Am. Soc. Hort. Sci.* 81, 369-370.

Echebarria, P.H. 2001. Influence of different rootstocks on the yield and quality of greenhouse grown cucumbers. *Acta Hort.* 559, 139-143.

Edelstein, M., Cohen, R., Burger, Y., Shriber, S., Pivonia, S. and Shtienberg, D. 1999. Integrated management of sudden wilt in melons, caused by *Monosporascus cannonballus*, using grafting and reduced rates of methyl bromide. *Plant Dis.* 83 (12), 1142-1145.

EPA. 1997. Methyl bromide alternative case studies. Vol. 3, 18 Sept. 2007. <http://www.epa.gov/spdpublc/mbr/casestudies/volume3/index.html>

Ferrari, V. 1998. Fusarium and root knot nematodes, two adversaries of melon which are difficult to control chemically. Informatore-Agrario-Supplemento 54 (3), 48-50.

Friedlander, M., Atsmon, D. and Galun, F. 1977. The effect of grafting on sex expression in cucumber. *Plant Cell Physiol.* 18, 1343-1350.

Gambley, C., Harrison, D., Teasdale, R., Goodman, P. and Hinsch, A. 1997. The development of the VITRON ™ robot for automated micropropagation. *Acta Hort.* 447, 637-642.

Giannakou, I.O. and Karpouzas, D.G. 2003. Alternatives to methyl bromide for root-knot nematode control. *Pest Management Sci.* 59, 883-892.

Gomi, K. and Masuda, M. 1981. Studies on the characteristics of nutrient absorption of rootstocks on grafting fruit vegetables. I. Magnesium deficiency of leaves of cucumber as affected by a rootstock, *C. ficifolia* and potassium concentration in culture solution. Bul. Fac. Agr. Miyazaki Univ., Miyazaki, Japan 27 (2), 179-186.

Grigoriadis, I., Nianiou-Obeidat, I. and Tsaftaris, A.S. 2005. Shoot regeneration and micrografting of micropropagated hybrid tomatoes. *J. Hort. Sci. Biotechnol.* 80 (2), 183-186.

Gruscenko, I.E. and Drobkov, A.A. 1952. Introduction and distribution of radioactive elements in grafted plants and their effect on the development of tomato. Izv. Akad. Nauk. S.S.R.R. Ser. Biol. 6, 62-66.

Hartmann, H.T., Kester, E.D., Davies, T.F. and Geneve, R.L. 2002. Plant propagation: Principles and practices (7th ed.). Prentice-Hall, NJ., 880 pp.

He, Z.L. and Fu, X.Y. 1988. Trials on watermelon grafting for resistance to wilt. *Fujian Agric. Sci. Tech.,* 3, 4.

Hirai, G., Nakazumi, H., Yagi, R. and Nakano, M. 2002. Fusarium wilt (race 1,2y)-resistant melon (*Cucumis melo*) rootstock varieties 'Dodai l' and 'Dodai 2'. *Acta Hort.* 588, 155-160.

Hong, M.S. 1710. Forest Economics 1, 207-211.

Horvath, I., Vigh, L., van Hasselt, P.R., Woltjecs, J. and Kuiper, P.J.C. 1983. Lipid composition in leaves of cucumber genotypes as affected by different temperature regimes and grafting. *Physio. Plant.* 57, 532-536.

Hoyos, P. 2001. Influence of different rootstocks on the yield and quality of greenhouse grown cucumbers. *Acta Hort.* 559, 139-143.

Hwang, H., Kim, S.C. and Ko, K.D. 1999. Development of prototype automatic grafting system for fruit-bearing vegetables. *Kor. J. Agr. Machinery* 24 (3), 217-224.

Ikeda, H., Okitsu, S. and Arai, K. 1986. Comparison of magnesium deficiency of grafted and non-grafted cucumbers in water culture and soil culture and the effect of increased magnesium application on the prevention of magnesium deficiency disorder (in Japanese with English summary). *Bul. Natl. Res. Inst. Veg. Ornam. Plants and Tea*, C9, 31-41.

Itai, C. and Birnbaum, H. 1991. Synthesis of plant growth regulators by roots. In: Y. Waisel, A. Eshel and U. Kafkati, (Eds), Plant Roots, the Hidden Half. Marcel Dekker, New York, pp 163-178.

Janick, J. 1986. Horticultural science, 4th edition. W.H. Freeman and Co., New York. p 339-346.

Jebari, H. 1994. Grafting early melon in cold plastic houses. Annales de l' Institute Nat. Rech. Agron. Tunisie 67 (1-2), 165-176.

Jenkins, S.F. and Wehner, T.C. 1983. Occurrence of *Fusarium oxysporum* f.sp *cucumerinum* on greenhouse-grown *Cucumis sativus* seed stock in North Carolina. *Plant Dis.* 67, 1024-1025.

Kang, C.H. 2000. Status of vegetable grafting machine development in Korea and practical problems. Problems and counter plants of vegetable seedling production. *Kor. Res. Soc. Prot. Hort.* 12, 91-110.

Kato, T. and Lou, H. 1989. Effect of rootstock on the yield, mineral nutrition and hormone level in xylem sap in eggplant. *J.Japan. Soc. Hort. Sci.* 58, 345-352.

Kim, S.E. and Lee, J.M. 1989. Effect of rootstocks and fertilizers on the growth and mineral contents in cucumber (*Cucumis sativus*). Res. Collection, *Inst. Food Develop.* 10, 75-82. Kyung Hee Univ., Korea.

Kim, D.H. and Lee, J.M. 2000. Seed treatment for cucumber green mottle mosaic virus (CGMMV) in gourd (*Lagenaria siceraria*) seeds and its detection. *J. Kor. Soc. Hort. Sci.,* 41, 1-6

Ko, K.D. 1999. Response of cucurbitaceous rootstocks species to biological and environmental stresses. Ph.D. Diss., Seoul Natl. Univ., Suwon, Korea (in Korean with English summary).

Kobayashi, K. 1991. Development of grafting robot for the fruit vegetables. *Plant Cell Techn.* 3 (6), 477-482.

Kobayashi, K., Onoda, A., Suzuki, M. and Otsuka, H. 1996. Development of grafting robot for cucurbitaceous vegetables. Part 4. Test for practical use. *J. Jap. Soc. Agr. Machinery* 58 (3), 59-68.

Konstas, J. and Kintzios, S. 2003. Developing a scale-up system for the micropropagation of cucumber (*Cucumis sativus* L.): the effect of growth retardants, liquid culture and vessel size. *Plant Cell Rep.* 21, 538-548.

Kouvelos, I. 2005. Evaluation of resistance five watermelon rootstocks in soil borne fungi *Fusarium oxysporum* f.sp. *niveum* and *Verticillium dahliae*. MS dissertation, Aristotelian University, Thessaloniki, Greece, 173 pp.

Kubota, C. and McClure, M.A..2008. Vegetable grafting: History, use, and current technology status in North America. *HortScience* 43 (6), 1664-1669.

Kurata, H. 1976b. Studies on the sex expression of flowers induced by day-length and temperature in pumpkin and watermelon. Mem. Fac. Agr., Kagawa Univ. 29, 1-49.

Kurata, K. 1994. Cultivation of grafted vegetables. 2. Development of grafting robots in Japan. *HortScience* 29, 240-244.

Lee, J.M. 1989. On the cultivation of grafted plants of cucurbitaceous vegetables. *J. Kor. Soc. Hort. Sci.* 30, 169-179.

Lee, J.M. 1994. Cultivation of grafted vegetables. I. Current status, grafting methods, and benefits. *HortScience* 29, 235-239.

Lee, J.M., Bae, E.J., Lee, C.W. and Knon, S.S. 1999b. Seedling growth and fruit set and quality of cucumber as affected by triazole chemicals. *Acta Hort.* 483, 125-132.

Lee, J.M., Bang, H.J. and Ham, H.S. 1998. Grafting of vegetables. *J. Japan. Soc. Hort. Sci.* 67, 1098-1114

Lee, J.M., Bang, H.J. and Ham, H.S. 1999a. Quality of cucumber fruit as affected by rootstocks. *Acta Hort.* 483, 117-123.

Lee, J.M., Kwon, W.S., Lee, T.J., Ham, H.S. and Bang, H.J. 1999c. Qualification of cytokinins by HPLC after purification with immunoaffinity chromatography and application to crop production. *J. Kor. Soc. Hort. Sci.* 41, 13-18.

Lee, J.M. and Oda, M. 2003. Grafting of herbaceous vegetables and ornamental crops. *Hort. Rev.* 28, 62-124.

Lee, J.M., Schreiber, B.M.N. and Brenner, M.L. 2001. Cytokinin contents in bleeding xylem sap of cucurbitaceous crops. 2nd Int. Symp. Cucurbits. Tshukuba, Japan.

Lee, W.H., Kim and Kwack, B.H. 1991. The characteristics of *Sicyos angulatus* L. growing wild in the Andong area and its potential as a rootstock for cucurbitaceous crops (Abstract). J. *Kor. Soc. Hort. Sci.* 32, 299-304.

Lee, W.Y. 1992. Use of *Sicyos angulatus* cv. Andong as rootstock for cucurbits. RDA Research and Extension 33 (4), 69-73.

Leonardi, C. and Romano, D. 2004. Recent issues on vegetable grafting. *Acta Hort.* 631, 163-174.

Leoni, S., Grudina, R., Cadinu, M., Madeddu, B. and Garletti, M.G. 1990. The influence of four rootstocks on some melon hybrids and a cultivar in greenhouse. *Acta Hort.* 287, 127-134.

Liu, H.C., Lin, Z., Tian, Y.G., Yu, A.M., Liu, H.C., Lin, Z.J., Tian, Y.G. and Yu, A.M. 1995. Control of Fusarium in watermelon by grafting in successive seasons. *China Veg.* 1, 12-14 (Abstract).

Liu, H.Y., Zhu, Z.J., Lu, G.H. and Qian, Q.Q. 2003. Study on relationship between physiological changes and chilling tolerance in grafted watermelon seedlings under low temperature stress. *Sci. Agric. Sinica* 36, 1325-1329.

Liu, Y.Q., Liu, S.Q. and Wang, H.B. 2004. Effect of salt-tolerant stock on growth, yield, and quality of watermelon, Shandong. *Agri. Sci.* 4, 30-31.

Lopez-Galarza, S., San Bautista, A., Perez, D.M., Miguel, A., Baixauli, C., Pascual, B., Maroto, J.V. and Guardiola, J.L. 2004. Effects of grafting and cytokinin-induced fruit setting on colour and sugar-content traits in glasshouse-grown triploid watermelon. *J. Hort. Sci. Biotech.* 79, 971-976.

Louvet, J. 1974. The use of grafting in market gardening. P. H. M. 152, 13-16.

Lu, S.F., Yang, D.T., Song, J.Y., Liu, M.Q. and Yang, S.J. 1996. Preliminary studies on controlling graft union through plant hormones (in Chinese with English summary). *Acta Bot. Sinica* 38, 307-311.

Masuda, M. and Gomi, K. 1984. Mineral nutrition and oxygen consumption in grafted and non-grafted cucumbers. J. Japan. Soc. Hort. Sci., 52, 414-419 (in Japanese with English summary).

Matsubara, S. 1989. Studies on salt tolerance of vegetables. 3. Salt tolerance of rootstocks. Bul. Okayamma Univ. Agr. 73, 17-25.

Matsuda, T. and Honda, H. 1981. Studies on physiological disorder in the fruit of melon. (1) Influence of grafting and plant-and fruit pruning in 'Prince' melon (in Japanese with English summary). *Bul. Natl. Res. Inst. Veg. Ornam. Plants Tea.* C5, 31-50.

Miguel, A. 1997. Injerto de hortalizas. Serie Divulgación Técnica. Consellería de Agricultura, Pescay Alimentaci aculeón, Genaralitat Valenciana, Valencia, 50-52.

Miguel, A., Marsal, J.I., López-Galarza, S., Maroto, J.V., Tarazona, V. and Bono, M. 2005. Comportamiento de portainjertos de sandía frente a nematodos. Phytoma-España 175, 20-24.

Mijuskovic, M. and Vucinic, Z. 1977. Fusarium wilt of watermelon in Montenegro and the possibility of its conrol. *Glasnik Odjelenja Prirodnih Nauka* 2, 229-250 (Abstract).

Mockaitis, J.M. and A. Kivilaan. 1964. Graft-induced sex changes in *Cucumis melo* L. Nature 202, 216.

Mondal, S.N., Hossain, A.K.M.A., Hossain, A.E., Islam, M.A. and Bashar, M.A. 1994. Effect of various rootstocks in the graft culture of watermelon in Bangladesh. *Punjab Veg. Grower* 29, 15-19 (Abstract).

Morel, G. and Martin, C. 1952. Guerison de dahlias atteints d'une maladie a virus. *C.R. Acad. Sci.* 235, 1324-1325.

Murashige, T. and Skoog, F. 1962. A revised medium for rapid growth and bioassays with tobacco tissue cultures. *Physiol. Plant.* 15, 473-497.

Nobuoka, T., Oda, M. and Sasaki, H. 1996. Effect of relative humidity, light intensity and leaf temperature on transpiration of tomato scions (in Japanese with English summary). *J. Japan. Soc. Hort. Sci.* 64, 859-865.

Nobuoka, T., Oda, M. and Sasaki, H. 1997. Effects of wind and vapor deficit on transpiration of tomato scions (in Japanese with English summary). *J. Japan. Soc. Hort. Sci.* 66, 105-112.

Obeidy, A.A. and Smith, M.A.L. 1991. A versatile new tactic for fruit tree micrografting. *HortTechnology* 1 (1), 91-95.

O'Brien, M. 1983. Evaluation of eggplant accessions and cultivars for resistance to Verticillium wilt. *Plant Dis. Rep.* 67, 763-764.

Oda, M. 1994. Effects of uniconazole and gibberelic acid application on elongation of hypocotyl and internodes in figleaf gourd for rootstock. *Japan Agric. Res. Quart.* 28, 195-199.

Oda, M. 1995. New grafting methods for fruit-bearing vegetables in Japan. Jarq 29 (3), 187-194.

Oda, M. 1999. Grafting of vegetables to improve greenhouse production. Food and Fertilizer Technology Center, Extension Bulletin 480, Taipei city, Republic of China on Taiwan.

Oda, M., Dsai, M., Ikeda, H. and Furukawa, H. 2001. Causes of low survival in cucumber (*Cucumis sativus*) plants grafted onto pumpkin (*Cucurbita moschata* Duch.) rootstocks by horizontal cut grafting at the center of the hypocotyls. Sci. Rep. Coll. Agric. Osaka Prefecture University 53, 1-5.

Oda, M., Nagaoka, M. and Tsuji, K. 1992. Inter-generic, inter-specific and inter-varietal grafting in *Cruciferae*. Acta Hort. 319, 425-430.

Oda, M., Tsuji, K. and Sasaki, H. 1993. Effects of hypocotyl morphology on survival rate and growth of cucumber seedlings grafted on *Cucurbita spp. Japan. Agr. Res. Quarterly* 26, 259-263.

Oda, M., Tsuji, K., Ichimura, K. and Sasaki, H. 1994. Factors affecting the survival of cucumber plants grafted on pumpkin plants by horizontal grafting at the hypocotyl level. *Bul. Natl. Res. Inst. Veg. Ornam. Plants Tea* A9, 51-60.

Oh, S.Y. 2000. Graft compatibility and xylem sap composition in cucurbitaceous vegetables (Abstract). MS Diss., Kyung Hee Univ., Suwon, Korea.

Okimura, M., Matsuo, S., Arai, K. and Okitsu, S. 1986. Influence of soil temperature on the growth of fruiting vegetables grafted on different rootstocks (In Japanese with English summary). *Bul. Natl. Res. Inst. Veg. Ornam. Plants Tea* C 9, 43-58.

Olympios, C.M. 2001. The Technique of Vegetable Cultivation in Greenhouses (in Greek). Stamoulis, Athens, 630 pp.

Onoda, S.A., Kobayashi, K. and Suzuki M. 1992. The study of the grafting robot. *Acta Hort.* 319, 535-540.

Owen, J.H. 1956. Cucumber wilt, caused by *Fusarium oxysporum* f.sp *cucumerinum*. Phytopathology 46, 153-157.

Papadaki, A.M. 2006. The effect of grafting on growth, production and resistance in adromicosis of tomato by *Verticillium dahliae* and cucumber by *Fusarium oxysporum* f.sp. *cucumerinum*. MS dissertation, Aristotelian University of Thessaloniki, Greece, 85 pp.

Papadaki, A.M., Bletsos, F.A. and Lagopodi, A.L. 2005. The effect of cucumber grafting onto commercial rootstocks on *Fusarium* development. Proc. of 22nd Scientific Congress of the Panhellenic Society of the Horticultural Science, 19-21 October 2005, Patra, Greece, 12 (B), 559-562.

Paplomatas, E.J., Elena, K., Tsagkarakou, A. and Perdikaris, A. 2002. Control of Verticillium wilt of tomato and cucurbits through grafting of commercial varieties on resistant rootstocks. *Acta Hort.* 579, 281-284.

Pardossi, A. and Tognoni, F. 1985. Possibility of controlling soil-borne pathogens in melon-growing. Colture-Protette 14 (12), 27-31.

Paroussi, G., Bletsos, F.A., Bardas, G.A., Kouvelos, J.A. and Klonari, A. 2007. Control of *Fusarium* and *Verticillium* wilt of watermelon by grafting and its effect on fruit yield and quality. *Acta Hort.* 729, 281-285.

Pavlou, G.C., Vakalounakis, D.J. and Ligoxigakis, E. K. 2002. Control of root and stem rot of cucumber, caused by *Fusarium oxysporum* f.sp. *radicis-cucumerinum*, by grafting onto resistant rootstocks. *Plant Dis.* 86 (4), 379-382.

Pontiki, M. 2005. Combination of micrografting with micropropagation in melon. MS dissertation, Aristotelian University of Thessaloniki, Greece, 53 pp.

PSNCK. 1982. Forest economics (Rev. Korean edition of original writings of M. S. Hong). Proc. Soc. Nat. Culture of Korea.

Pulgar, G., Villora, G., Moreno, D.A. and Romero, L. 2000. Improving the mineral nutrition in grafted watermelon plants: nitrogen metabolism. *Biologia Plantarum* 43 (4), 607-609 (Abstract).

Qi, H.Y., Li, T.L., Liy, Y.F. and Li, D. 2006. Effects of grafting on photosynthesis characteristics, yield, and sugar content in melon. J. Shenyang Agr. Univ. 37, 155-158.

Ruiz, J.M., Belakbir, A., Lopez-Cantarero, I. and Romero, L. 1997. Leaf-macronutrient content and yield of grafted melon plants. A model to evaluate the influence of root-stock genotype. *Scientia Hort.* 71, 227-234.

Ryu, J.S., Choi, K.S. and Lee, S.S. 1973. Effects of grafting stocks on growth, quality and yields of watermelon. *J. Kor. Soc. Hort. Sci.* 13, 45-49.

Sakata, Y., Ohara, T. and Sugiyama, M. 2008. The history of melon and cucumber grafting in Japan. *Acta Hort.* 767, 217-228.

Sakata, Y., Takayoshi, O. and Mitsuhiro, S. 2007. The history and present state of the grafting of cucurbitaceous vegetables in Japan. *Acta Hort.* 731, 159-170.

Sapountzakis, G. and Tsaftaris, A.S. 2002. Combination of micropropagation with micrografting in cucumber. *Acta Hort.* 579, 233-236.

Sarowar, S., Oh, H.Y., Hyung, N.I., Min, B.W., Harn, C.H., Yang, S.K., Ok, S.H. and Shin, J.S. 2003. *In vitro* micropropagation of a *Cucurbita* interspecific hybrid cultivar-a rootstock plant. *Plant Cell Tiss. Org. Cult.* 75, 179-182.

Satoh, S. 1996. Inhibition of flowering of cucumber grafted on rooted squash stocks. *Physiol. Plant.* 97, 440-444.

Shimada, N. and Moritani, M. 1977. Nutritional studies on grafting of horticultural crops. II. Absorption of minerals from various nutrient solutions by grafted cucumber and pumpkin plants. *J. Japan. Soc. Soil Sci. Plant Nutr.* 48, 396-401.

Siguenza, C., Schochow, M., Turini, T. and Ploeg, A. 2005. Use of *Cucumis metuliferus* as a rootstock for melon to manage *Meloidogyne incognita*. *J. Nematology* 37, 276-280.

Singh, P.K., Dasgupta, S.K and S.K. Tripathi. 2004. Hybrid vegetable development. Food Products Press, Binghampton, USA, 442 pp.

Suge, H. 1984. Nature of floral stimulus in *Perilla* as studied by grafting. I. Method of evaluation and the movement of floral stimulus as affected by N^b –benzylaminopurine. *Japan Soc. Crop Sci.* 53, 423-329.

Suzuki, M., Kobayashi, K., Inoku, K., Miura, K. and Hirata, K. 1995. Development of grafting robot for cucurbitaceous vegetables. *J. Jap. Soc. Agric. Machinery* 57 (2), 67-75.

Tachibana, S. 1982. Comparison of root temperature on the growth and mineral nutrition of cucumber cultivars and figleaf gourd. *J. Jap. Soc. Hort. Sci.* 51, 299-308.

Tachibana, S. 1986. Effect of root temperature on lipid and its fatty acid composition in cucumber and figleaf gourd roots. *J. Jap. Soc. Hort. Sci.* 55, 187-193.

Tachibana, S. 1988. Cytokinin concentrations in roots and root xylem exudate of cucumber and figleaf gourd as affected by root temperature. *J. Jap. Soc. Hort. Sci.* 56, 417-425.

Tachibana, S. 1989. Respiratory response of detached roots to lower temperature in cucumber and figleaf gourd grown at 20°C root temperature. *J. Jap. Soc. Hort. Sci.* 58, 333-337.

Takahashi, H., Saito, T. and Siuge, H. 1982. Intergeneric translocation of floral stimulus across graft union in monoecious Cucurbitaceae with special reference to the sex expression of flowers. *Plant Cell Physiol.* 23, 1-9.

Tjamos, E.C., Antoniou, P.P., Tjamos, S.E., Fatouros, N.P. and Giannakou, J. 2002. Alernatives to methyl bromide for vegetable production in Greece. *Proc. Fifth Int. Conf.* on Alternatives to methyl bromide, Lisbon, 168-171.

Torii, T., Kasiwazaki, M., Okamoto, T. and Kitani, O. 1992. Evaluation of graft-take using a thermal camera. *Acta Hort.* 319, 631-634.

Traka-Mavrona, E., Koutsika-Sotiriou, M., Bletsos, F. and Pritsa, Th. 1998. Performance *Cucurbita spp.* as rootstock of Greek melon (*Cucumis melo* L.) cultivars. Proc. 7th Scientific Congress in Genetics and Plant Breeding, 21-23 October 1998, Heraclion Kreta, Greece, p, 119-128.

Trionfetti-Nisini, P., Colla, G., Granati, E., Temperini, O., Crino, P. and Saccardo. F. 2002. Rootstock resistance to Fusarium wilt and effect on fruit yield and quality of two muskmelon cultivars. *Sci. Hort.* 93, 281-299.

Weng, Z.X., Li, B.D. and Feng, D.X. 1993. Study on enchancement of cucumber resistance and yield by grafting on *Cucurbita ficifolia*. China Veg. 3, 11-15.

Xu, C.Q., Li, T.L., Qi, H.Y. and Wang, H. 2005b. Effects of grafting on growth and development, yield, and quality of muskmelon. China Veg. 6, 12-14.

Xu, C.Q., Li, T.L., Qi, H.Y. and Qi. M.F. 2006. Effects of grafting on development and sugar content of muskmelon fruit. *J. Shenyang Agr. Univ.* 1, 378-381.

Xu, S.L., Chen, Q.Y., Chen, X.Q., Gao, J.S. and Li, S.H. 2005c. Effects of grafting on 'Jiashi' muskmelon yield and its resistance to melon fusarium wilt. *Acta Hort. Sinica* 32, 521-523.

Xu, S.L., Chen, Q.Y., Li, S.H., Zhang, L.L., Gao, J.S. and Wang, H.L. 2005d. Role of sugar-metabolizing enzymes and GA_3, ABA in sugar accumulation in grafted muskmelon fruit. *J. Fruit Sci.* 22, 514-518.

Yamakawa, K. 1982. Grafting. pp 141-153. In: S. Nishi (ed), Handbook of Vegetable Production. Yokendo, Tokyo.

Yamamoto, Y., Hayashi, M., Kanamaru, T., Watanabe, T., Mametsuka, S. and Tanaka, Y. 1989. Studies on bloom on the surface of cucumber fruits. 2. Relation between the degree of bloom occurrence and contents of mineral elements (in Japanese with English summary). *Bull. Fukuoka Agr. Res. Cent.* B9, 1-6.

Yamasaki, A., Yamashita, M. and Furuya, S. 1994. Mineral concentrations and cytokinin activity in the xylem exudate of grafted watermelons as affected by rootstocks and crop load. *J. Jap. Soc. Hort. Sci.* 62, 817-826.

Yetisir, H. and Sari, N. 2004. Effect of hypocotyl morphology on survival rate and growth of watermelon seedlings grafted on rootstocks with different emergence performance at various temperatures. *Turk. J. Agric. For.* 28, 231-237.

Zaiter, H.Z., Coyne, D.P. and Clark, R.B. 1987. Temperature, grafting method, and rootstock influence on iron-deficiency chlorosis of bean. J. Am. Soc. Hort. Sci. 112, 1023-1026.

Zheng, G.F. 1995. A fine watermelon stock, ChaoFeng F_1, and its cultivation. *Henan Nongye Kexue* 3, 24-25 (Abstract).

Zijlstra, S., Groot, S.P.C. and Jansen, J. 1994. Genotypic variation of rootstocks for growth and production in cucumber: possibilities for improving the root system by plant breeding. Sci. Hort. 56, 185-196.

Ziv, M., Ronen, G. and Ravin, M. 1998. Proliferation of meristematic clusters in disposable presterillized plastic bioreactors for large-scale micropropagation of plants. *In vitro Cell Dev.Biol. Plant* 34, 152-158.

This paper has been reviewed by Dr. D. Savvas, Assistant Professor of Horticulture, Agricultural University of Athens.

In: Horticulture in the 21st Century
Editors: Abigail N. Sampson, pp. 121-142
ISBN 978-1-61668-582-9

Chapter 4

APPLICATION OF BIOTECHNOLOGY TOOLS IN THE CONSERVATION AND CHARACTERIZATION OF CRITICALLY ENDANGERED ANCIENT MOROCCAN SPECIES ARGANIA SPINOSA (SAPOTACEAE)

***Pedro Martínez-Gómez*[1 1]*, Abdellatif Hafidi*[2]*, Abel Piqueras*[1]*, and Khalid Majourhat*[3]**

[1] Department of Plant Breeding, CEBAS-CSIC. PO Box 164.
E-30100 Espinardo, Murcia, Spain
[2] Faculté des Sciences Semlalia, University Cadi-Ayyad of Marrakech,
BP 2390, Marrakech 400001, Morocco
[3] Faculté des Sciences, University Ibn Zohr of Agadir, BP 32/S Agadir, Morocco

ABSTRACT

Argan tree [*Argania spinosa* (L.) Skeel] is an endemic species from Southwest Morocco being the unique representative of the tropical Sapotaceae in this area. This species plays a great socio-economical and ecological role in these arid zones due to the prized oil from its kernels and its foliage to feed cheep and goat herds. However, a great zoo-anthropologic action led to the absence of natural regeneration of this species. In this work, the application of recent biotechnology tools for conservation and characterization of these critically endangered species is exposed. Promising propagation methods included *in-vitro* protocols well adapted to this species that allowed the introduction, micropropagation and rooting of plant material. On the other hand, a standard karyotyping protocol has been described as preliminary tool to start molecular studies. In addition, molecular markers based on PCR amplification of the DNA have also become an essential tool for the characterization and conservation of argan tree. Regarding this PCR amplification of the DNA, three main strategies, RAPD (random amplification of the DNA), SSR (using sequences from related species) and SCAR (sequence-characterized amplified regions) markers are being assayed to date. These molecular markers have been applied in the genetic characterization of argan germplasm, the

1 pmartinez@cebas.csic.es.

establishment of genetic relationships between morphotypes, and the future construction of genetic maps. In addition, these initially DNA research open the door to the future DNA sequencing of this species.

INTRODUCTION

In arid and semiarid regions of south-western Morocco, argan tree [*Argania spinosa* (L.) Skeels (synonyms: *Argania sideroxylon* Roem. and Schult., *Sideroxylon spinosum* L.)] has proven valuable for production of prized seed oil, forage and firewood, being also characterized by its drought resistance and fighting against desertification (Bechakroun and Buttoud, 1989). Argan tree belonging to the *Sideroxyleae* tribe is the only representative species of the tropical family *Sapotaceae* in Morocco (Emberger, 1939). *Argania* is a monospecific genus with *A. spinosa* as the unique representative (Pennington 1991; Swenson and Anderberg, 2005). This species develops under the hot arid Mediterranean climate and has very different distribution and ecological preferences compared to the other species of the *Sideroxylon* tribe (Benchakroun, 1990).

Argan is an endemic species to Southwest Morocco where it occupies a wide distribution area around 800,000 ha. It plays great socio-economical and ecological roles in these arid zones. Indeed, thanks mainly to the prised oil extracted from its kernels and its foliage that feed cheep and goat herds, 90% of the local population incomes depend on an agroforestry system based on this multiuse tree (Benchkroun, 1990) (Figure 1). However, the argan forest undergone and excessive degradation caused by overgrazing, wood removal, urban expansion and land clearing for intense agriculture installation, which make it threatened of disappearance (Benchekroun and Buttoud, 1989). The strong zoo-anthropic pressure on argan groove generates ecological and social problems whose signs are already announced like the projection of the desert. Soils erosion and rural exodus are also among these direct consequences that require an urgent intervention to preserve and safeguard this fragile ecosystem (Benchekroun and Buttoud, 1989).

This state of affairs underlines the need for launching massive afforestation programs in these regions. Currently, reforestation programs are based on conventional techniques by sexual and asexual propagation from cuttings and grafting (Benismail et al., 2002, Nouaim et al., 2002). Nevertheless, they fail to response to high demand and to overcome difficulties of field establishment due mainly to the poor quality of root of produced plantlets. In this sense, *in vitro* culture could be a very interesting alternative to the traditional progation of argan tree.

To protect argan groves from present overexploitation, it was classified in 1998 as a "Biosphere reserve" by UNESCO, and many efforts were undertaken by the Moroccan government for its rehabilitation. This species is considered as "underutilized plant species" with a potential to contribute to food and nutritional security and poverty alleviation. In this sense, the chronic food shortstages present in many developing countries demonstrate clearly the fragility of food security systems based on a few staples.

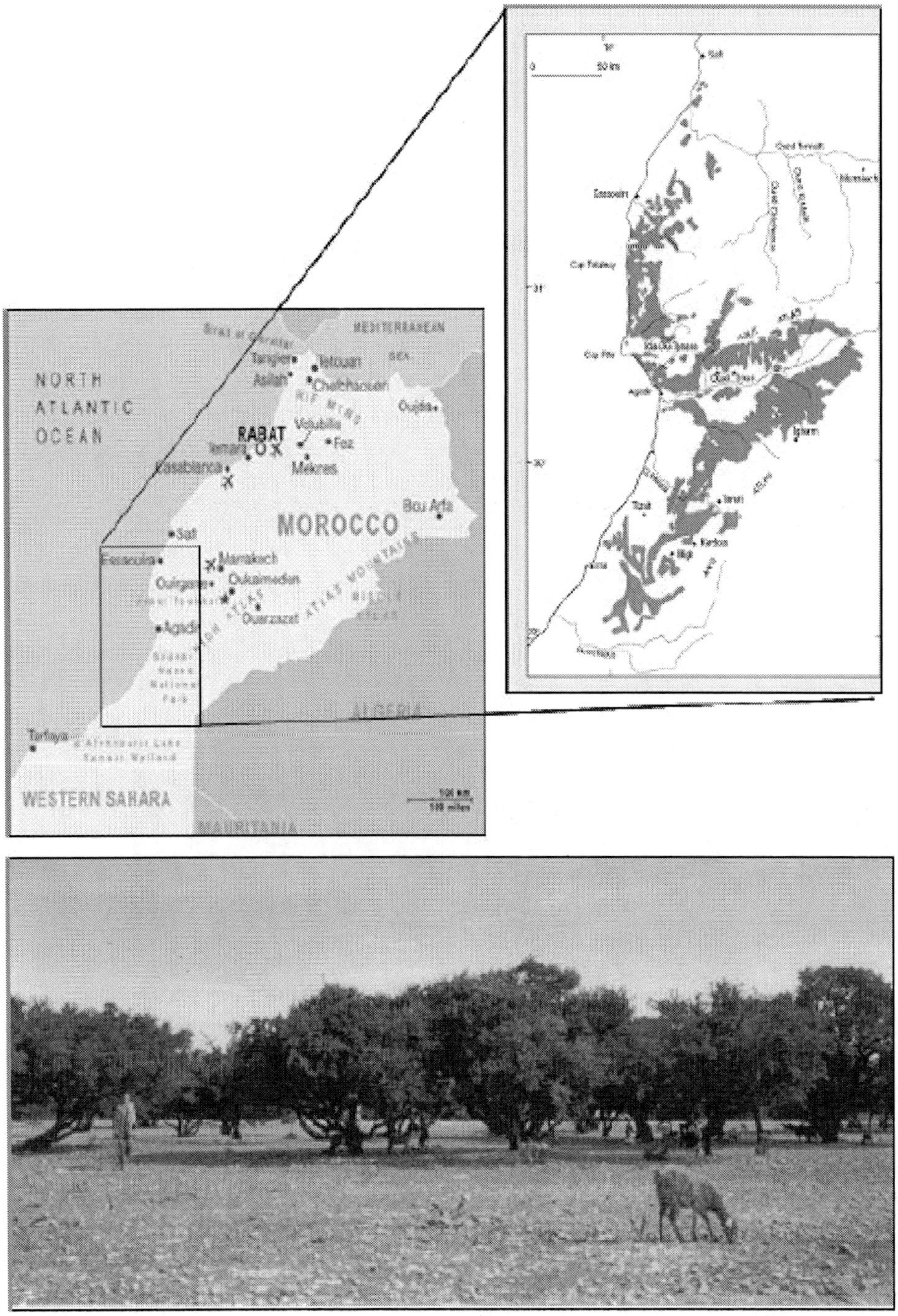

Figure 1. Location of argan populations in Southwest of Morocco and detail of tree population in the north of Essaouira city.

Agrobiodiversity is them a necessity for the survival of poor communities and the stabilization of agroecosystems. Many of these "underutilized plant species" such as argan tree are particularly useful in marginal lands, where they have been selected to withstand stress conditions and where they contribute to sustainable production with few inputs (IPGRI, 2006a; Martínez-Gómez et al., 2007). However, because they have been neglected by researchers in the past not enough is known about their agronomy and yield improvement potential or how to best take advantages of their inherent qualities (IPGRI, 2006a; 2006b; 2006c).

This species is genetically and molecularly very poorly known (Majourhat et al., 2007a). Traditionally, characterization and identification of variability has been based on morphological (Bani-Aameur, 2004; Chernane et al., 2000). It has been suggested to use fruit form as criterion for varietal selection in this species (Bani-Aameur, 2004). The three most commonly identified morphotypes of argan tree were sampled, oval, spherical and spindle fruit types (Figure 2). However, such morphological markers are not always available for analysis, and are affected by environmental conditions. DNA marker technology offers several advantages over the sole use of conventional markers. In the case of argan, only one work studying the genetic variability of argan populations using DNA amplification and restriction was published until now (El Mousadik and Petit, 1996b).

On the other hand, it is also necessary to note that although the initial aim of argan culture focuses mostly on oil production, forage and wood, recently the scope of this culture has been extended for use in bioprocess technology for production of high value chemicals of immense importance in the cosmetics, pharmaceutical and nutraceutical sectors making that 90 % of local population incomes depend on based on this multiuse tree (Benchkroun, 1990; Martínez-Gómez et al., 2007).

Figure 2. Detail of argan tree, most common identified fruit morphotypes: oval, spherical and spindle, and elaboration of argan oil in a cooperative farm.

This work offers an overview of the current biotechnological approaches being developed to optimize the conservation and characterization of critically endangered ancient Moroccan species *Argania spinosa* (Sapotaceae). These approaches include improved propagation techniques such as *in vitro* culture, karyotype characterization and molecular marker (RAPDs, SSRs and SCAR) application.

IN VITRO CULTURE

Conservation of genetic resources, mainly of the underutilized species such as argan tree, can be one of the most important aspects in the control of food security problems being propagation techniques an important component for this purpose. Developing adequate protocols for sexual and/or vegetative plant multiplication is then the first prerequisite in these conservation programs (IPGRI, 1994; 1996; Phartyal et al., 2002).

Beside sexual and conventional vegetative propagation techniques, *in vitro* tissue culture provide complementary protocols for rare fruit species propagation conservation. There are used mainly when problems are encountered with sexual propagation (seeds not available, problems with germination). *In vitro* culture can have major advantages over conventional propagation techniques in the management of some of these species. Rapid multiplication under controlled, pathogen-free conditions can be achieved by the inclusion of plant growth regulators or hormones in culture media. In addition, existing meristematic materiel is the explant of choice for vegetative propagation of rare plant and endangered species (Bornman, 1993).

The possibility of growing plants and even isolated plant cells in a test-tube under controlled *in vitro* conditions offers unique opportunities for micropropagation and multiplication. Advantages include minimizing environment influences, the potential to handle large numbers of individuals in a very small space, and accelerated growth and development. Vegetative micropropagation through tissue culture has many advantages over conventional vegetative propagation including minimizing environment influences, the potential to handle large numbers of individuals in a very small space, and the accelerated growth and development. It is therefore a promising technique for mass-producing genetically identical, physiologically uniform, developmentally normal, and pathogen-free plantlets, which can be acclimatized in a reduced time period and at a lower cost (Datée and Branchard, 1986).

At this moment, different methods were described for argan micropropagation (Bousselmame et al., 2001; Nouaim et al., 2002; Bekkali-Kacimi, 1997). These methods, however, are often suboptimal and need to be improved. In fact, these micropropagation systems are confronted to many problems. In stage II micropropagation phase, *in vitro* propagated shoots manifested diminished elongation through repeated subcultures; low laterals produced in response to cultural treatments, high rate of apical and leaf necrosis and hyperhydricity. In addition, in stage III, poor adventitious roots are formed (Bousselmame et al., 2001; Nouaim et al., 2002; Bekkali-Kacimi, 1997). All these problems could be a consequence of the inadequacy of the culture medium during the initiation and multiplication phase, which could be detrimental on the rooting phase which constitutes an important step for plant acclimation and outdoor transplantation success. In literature, Ca^{+}, NO_3^{-} , NH_4^{+} (Sha et al., 1985, McLaughlin and Karnosky, 1989) and the hormonal composition of the medium mainly cytokinin type and concentration were reported to play and important role in the reduction of hyperhydricity, apical necrosis and stimulation of shoots growth (Sharma and Ramamurthy, 2000; Niedz and Evens, 2007).

In our recent experiments efficient *in-vitro* protocols well adapted to argan tree species have been developed allowing the introduction, multiplication and elongation of plant material (Majourhat et al., 2007b) (Table 1; Figure 3).

Table 1. Mediums assayed for *in vitro* argan micropropagation

Seed Germination		Introduction + Multiplication	
MS macro and micronutrients		MS including vitamins	
Sucrose	20 g	Sucrose	30 g
GA3	0.05 mg	BAP	5 mg
Agar	8 g	NAA	0.01 mg
		MES	0.200 g
		GA3	0.5 mg
		L. glutamine	0.120 g
		Sequestrene	0.010 g
		Agar	8 g
Water qsp 1 liter. pH of media was adjusted to 5.8 before sterilization			
Elongation		Rooting	
MS including vitamins		MS including vitamins	
Sucrose	30 g	Sucrose	15 g
BAP	2,5 mg	IBA	0,1 – 1* mg
GA3	0.075 mg	Agar	8 g
L. glutamine	0.120 g		
Sequestrene	0.010 g		
Agar	8 g		
Water qsp 1 liter. pH of media was adjusted to 5.8 before sterilization			

These results open the possibilities of more efficient conservation strategies for the argan germplasm. The experiments undertaken until now show that argan explants have great plasticity regarding the mineral composition of the media during the multiplication phase. Nevertheless during the elongation phase Murashige and Skoog (1962) MS medium gives better results than the others.

Figure 3. In vitro propagation of argan. Detail of the seed germination inf in vitro culture of different ecotypes of argan tree (A) and the elongation madium (B) .

Young shoots each with 3–4 nodes, were collected from six month old seedlings of *A. spinosa* and washed in running tap water. Surface disinfection was carried out with 5 % sodium hypochlorite containing a drop of detergent for 10 min after a brief rinse (30 seconds) in 70% (v/v) water/ethanol. The material was then thoroughly washed with sterilized distilled water to eliminate any remaining traces of disinfectants. To induce proliferation, single node explants were then excised and cultured on a Murashige and Skoog medium (MS) (Murashige and Skoog, 1962) containing different concentrations (1.5, 3 and 5 mgl^{-1}) of 6-benzyl aminopurine (BA) in combination with 3 % sucrose, 0.01 mgl^{-1} of indole butyric acid (IBA) and 0.075 mgl^{-1} of giberillic acid (GA3). The medium was solidified with 0.7 % agar and the pH adjusted to 5.8 with NaOH 1N or HCl 1N solutions before autoclaving. After 4 weeks of culture, number of shoots per culture and shoot length (cm) were recorded.

After four subcultures on the initiation medium, explants were transferred to an MS basal medium for two weeks to eliminate any carry over from the initiation medium then transferred to multiplication mediums (Figure 4). To optimize the multiplication medium, the influence of macronutrients on *in vitro* morphogenesis was assessed by placing the individual shoots (approximately 0.5 to 1 cm in length) on 6 different culture media varying only in their macronutrient composition. These consisted of "M_1" for Murashige and Skoog (1962), "M_2" for Driver and Kuniyuki, (1984), "M_3" for Lloyd and McCown (1980), "M_4" for Anderson (1980), "M_5" for Quoirin and Lepoivre (1977) and "M_6" for Quoirin and Lepoivre (1977) completed with 220 mgl^{-1}calcium chloride. All these macronutrient formulations were supplemented with Murashige and Skoog (1962) micronutrients and vitamins as described in Table 1. These medium were supplemented with 3 % sucrose, 0.9 mgl^{-1} BAP, 0.01 mgl^{-1} IBA, 0.075 mgl^{-1} GA3 and 0.7 % of agar. The pH was adjusted to 5.8 with NaOH 1N orHCl 1N solutions, before autoclaving. After 30 days of culture, offshoot length, stem diameter and percent of regeneration were recorded.

Finally, rooting medium consisted of half strength M_6 medium supplemented with 220 $mg.l^{-1}$ $CaCl_2$ $2H_2O$, 1.5 % sucrose and different concentration of IBA (0, 0.01, 0.1, 0.2, 0.4, 0.6, 0.8 and 1 $mg.l^{-1}$). The medium was solidified with 0.7 % agar and the pH adjusted to 5.8 before autoclaving. After 30 days of culture, percentage of rooting and number of root per explant were recorded.

On the other hand, *in vitro* cryopreservation of clonal materials at ultra low temperatures (e.g. liquid nitrogen) offers several advantages including indefinite storage life with reduced space requirements, and freedom from labor-intensive subculturing and other maintenance tasks. It consists of suspending all metabolic processes until the revival of the tissue. To ovoid freeze injuries caused by intracellular ice crystals formation different techniques were developed to allow tissue acquiring cold hardiness. The two steps (or slow cooling) method is the must used technique. It consist of a preconditioning of the tissue in presence of low amount of cryoprotectant and a progressive cooling before immersion in liquid nitrogen (Ryynänen and Haggman, 2001; Harvengt et al., 2004). Other methods are used like vitrification (or rapid cooling) and encapsulation/dehydration (Engelmann, 1997).

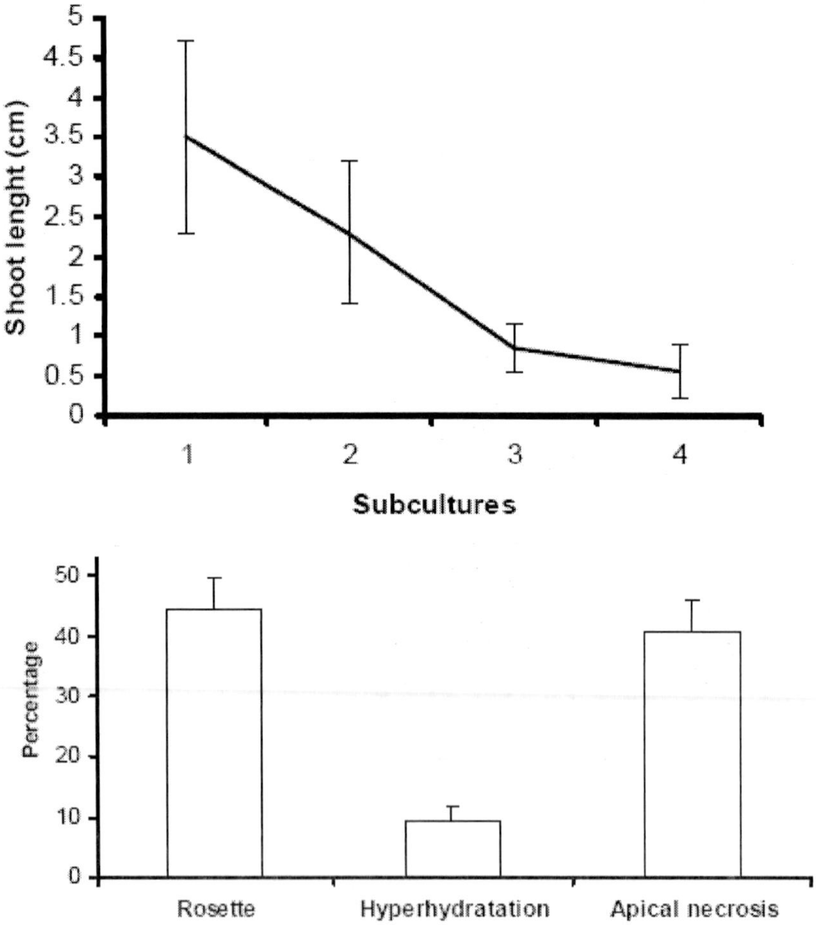

Figure 4. Effect of subsequent subcultures on initiation medium of *A. spinosa* nodal explants and symptoms of apical necrosis hyperhydricity and diminished elongation of nodal explants after four subcultures on initiation medium.

The later methods are bases on controlled application of concentrated cryoprotectants (eg. sorbitol, glycerol, polyethylenglycol, sucrose, etc.) that facilitate tissue desiccation at nonfreezing temperatures, leading to increased viscosity of the cytosol and, upon rapid cooling in liquid nitrogen. Cryoconservation of leaf, petiole, dormant buds, stem pieces, shoot-tips or somatic embryos was investigated as a means for long-term maintenance of germplasm of rare and endangered species. *Grevillea scapigera* an endangered species from Western Australia was conserved liquid nitrogen using four week old in vitro shoot tips as explants. The cryopreserved material was able to generate new shoot from callus that was successfully transferred to soil. However the problem encountered here is the low the low survival of thawed explants (Touchell et al., 1992). To date, this alternative is not possible in argan tree although future studies in this line could be of great interest.

Karyotype Studies

As we indicated above, argan tree is an specie cytogenetically very poorly known. Data available about this topic is scarce and controversial. Johnson (1991), in his review about the cytology of *Sapotaceae*, reported that the basic chromosome number of tribe *Sideroxyleae*, to which argan tree belongs, is suggested to be x = 11, whereas, the most common chromosome number in *Sapotaceae* family is x = 13 (with 39 counts) followed by x = 12 and 14 (with 27 and 25 counts respectively). This author also reported that polyploidy in this family is rare with only 5 counts mainly occurring in the tribe *Chrysophylaceae*. However, previous work on the cytology of *A. spinosa* failed to establish a common chromosome number. In fact, Miege in 1954 reported a chromosome number of 20 in somatic cells, whereas Humphries et al. (1978), in meiotic cells reported 24. Such information is of great interest for the advance of genetic studies and to identify numerical and structural abnormalities during tissue culture protocols and selection programs in this species. In addition, this information deals with physiological process in plants including plant morphology and fertility (Pegtel, 1999; Pagliarini, 2000; Kermani et al., 2003).

In general, chromosome studies are an important prerequisite for fruit genetic and breeding studies (Lespinasse et al., 1976; Darlington and Lacour, 1979; Schuster, 1996). These techniques have been used in tree species in the study of interspecific crosses (Salesses and Bonherta, 1993), the characterization of germplasm with an unknown ploidy level (Hesse, 1971; Toyama, 1974), and the establishment of taxonomic and phylogenetic relationships (Singh et al., 1984; Soodan et al., 1988; Harandi and Ghaffari, 2001). In addition, chromosome studies are a preliminary tool to start DNA studies including molecular characterization of the species and development of linkage maps (Majourhat et al., 2007a; Martínez-Gómez et al., 2007).

A standard karyotyping protocol has been developed at CEBAS-CSIC of Murcia as preliminary tool to start molecular studies. Our results indicated that the following protocol was successfully applied in argan. Samples are first placed in cold water for 4 hrs at 0°C, and later treated with 0.2% colchicine for 3 hrs at 5°C; colchicine treatment effectively stopped cell mitosis. Samples are fixed for 24 hrs at 5°C in [methanol, propionic acid, chloroform (6:3:2)] and then stored in 70% ethanol. Samples are then hydrolyzed in 1N HCl at 60°C for 20 min. Longer periods produce an excessive breaking of cells. Finally, staining of samples is performed in acetic acid/orcein 45% for 2 hrs (Majourhat et al. 2007a). This protocol has been also successfully assayed in other tree species such as almond (Martínez-Gómez et al., 2005).

In the case of argan, the diploid number determined in the somatic cells from the argan root tips analyzed was of 2n = 20. Our result confirms the information reported previously by Miège (1954) but differs from that of Humphries et al. (1978). The samples analyzed showed a karyotype constituted for ten pairs of chromosomes and the putative karyotype proposed has been of four sub-metacentric and six metacentric pairs. The four sub-metacentric pairs were the longest with a mean total length between 1.14 and 1.69 µm and the total length of six metacentric pairs were between 0.59 and 1.03 µm (Figure 5) (Majourhat et al., 2007a).

This study showed that *A. spinosa* has a different chromosome number with respect to the other genus in the Sideroxyleae (basic chromosome number of x=11), and in the Sapotaceae (basic chromosome number of x=13, x=12 or x=14) (Johnson, 1991).

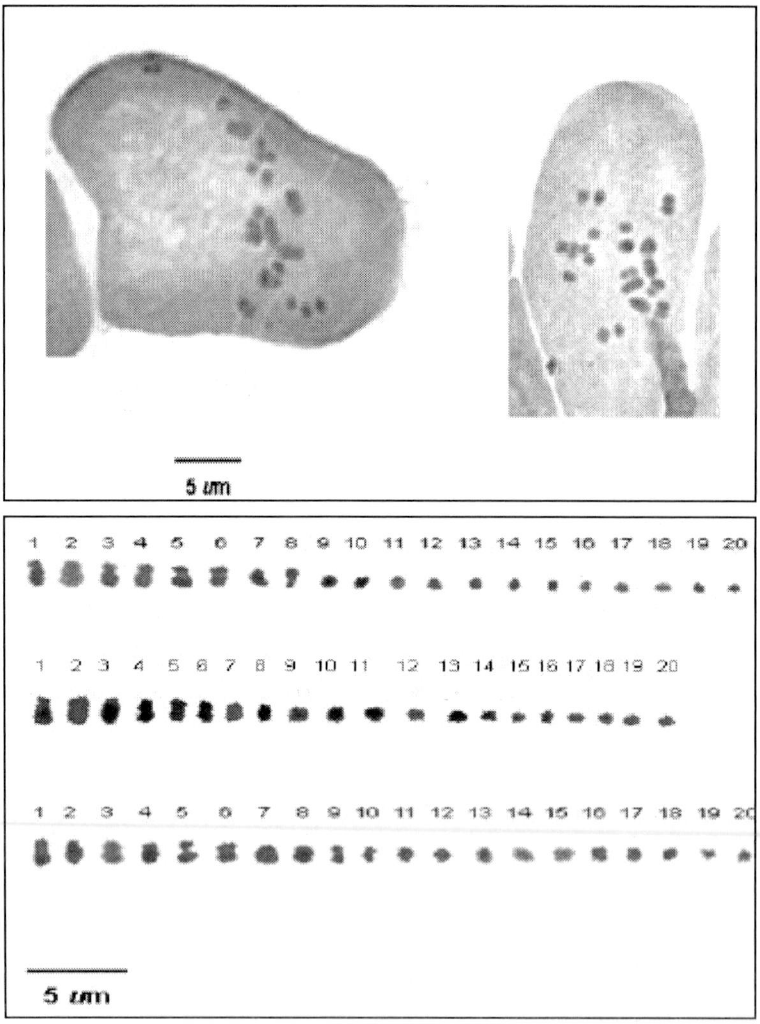

Figure 5. Karyotypes studies in argan tree. Metaphasic plaques and karyotypes of different mitotic root cells from argan tree.

An ascending or descending dysploidy process could be the hypothesis explaining this situation. In the case of descending dysploidy (meroaneuploid reduction), the decrease of parts of normal parental chromosomes can occur in plants adapted to a more rigorous environments than the ancestral species and this could be the case of Argania, the only species of the tropical family Sapotaceae adapted to the arid environment of Southwest Morocco (Pennington, 1991). *Argania* species is an extremely drought resistant plant being one of the few nontropical species of the Sapotaceae family (Schnell, 1970; 1971). Such reduction in chromosome number by multiple translocations was reported in other species like *Eleocharis subarticulata* (DaSilva et al., 2005).

A. spinosa develops under the hot arid Mediterranean floor and have very different distribution and ecological exigencies in comparison with the rest of Sideroxylon species (Emberger, 1939). On the other hand, Quezel and Barbero (1993) hypothesized that argan groove had taken place during the Mio–Pliocene area. In addition, *Arganioxylon sardum*, a fossil plant encountered in Sardinia (Italy) (Biondi, 1981), was similar to *A. spinosa*. This evidence can be the element to demonstrate the old origin of *A. spinosa* and its old distribution out of the tropical zone.

Application of Molecular Markers

One of the goals of argan conservation programmes must be to maintain existing level of genetic variation and genetic resources. In this sense, the molecular characterization of the accessions studied is of great interest. Characterization and identification of argan tree has been traditionally based on morphological traits (Bani-Aameur, 2004; Chernane et al., 2000) and biochemical (protein and isoenzyme) markers (Chernane et al., 2000; El Mousadik and Petit, 1996a). Even, it has been suggested to use fruit form as criterion for varietal selection in this species (Bani-Aameur, 2004). However, such markers are not always available for analysis, and are affected by environmental conditions. Molecular DNA marker technology offers several advantages over the sole use of convencional markers. Several types of molecular markers, including RFLPs (Restriction Fragment Length Polymorphysms) or AFLPs (Amplification Fragment Length Polymorphysms) are being employed for the genetic characterization of underutilized fruit germplasm, the establishment of genetic relationships between cultivars and species, and the future construction of genetic maps of these rare fruit species. However, the more recent utilization of PCR-based markers has increased the opportunities for molecular characterization and mapping of populations in a wider range of species with unknown DNA sequences (using Random amplified polymorphic DNA, RAPDs), and known DNA sequences (using simple sequence repeat, SSRs, microsatellites). In the case of argan, only one work studying the genetic variability of argan populations using DNA amplification and restriction was published until now (El Mousadik and Petit, 1996b).

DNA Extraction

For developing countries such as Morocco, the main obstacle for molecular marker use is the cost of the routinely used products. This situation is the same for the different part of the process, DNA extraction, DNA multiplication (if you are using a PCR based strategy), and amplified DNA electrophoresis.

Looking for a rapid and cheap DNA minipreparation protocol, we assayed several protocols for DNA extraction and quantification. A cheap protocol based on CTAB or SDS and using few amount of plants material (40 mg) was optimized. Total genomic DNA was isolated using the CTAB procedure described by Doyle and Doyle (1987) and the DNA isolation procedure described by Gepts and Clegg (1989) with minor variations using SDS extraction buffer.

Our results indicated that DNA average yield, range from 1 to 1.5 μg/mg of fresh leaf tissue and 260/280 ratio range from 1.61 to 1.85. Even without using liquid nitrogen this extraction protocol gave rise to high yield and quality DNA from different species of great economical and ecological interest. DNA extracted with this protocol was useful for RAPD-PCR reaction. However, using Qiagen Kit, we were not able to extract DNA with high molecular weight and quality from *Argania* species (Figure 6).

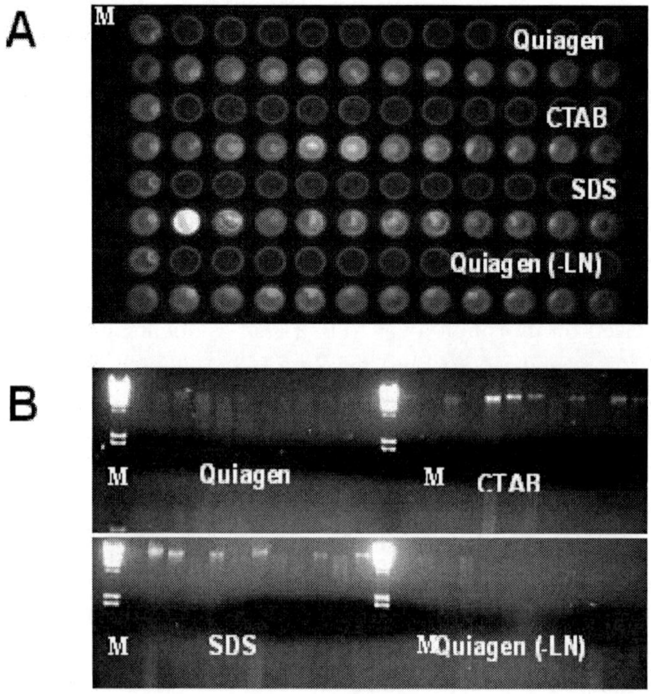

Figure 6. Protocols assayed for DNA extraction (CTAB, SDS, Quiagen membranes withouh liquid nitrogen and Quiagen membranes with liquid nitrogen) and quantification of argan DNA using Etidium Bromide in microplates (A) and agarose gel electrophoresis (1.5%) (B). M molecular marker λ Hind II.

Application of RAPD Markers

Random amplified polymorphic DNA (RAPD) markers are based on the PCR amplification of random locations in the genome (Welsh and McClelland, 1990; Messaoud et al., 2007). RAPDs are characterized by using arbitrary primers and permit the quick construction of genetic maps and the saturation of specific genomic regions with molecular markers. A single oligonucleotide is utilized for the amplification of genomic DNA. In contrast to isoenzymes and RFLPs, RAPDs are dominant markers. These markers are especially used in rare species including argan tree where there are not known sequences. The most crucial factor for this technique is the DNA quality and concentration offering the advantage of rapid screen for polymorphisms (Dax et al., 1993; Gérard et al., 2006).

These RAPD markers are also being used in the molecular characterization of argan tree species a fruit species which is poorly known from the molecular point of view. In our assays, nineteen RAPD universal primers purchased from Operon Technologies Huntsville, USA) were assayed. Amplifications were carried out in 20 μL total volume containing 1× Buffer (New England, Biolabs, Ipswich, USA), 1 mM MgCl2, 0.16 mM of dNTP, 0.4 μmol of primer, 1.0 unit of *Taq* DNA polymerase (Biolabs), and 4 ng templates DNA. The amplification program consisted of a step of DNA melting of min at 94 ◦C, followed by 35 cycles of 94 ◦C for 1 min, 35 ◦C for 1 min, and 72 ◦C for 1 min, and a final elongation step of 72 ◦C for 10 min. The amplified products were resolved in 2% agarose gels stained with ethidium bromide and visualized with UV transmitted light. A 1 Kb DNA Ladder (Invitrogen Life Technologies, Barcelona, Spain) was used as molecular size standard (Majourhat et al., 2008). RAPD amplifications were repeated at least twice in order to check the reproducibility of bands.

The analysis of the 19 RAPD primers showed that all primers generated polymorphic and reproducible patterns. A total of 140 polymorphic RAPD bands were detected out of 146 bands with a size range between 322 and 2009 bp. The number of RAPD bands detected by each primer depends on primer, sequence and the extent of variation in specific argan genotype. The number of bands varied from 2 (for OPA-16) to 14 (for OPW-13), with an average of 7.8 bands per primer. The mean PIC score over all loci was 0.84, ranging from 0.35 in OPA-16 to 0.96 in OPW-13 (Majourhat et al., 2008) (Table 2 and Figure 7).

Problems with DNA quality and a general sensitivity to changes in the reaction conditions can hamper the routine utilization of RAPD markers. This feature as well as the codominant nature, the variable degree of repeatability and the problems in transferring across populations, limits their utilization primarily to map construction. These difficulties can be overcome by converting RAPDs to sequence-characterized amplified regions or SCARs (Paran and Michelmore 1993). In contrast to RAPD and AFLP methods, SCAR is a PCR-based method that employs specific primers. These primers amplify single bands corresponding to genetically defined loci. SCARs can potentially be converted into codominant markers and are less sensitive to reaction conditions. To date, this alternative is not possible in argan tree although new studies are being performed at CEBAS-CSIC of Murcia (Spain).

Table 2. Universal primers (decamers) purchased from Operon Technologies Inc. (California, USA) used as RAPD markers and SSR markers obtained in *Manilkara huberi* and *Vitellaria paradoxa* assayed in the molecular characterization of the argan accessions

Marker	Primers sequence	Origin	Reference	T[a] annealing	polymorphic bands	Size (bp)
RAPD markers						
OPA-08	GTGACGTAGG	--	Operon Technologies	36	7	447-1413
OPA-10	GTGATCGCAG	--	Operon Technologies	36	7	620-1839

Table 2 (Continued)

Marker	Primers sequence	Origin	Reference	T[a] annealing	polymorphic bands	Size (bp)
OPA-11	CAATCGCCGT	--	Operon Technologies	36	6	636-2514
OPA-16	AGCCAGCGAA	--	Operon Technologies	36	2	1000-2009
OPB-07	GGTGACGCAG	--	Operon Technologies	36	4	392-669
OPB-11	GTAGACCCGT	--	Operon Technologies	36	7	589-2090
OPG-13	CTCTCCGCCA	--	Operon Technologies	36	9	373-2553
OPN-14	TCGTGCGGGT	--	Operon Technologies	36	7	548-1363
OPN-15	CAGCGACTGT	--	Operon Technologies	36	8	349-2009
OPR-15	GGACAACGAG	--	Operon Technologies	36	7	615-1381
OPR-16	CTCTGCGCGT	--	Operon Technologies	36	9	317-1844
OPW-09	GTGACCGAGT	--	Operon Technologies	36	7	478-1075
OPW-12	TGGGCAGAAG	--	Operon Technologies	36	5	521-1244
OPW-13	CACAGCGACA	--	Operon Technologies	36	14	493-2001
OPX-03	TGGCGCAGTG	--	Operon Technologies	36	4	586-1875
OPX-06	ACGCCAGAGG	--	Operon Technologies	36	11	444-2098
OPY-06	AAGGCTCACC	--	Operon Technologies	36	8	447-1798
OPY-13	GGGTCTCGGT	--	Operon Technologies	36	12	322-1683
OPZ-09	CACCCCAGTC		Operon Technologies	36	7	531-1485
SSR markers						
Mh04	F: GCACTCTCCATG GTTCCAGT R:AAAGAGTCAATG GCGTGAGC	*Manilkara huberi*	Azevedo et al. 2005	52	4	161-197
Mh07	F: ATTGCAGCATAT CCACACCA R: GCAAAGGGTGATGG GTTAGA	*Manilkara huberi*	Azevedo et al. 2005	56	3	138-143
Mh12	F: TGCGGAACTGTGGA AAGAGT R: ATCCACAGCAATGA CTGACG	*Manilkara huberi*	Azevedo et al. 2005	56	1	200
Mh22	F: CCCATTATAGCCCTC CACCT R: AGAGAGCACATGCAA GCTCA	*Manilkara huberi*	Azevedo et al. 2005	56	3	190-206
mVpCIRB03	F: GCACTCTCCATGGTT CCAGT R:AAAGAGTCAATGGC	*Vitellaria paradoxa*	Cardi et al. 2005	52	4	143-151

	GTGAGC					
mVpCIRB05	F: TATCCGTAGTGTGC GTGT R: TTTGGGGTCTATTT CATCT	*Vitellaria paradoxa*	Cardi et al. 2005	48	2	183-197
mVpCIRE04	F: CGGCATTTGGTCTT ACTT R: TTTTCCCCTTTGGTCA	*Vitellaria paradoxa*	Cardi et al. 2005	48	1	129
mVpCIRE05	F: GTTTGTATGGTTT CGGTT R: CTTCGTTTTCAGTAG GTCTC	*Vitellaria paradoxa*	Cardi et al. 2005	52	3	244-257
mVpCIRE11	F: TTGCTATTTGCCT GTT R: TTCATCACCTTCC TCTC	*Vitellaria paradoxa*	Cardi et al. 2005	48	4	225-235
mVpCIRF05	F: AGGACGGCGAATG GTG R: GCAGGGCAACTACAA AGACA	*Vitellaria paradoxa*	Cardi et al. 2005	52	3	223-230
mVpCIRH04	F: CTCTGTGCTTTATG TCGT R: ACTCACCATTGTTG CTT	*Vitellaria paradoxa*	Cardi et al. 2005	48	5	234-278

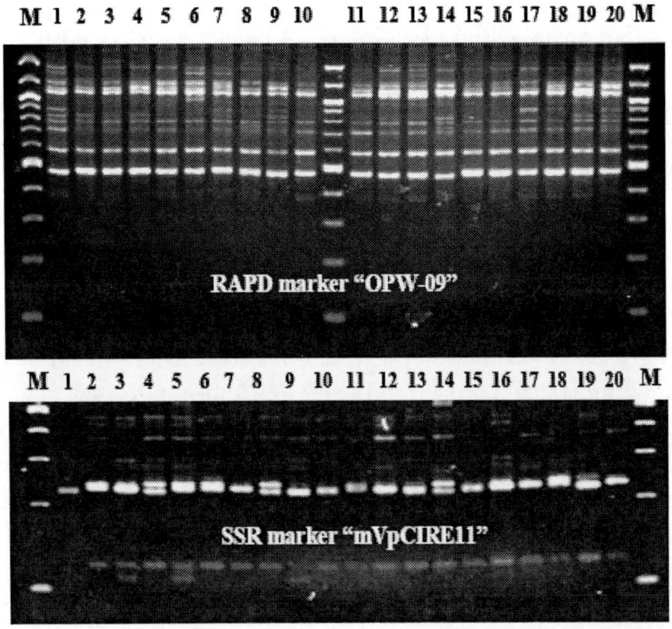

Figure 7. Molecular characterization of 20 accessions of *Argania spinosa* using RAPD and SSR markers. Agarose gel (1.5%) showing the allelic segregation of the OPW-09 RAPD marker and Metaphor® agarose gel (3%) showing the allelic segregation of the mVpCIRE11 nuclear SSR marker. M, 1Kb DNA Ladder.

APPLICATION OF SSR MARKERS

Simple sequence repeat (SSR) markers (microsatellites), also based on the PCR technique, are currently becoming the markers of choice for genetic fingerprinting studies for a wide range of plants. Because of their high polymorphism, abundance, and codominant inheritance, they are well suited for the assessment of genetic variability within crop species and closely related species (Gupta et al., 1996; de-Lucas et al., 2008). Simple sequence repeat markers generated include highly informative markers which have been applied in the molecular characterization of almond and related wild and cultivated species. In addition, these transferable markers were used in several genetic and taxonomic studies (Gupta, 1996; Martínez-Gómez et al., 2003). However, to date, none of these markers have been described in argan although the ability to use the same microsatellite primers in different plant species from the same family, called transportability or transferability, has been well described (Decroocq et al., 2003; Martínez-Gómez et al., 2003).

In our assays, twenty SSRs developed in two different species inside the *Sapotaceae* family, 12 primer-pairs in *Manilkara huberi* (Ducke) Standl (Azevedo et al., 2005) and 8 additional ones in *Vitellaria paradoxa* C.F. Gaertn (Cardi et al., 2005) were tested for their transferability to amplify microsatellite loci in the 38 argan samples studied. PCRs were performed in 20 µL mix containing 0.18 µM each primer, 1 × *Taq* buffer (Biolabs), 2.5 mM MgC12, 0.1 mM of each dNTP, 1 unit of *Taq* DNA polymerase (Biolabs), and approximately 5 ng of DNA. The PCR program consisted of an initial melting step (94 °C, 4 min) followed by 30 cycles (94 ◦C, 30 s, annealing temperature, 30 s and 72 ◦C, 1 min) and a final extension step (72 ◦C, 7 min). PCR products were separated using 3% Metaphor Agarose gel electrophoresis (Biowittaker, Maine, USA). To confirm the results, SSR products smaller than 5 bp were also separated using 10% polyacrylamide gel electrophoresis (PAGE). A 1 kb DNA Ladder was also used as molecular size standard.

In our studies, amplification was successful with 4 of the 12 (33%) SSR markers developed in *Manilkara huberi* and 7 of the 8 (87%) SSR markers developed in *Vitellaria paradoxa*. A total of 32 polymorphic bands were scored for argan accessions. The number of presumed alleles revealed by the SSR analysis ranged from one to five with a size range of amplified bands between 129 and 278 bp (Majourhat et al., 2008) (Table 2 and Figure 7).

The level of polymorphism was slightly smaller in argan to that reported by Azevedo et al. (2005) and Cardi et al. (2005) in the original species in which the SSR markers were developed, i.e. *Manilkara huberi* and *Vitellaria paradoxa* respectively. However, the range of the amplified band sizes in argan was similar to those reported in the original species in which the SSR markers were developed. Overall, the results of the obtained polymorphism suggest a high degree of transportability of SSR in *Sapotaceae* family. These results agree with reports on the successful utilization of these markers in different species from the same genus (Martínez-Gómez et al., 2003) and the successful utilization of these markers in different species from the same family (Decroocq et al., 2003; de-Lucas et al., 2008).

Differences in amplification success for SSR markers observed among species are due to the genetic variability between the different species in *Sapotaceae* family where the SSRs were developed. A decrease in the amount of polymorphic SSR markers would be expected as genetic distance increases from the designated anchor species, as has been previously reported in *Prunus* by Martínez-Gómez et al. (2003). It has been shown that closely related species are more likely to share microsatellite priming sites than more distantly related ones, but it is possible to transfer functional microsatellite primers from more distantly related species in some cases (Lorieux et al., 2000). In this sense, these results also indicate that *Vitellaria paradoxa* is genetically closer to *Argania spinosa* than *Manilkara huberi*.

CONCLUSION

In the case of argan germplasm the typical long generation time along with the extensive space requirements and other limitations of this tree species have frustrated the development and testing of these species in their regions of origin in developing countries and in other regions. These limitations, however, allow biotechnological tools (including *in vitro* culture, karyotype characterization and molecular marker application) to improve conservation, particularly applicable to this species. Argan tree crop include a large number of intercompatible species which provide an enormous gene pool available for breeding. Little use has been made of this variability because the slowness of classical breeding methods. Promising propagation methods included *in-vitro* protocols well adapted to argan tree can accelerate the introduction, micropropagation and rooting of new plant material. On the other hand, a standard karyotyping protocol has been described as preliminary tool to start DNA studies. In addition, molecular DNA markers based on PCR amplification of the DNA have also become an essential tool for the characterization and conservation of argan species. Regarding this PCR amplification of the DNA, three main strategies, RAPD (random amplification of the DNA), SSR (using sequences from related species) and SCAR (sequence-characterized amplified regions) markers are being developed. These molecular markers have been applied in the genetic characterization of argan germplasm, the establishment of genetic relationships between morphotypes, and the future construction of genetic maps of this rare woody species. In addition, these initially DNA research open the door to the future DNA sequencing of this species. Additional advantages encouraging the utilization of these new biotechnologies to tree crop conservation and improvement include high levels of synteny between genomes and a well-established international network of cooperation among researchers.

ACKNOWLEDGMENTS

This work has been supported partially by the research projects A/6210/06 "Caractérisation biochimique et génomiques des morphotypes de l´arganier (*Argania spinosa* L. Skeels)" and B/020580/08 ("Selección de clones de argania de interés y caracterización molecular mediante marcadores específicos tipo SCARs") from the AECI (Agencia Española de Cooperación Internacional) of Spain.

References

Anderson, W.C. (1980). Tissue culture propagation of red and black raspberries, *Rubus idaeus* and *R. occidentalus*. *Acta Horticulturae*, 112, 13–20.

Azevedo, V.C.R., Vinson, C.C., Ciampi, Y. (2005). Twelve microsatellite loci in *Manilkara huberi* Standl. (*Sapotaceae*), an Amazonian timber species. *Molecular Ecology Notes*, 5, 13–15.

Bani-Aameur, F. (2004). Morphological diversity of argan (*Argania spinosa* (L.) Skeels) populations in Morocco. *Forest Genetic*, 11, 311– 316.

Bekkali-Kacimi, M. (1997). *Organogenese in vitro de l'arganier (Argania spinosa L. Skeels): relations entre les elements mineraux des tissues et des millieux de culture et l'aptitude a la regeneration des pousses*. PhD dissertation, University Caddi Ayyad, Faculty of sciences of Marrakech, Morocco, 97 p.

Benchakroun, F. (1990). Un systeme typique d'Agroforesterie au Maroc: L'arganerai. *Séminaire Maghrébin D'Agroforesterie*, Jebel Oust, Tunisie, October 1990, 150-156.

Benchakroun, F., Buttoud, G. (1989). L'arganeraie dans l'économie rurale du sud-ouest marocain. *Forêt Méditerranéenne*, 2, 127-136.

Benismail, M.C., Harrouni, M.C., Mokhtari, M. (2002). Production rapide de plants d'arganier aptes a la transplantation. *Transfer de Technologie en Agriculture*, 95, 1-4.

Biondi, E. (1981). Arganioxylon sardum n. gen., n. sp. Et Sclerocaryoxylon chiarugii n. gen., n. sp,: bois fossiles du Miocène de la Sardaigne (Italie). *Revue de Palaeobotany Palynology*, 34, 301–320.

Bornman, C.H. (1993). Micropropagation and somatic embryogenesis. In: Hayward MD, Bosemark NO, Romagosa I (Eds) *Plant Breeding. Principles and Prospects*, Chapman and Hall, London, UK, pp 246-260.

Bousselmame, F., Kenny, L., Chlyah, H. (2001). Optimisation des condition de culture pour l'enracinement in vitro de l'arganier (*Argania spinosa* L.). *C.R. de l'Academie des Sciences de Paris*, 324, 995-1000.

Cardi, C., Vaillant, A., Sanou, H., Kelly, A., Bouvet, J.M. (2005). Characterization of microsatellite markers in the shea tree (*Vitellaria paradoxa* Gaertn.) in Mali. *Molecular Ecology Notes*, 5, 524-526.

Chernane, H., Hafidi, M., El Hadrami, I., Ajana, H. (2000). Evolution des paramètres biométriques et de la composition en acides gras de l'huile de quatre types de fruits d'Arganier (*Argania spinosa* L. Skeels) durant la maturation. *Agrochimica*, 44, 180–196.

Darlington, C.D., Lacour, L.F. (1979). *The Handling of Chromosomes*, George Allen and Unwin Ltd., London, 189 pp.

DaSilva, C.M., Gonzalez-Elizondo, M.S., Vanzela, L.L. (2005). Reduction of chromosome number in Eleocharis subarticulata (Cyperaceae) by multiple translocations. *Botanical Journal of the Linnean Society*, 149, 457–464.

Datée, Y., Branchard, M. (1986). *In* vitro culture and cellular biotechnologies applied to plant breeding. Present state and perspectives. *Acta Horticulturae*, 193, 17-28.

Dax, E., Livneh, O., Edelbaum, O., Kedar, N., Gavish, N., Karchi, H., Milo, J., Sela, I., Rabinowit, H.D. (1993). A random amplified polymorphic DNA (RAPD) molecular marker for the Tm-2agene in tomato. *Euphytica*, 74, 159–163.

Decroocq, V., Fave, M.G., Hagen, L., Bordenave, L., Decroocq, S. (2003). Development and transferability of apricot and grape EST microsatellite markers across taxa. *Theoretical and Applied Genetics*, 106, 912–922.

De-Lucas, A.I., Santana, J.C., Recio, P., Hidalgo, E. (2008). SSR-based tool for identification and certification of commercial *Populus* clones in Spain. *Annals of Forest Science*, 65, 107.

Doyle, J.J., Doyle, J.L. (1987). A rapid DNA isolation procedure for small quantities of fresh leaf tissue. *Phytochemistry Bulletin*, 19, 11-15.

Driver, J.A., Kuniyuki, A.H. (1984). *In Vitro* propagation of paradox walnut rootstock. *HortScience*, 19, 507-509

El Mousadik, A., Petit, R.J. (1996a). High level of genetic differentiation for allelic richness among population of argan tree (*Argania spinosa* (L.) Skeels) endemic to Morocco. *Theoretical and Applied Genetics*, 92, 832–839.

El Mousadik, A., Petit, R.J. (1996b). Chloroplast DNA phylogeography of argan tree of Morocco. *Molecular Ecology*, 5, 547–555.

Emberger, L. (1939). *Aperçu général sur la végétation du Maroc. Commentaire de la carte phytogéographique du Maroc 1:1500000.* Rubel E. et LfidiW. (Eds.), Ed. Veröff. Geobot. Inst. Rfibel Zfirich, 14, pp. 40–157.

Engelmann, F. (1997). Importance of desiccation for the cryopreservation of recalcitrant seed and vegetatively propagated species. *Plant Genetic Resources Newsletter*, 112, 9-18.

Gepts, P., Clegg, M. (1989). Genetic diversity in pearl millet (*Penisetum glaucum*) at the DNA sequence level. *Journal of Heredity*, 80, 302-208

Gérard, P.R., Fernández-Manjarrés, J.F., Bertolino, P., Dufour, J., Raquin, C., Frascaria-Lacostem, F. (2006). New insights in the recognition of the European ash species *Fraxinus excelsior* L. and *Fraxinus augustifolia* Vahl. as useful tools for forest management. *Annals of Forest Science*, 63, 733–738.

Gupta, P.K., Balyan, H.S., Sharma, P.C., Ramesh, B. (1996). Microsatellites in plants: a new class of molecular markers. *Current Science*, 70, 45-54.

Harandi, O.F., Ghaffari, M. (2001). Chromosome studies on pistacio (*Pistacia vera* L.) from Iran. *Option mediterraneenne*, 56, 35-40

Harvengt, L., Meier-Dinkel, A., Dumas, E., Collin, E. (2004). Establishment of a cryopreserved gene bank of European elms. *Canadian Journal of Forest Research*, 34, 43–55.

Hesse, C.O. (1971). Monoploid peaches, *Prunus persica* Basch: Description and meiotic analysis. *Journal of the American Society of Horticultural Science*, 96, 326-330.

Humphries, C.J., Murray, B.G., Bocquet, G., Vasudevan, K.N. (1978). Chromosome numbers of phanerogams from Morocco and Algeria. *Botany Notiser*, 131, 391-406.

IPGRI - International Plant Genetic Resources Institute. (1994). *Genebank Standards*. Food and Agriculture Organization of the United Nations, Rome, International Plant Genetic Resources Institute, Rome. 13 pp.

IPGRI - International Plant Genetic Resources Institute. (1996). *Central crop databases: Tools for plant genetic resources management*. Ed. IPGRI Rome, Italy. 100 pp.

IPGRI - International Plant Genetic Resources Institute. (2006a). *The global facilitation unit for underutilized species*. Ed. IPGRI Rome, Italy. 8 pp.

IPGRI - International Plant Genetic Resources Institute. (2006b). *Strategic framework for underutilized plant species. Research and development*. Ed. IPGRI Rome, Italy. 33 pp.

IPGRI - International Plant Genetic Resources Institute. (2006c). *Approaches and decision steps for the promotion and development of underutilized plant species*. Ed. IPGRI Rome, Italy. 4 pp.

Johnson, M. (1991). Cytology. In: Pennington TD (Ed) *The Genera of Sapotaceae,* Royal Botanic Gardens, Kew, UK, pp 15-22.

Kermani, M., Sarasan, J., Roberts, A.V., Yokoya, K., Wentworth, J., Sieber, V.K. (2003). Oryzalin-induced chromosome doubling in *Rosa* and its effect on plant morphology and pollen viability. *Theoretical and Applied Genetics*, 107, 1195-1200.

Lespinasse, Y., Alston, F.H., Watkins, R. (1976). Cytological techniques for use in apple breeding. *Annals of Applied Biology*, 82, 349-349

Lloyd, G., McCown, B. (1980). Commercially feasible micropropagation of mountain laurel, *Kalmia latifolia*, by use of shoot tip culture. *Proceedings of the International Plant Propagation Society*, 30, 421–427.

McLaughlin, J., Karnosky, D.F. (1989). Controlling vitrification in Larix decidua via culture media manipulation. *Canadian Journal of Forest Research*, 19, 1334–1337.

Majourhat, K., Araneda, L., Zeinalabedini, M., Hafidi, A., Martínez-Gómez, P. (2007a). Chromosome counting and karyotype characterization of *Argania spinosa* Skeel. South African Journal of Botany, 73, 661–663.

Majourhat, K., Hafidi, A., Piqueras, A., Martínez-Gómez, P. (2007b). Multiplicación *in vitro* de la especie oleaginosa argania [*Argania spinosa* (L.) Sheel.] procedente de Marruecos. *XI Congreso Nacional de Ciencias Hortícolas*. April 2007, Albacete (Spain). *Actas de Horticultura*, 48, 646-649.

Majourhat, K., Jabbar, J., Hafidi, A., Martínez-Gómez, P. (2008). Molecular characterization and genetic relationships among most common identified morphotypes of critically endangered rare Moroccan species *Argania spinosa* (Sapotaceae) using RAPD and SSR markers. *Annals of Forest Science*, 65, 805-810.

Martínez-Gómez, P., Arulsekar, S., Gradziel, T.M. (2003). Relationships among peach and almond and related species detected by SSR markers. *Journal of the American Society of Horticultural Science*, 128, 667–671.

Martínez-Gómez, P., Sánchez-Pérez, R., Vaknin, Y., Dicenta, F., Gradziel, T.M. (2005). Improved technique for counting chromosomes in almond. *Scientia Horticulturae*, 105, 139-143.

Martínez-Gómez, P., Majourhat, K., Zeinalabedini, M., Erogul, D., Khayam-Nekoui, M., Grigorian, V., Hafidi, A., Piqueras, A., Gradziel, T.M. (2007). Use of biotechnology for preserving rare fruit germplasm. *Bioremediation, Biodiversity and Bioavailability*, 1, 31-40.

Messaoud, C., Afif, M., Boulila, A., Rejeb, M.N., Boussaid, M. (2007). Genetic variation of Tunisian *Myrtus communis* L. (Myrtaceae) populations assessed by isozymes and RAPDs. *Annals of Forest Science*, 63, 845– 853.

Miège, J. (1954). Nombres chromosomiques et repartition geographiques de quelques plantes tropicales et équatoriales. *Revue de Cytologie et Biologie Végétal*, 15, 312-348.

Murashige, T., Skoog, F. (1962). A revised medium for rapid growth and bioassays with tobacco tissue cultures. *Physiologia Plantarum*, 15, 473–497.

Niedz, R.P., Evens, T.J. (2007). the effects of mineral nutrients on callus growth and *in vitro* shoot multiplication. *In vitro Cellular and Developmental Plant Biology*, 43, 670-671.

Nouaim, R., Mangin, G., Breuil, M.C. and Chaussod R. (2002). The argan tree (*Argania spinosa*) in Morocco: propagation by seed, cutting and in vitro techniques. *Agroforestery Systems*, 54, 71-81.

Pagliarini, M.S. (2000). Meiotic behavior of economically important plant species: the relationship between fertility and male sterility. *Genetics and Molecular Biology*, 23, 997-1002.

Paran, I., Michelmore, R.W. (1993). Development of reliable PCR-based markers linked to downy mildew resistance genes in lettuce. *Theoretical and Applied Genetics*, 85, 985-993.

Pegtel, D.M. (1999). Effect of ploidy level on fruit morphology, seed germination and juvenile growth in scurvy grass (*Cochlearia officinalis* L. s.l., *Brassicaceae*). *Plant Species Biology*, 14, 201-215.

Pennington, T.D. (1991). *The Genera of the Sapotaceae*, Kew Publishing, UK, 307 pp

Phartyal, S.S., Thapliyal, R.C., Koedam, N., Godefroid, S. (2001). *Ex situ* conservation of rare and valuable forest tree species through seed-gene bank. *Current Science*, 83, 1351-1357.

Quezel, P., Barbero, M. (1993). Variations climatiques au Sahara et en Afrique sèche depuis le Pliocène: enseignement de la flore et de la végétation actuelles. *Bulletine Ecology*, 24, 191–202.

Quorin, M., Lepoivre, P. (1977). Improved medium for in vitro culture of *Prunus* sp. *Acta Horticulturae*, 78, 437–442.

Salesses, G, Bonherta, A. (1993). Meiotic behaviour of hybrids between *Prunus cerasifera*, *P. spinosa*, and *P. persica*: an approach to the peach-plum genome relationship. *Cytologia*, 58, 257-262.

Ryynänen, L.A., Häggman, H.M. (2001). Recovery of cryopreserved silver birch shoot tips is affected by the pre-freezing age of the cultures and ammonium substitution. *Plant Cell Reports,* 20, 354–360.

Schuster, M. (1996). Cytogenetics in fruit breeding. Preparation methods for mitotic chromosomes. *Gartenbauwissenschaft*, 61, 273-275.

Schnell, R. (1970). *Introduction à la phytogéographie des pays tropicaux*. Les problémes généraux. Gauthier-Villars, Paris vol. I: les flores, les structures, pp. 499.

Schnell, R. (1971). *Introduction à la phytogéographie des pays tropicaux*. Les problémes généraux. Gauthier-Villars, Paris, vol. II: les groupements végétaux, pp. 452.

Sha, L.; McCrown, B.H.; Peterson, L.A. (1985). Occurrence and cause of shoot-tip necrosis in shoot cultures. *Journal of the American Society of Horticultural Science*, 110, 631–634.

Sharma, S.K., Ramamurthy, V. (2000). Micropropagation of 4-year-old elite Eucalyptus tereticornis trees . *Plant Cell Reports*, 19, 511-518.

Singh, R., Singh, A., Koul, A.K., Wafai, B.A. (1984). Chromosome inventory of some Rosaceous fruits cultivars of Kashmir Valley. *Chromosome Information Service*, 36, 7-9.

Soodan, A.S., Koul, A.K., Wafai, B.A. (1988). Assessment of the germplasm of Rosaceous fruits under cultivation in Kashmir Valley II. Meiotic system and pollen production in almond, peach and their hybrids. *Cytologia*, 53, 665-670.

Swenson, U., Anderberg, A. (2005). Phylogeny, character evolution and classification of Sapotaceae (Ericales). *Cladistics*, 21, 101–130.

Touchell, D.H., Dixon, K.W., Tan, B. (1992). Cryopreservation of shoot-tips of *Grevillea scapigera* (Proteaceae): a rare and endangered plant from Western Australia. *Australian Journal of Botany*, 40, 305-310.

Toyama, T.K. (1974.) Haploidy in peach. *HortScience*, 9, 187-188.

Welsh, J., McClelland, M. (1990). Finger printing genomes using PCR with arbitraries primers. *Nucleic Acids Research*, 18, 7213-7218.

In: Horticulture in the 21st Century
Editors: Abigail N. Sampson, pp. 143-160
ISBN 978-1-61668-582-9

Chapter 5

IRRIGATING ORNAMENTAL HORTICULTURAL CROPS WITH ALTERNATIVE WATER SOURCES

Genhua Niu[1] and Raul I. Cabrera[2]
[1]Texas AgriLife Research and Extension Center at El Paso, Texas A&M University System, 1380 A&M Circle, El Paso, TX 79927
[2]Texas AgriLife Research and Extension Center at Dallas, Texas A&M University System, 17360 Coit Road, Dallas, TX 75252

ABSTRACT

Water shortages and poor water quality are critical issues in many regions of the world. With a rapidly increasing population and diminishing water supplies, the competition for fresh water among agriculture, industry, urban and recreational users has become intense. Use of alternative water sources, such as municipal reclaimed water and other poor-quality, non-potable saline waters, for irrigating horticultural crops and urban landscapes may be inevitable in the 21st century. The major concern of using these alternative water sources is their elevated salinity, which causes injury to tissues, growth or yield reductions on sensitive plants, and also causes salt accumulation in root zones if irrigation is not managed properly. Many horticultural commodities are high-value crops and are traditionally irrigated with high quality water. Recent research results have shown promising results on irrigating a variety of ornamental horticultural plants and crops with saline water at moderate salinity levels with little negative effects on plant growth and quality. This chapter presents and reviews the latest research results on salt tolerance for a number of ornamental horticultural species. Salinity tolerance of these plants varied widely with species, environmental conditions, substrate or soil type, and irrigation management. Tolerant plants are able to resist salt stress by limiting or reducing uptake of sodium (Na^+) and chloride (Cl^-) ions, confining these ions in roots or reducing transport of the ions accumulated in roots to shoots, selective ion uptake, and tolerating high osmotic stress.

INTRODUCTION

Soil salinity is a common problem encountered in arid and semi-arid regions and is generally caused by insufficient drainage and/or inappropriate irrigation management, even though irrigation water is seldom saline (Boland, 2008; Pasternak and Malach, 1994). More than 800 million hectares of land throughout the world are affected by salinity and sodic conditions (Munns and Tester, 2008; Tanji, 2002), accounting for more than 10% of the world's total arable land area. Furthermore, about 97% of the earths' water is saline, contained in oceans and seas, and less than 3% is continental, with only a small fraction of it being readily-manageable fresh water for agricultural and human needs (Tanji, 1990). As urban population grows, the demand for high-quality water increases, and the only water source that increases with population growth is municipal wastewater, which is treated for other beneficial uses, including irrigation. In fact, municipal reclaimed water has long been used for urban landscape and turfgrass irrigation in southwestern parts of the United States and other countries (Duncan et al., 2009; Dobrowolski et al., 2008; Tanwar, 2003), and more recently for vegetable production in California and citrus and ornamental crop production in Florida (Dobrowolski et al., 2008). As more alternative waters are used for irrigation, soils would become more saline if not managed properly. Therefore, it is important to identify horticultural crops that are salt tolerant.

Salt tolerance of food and fiber crops and plants not only varies considerably with genotype but also depends heavily on the environmental conditions and irrigation practices (Boland, 2008; Maas, 1990). Similar responses to salinity are also observed in ornamental horticultural crops (Fox et al., 2005; Niu et al., 2007a; Shannon et al., 1994). Therefore, successful use of alternative water sources to irrigate horticultural crops requires adequate understanding of the interaction among plant, soil/substrate, water, and environmental conditions. The intent of this chapter is to summarize current knowledge concerning the salt tolerance of ornamental horticultural plants and crops, including herbaceous plants (bedding plants, groundcovers, and perennials), garden and cut roses, and woody shrubs and trees. The biochemical or molecular aspects of the responses of the ornamental horticultural crops to salinity are beyond the scope of this chapter.

SALINITY PARAMETERS AND MEASUREMENT

Salinity is defined as the concentration of dissolved mineral salts found in waters or soil solutions (Farnham et al., 1985; Tanji, 2002). The individual chemical constituents that contribute to salinity in waters include potassium (K^+), calcium (Ca^{2+}), magnesium (Mg^{2+}), sodium (Na^+), chloride (Cl^-), bicarbonate (HCO_3^-), nitrate (NO_3^-), sulfate (SO_4^{2-}), boron (B), aluminum (Al^{3+}) and silica (SiO_2). While these individual constituents are reported in units of $mg \cdot L^{-1}$ (ppm), $mmol \cdot L^{-1}$ or $meq \cdot L^{-1}$ (mmol $charge \cdot L^{-1}$), overall salinity is reported as the concentration of total dissolved solids (TDS) and expressed in $mg \cdot L^{-1}$. TDS is obtained by evaporating a sample of water or soil solution down to dryness and weighing the residue. A more common expression, and simpler measurement, of overall salinity in agriculture and horticulture, however, is the determination of electrical conductivity (EC) of a water (EC_w) or soil/substrate extract (EC_e). The ability of a water, or solution, to conduct electricity is directly related to the concentration of salts contained in it, and the units to report EC are $mmhos{\cdot}cm^{-1}$ or $dS{\cdot}m^{-1}$ (same unitary value). While there is no exact relationship between TDS and EC, there are, however, some approximate conversion factors (Tanji, 2002). For waters or solutions with EC up to 5 $dS{\cdot}m^{-1}$, TDS is approximated by multiplying EC value (in $dS{\cdot}m^{-1}$) by 640, and by 800 for more saline solutions (>5 $dS{\cdot}m^{-1}$).

The measurement of EC in waters is straightforward, but becomes more involved when determining it in soil solution. The most widely used method to determine EC of soil solution involves its extraction from a saturated paste or media extract, requiring the collection (destructive procedure) of a representative sample of soil or substrate from the root zone (Cabrera, 1998; Lang, 1996). Other procedures to evaluate the EC of the soil solution in the root zone of plants and crops include 1:2 and 1:5 dilutions (soil to distilled or de-ionized water ratios, on a v/v basis), the pour-through technique and the use of various extraction devices like suction cups or samplers. The EC values obtained from each soil solution extraction method have to be diagnosed with their specific and respective interpretative scales or guidelines.

While still very practical and applicable in the production and maintenance of ornamental plants in production and landscape settings, these methods to determine soil solution EC are providing only a single reading, representative to that sample and point in time. Salinity, however, is not a static property of the soil solution, but it is strongly and dynamically influenced by soil moisture content, which changes both spatially and temporally (Tanji, 2002). The concentration of dissolved salts does not change in direct proportion to changes in soil moisture content due to solubility, cation exchange and ion association issues. Therefore, salinity is a dynamic property in the root zone, due to concentration of the soil solution by water extraction and selective uptake from plant roots, evaporation from the substrate and replenishment by irrigation or rainfall (Tanji, 2002).

General Information on Alternative Water Sources

The types of available alternative or non-potable water sources vary with regions or even locations. Generally, municipal reclaimed water or recycled water is the main alternative water. In some places, ground saline water or sallow aquifer saline water, and agricultural drainage water are available (Boland, 2008; Duncan et al., 2009). The salinity and composition of these alternative waters vary with locations and sources.

Reclaimed water is defined as domestic or municipal wastewater which has been treated to a quality suitable for beneficial use. Reclaimed water is also been called by other terms, including recycled water, urban wastewater, urban water reuse and urban effluent (Duncan et al., 2009). These latter, and broader, terms actually denote urban water sources that have not been treated, and include stormwater and rainwater. Reclaimed water is also different from greywater, which is untreated, non-toilet, and household water including water from sinks, showers, and baths. The salinity of reclaimed water, measured as EC_w, varies with its source water, treatment processes, location and time of the year, and typically range from 1.0 to 1.9 $dS{\cdot}m^{-1}$ (Schuch, 2005; Wu et al., 2001). While some water reclamation treatment facilities offer periodic laboratory test results, the data is, however, often incomplete for assessing irrigation quality because they are oriented to the "human impact factor" and less to the performance of plants irrigated with them (Duncan et al., 2009). The main chemical compounds of horticultural interest in reclaimed water, after going through all treatment processes, are K^+, Ca^{2+}, Mg^{2+}, Na^+, Cl^-, HCO_3^- and SO_4^{2-}. The typical concentrations of the latter four ions (Na^+, Cl^-, HCO_3^- and SO_4^{2-}) found in reclaimed water (Duncan et al., 2009) exceed, on average, 60-120% the thresholds recommended for ornamental plants (Farnham et al., 1985; Wu et al., 2001).

METHODS OF SCREENING SALT TOLERANCE FOR HORTICULTURAL PLANTS

Greenhouse studies. Compared to field trials, greenhouse studies for screening salt tolerance are less labor intensive, less costly, and the environmental conditions can be more easily controlled. Salt treatments in greenhouse studies are usually created by adding salts to irrigation water to mimic a composition of a target saline water source or saline soil. A large number of salinity studies have used NaCl as the sole salinizing agent (Cassaniti et al., 2009; Eom et al., 2007; Marosz, 2004; Fostad and Pedersen, 2000; Wahome, 2003; Wu et al., 2001). However, composition of salts in water and salt-affected soils varies widely across the globe (Grattan and Grieve, 1999). Saline solutions with a single salt may result in misleading and erroneous interpretations about plant response to salinity (Carter and Grieve, 2006; Grattan and Grieve, 1999). In our studies, NaCl + $CaCl_2$ (2:1 molar ratio) or NaCl + $CaCl_2$ + $MgSO_4$ (87:8:5 by weight) have been added to tap water or nutrient solutions to mimic the major ion composition of local poor-quality or reclaimed waters. A wide range of salinity levels, depending on the estimated salt tolerance of selected species being tested, are prepared to distinguish the differences and/or behavior in plant responses to increasing salinity.

Outdoor container and field studies. Since environmental conditions such as temperature, light intensity, humidity, and wind speed can considerably affect plant performance and therefore, their response to salinity, results obtained from greenhouse studies provide a relative order of salt tolerance among tested plant species. The absolute threshold of salinity for landscape plants needs, however, to be confirmed under field or outdoor conditions. For field studies, soils will be salinized after a salt tolerance study, and thus will require replacement prior to re-utilization in subsequent experiments in the same field. In our research, raised beds have been used and soils in the beds are replaced prior to each experiment (Niu et al., 2007a). Treatments are created similarly to greenhouse studies by adding salts at various ratios to irrigation water to generate various salinity levels. Like greenhouse container studies, salt accumulation in the field varies among beds and locations within the same bed due to differences in evaporation, plant water use and micro-environment. Outdoor studies using containers can provide results under standard outdoor environmental conditions, while eliminating the variation among adjacent plots and ease the control and disposal of salinized substrates or soils. Nevertheless, selection of methods for screening salt tolerance still depends on the biological characteristics and cultural practices for the target plant species, seasons, and availability of resources.

Assessment of salt tolerance. Depending on species or genotype, growth can be quantified non-destructively by recording plant height and/or canopy size periodically. All parts of a plant, including leaves, stems, roots, and fruit may be reduced in size under saline conditions. Usually, growth reduction in shoots is greater than in roots (Munns, 2002). For agricultural crop plants, salt tolerance is often assessed based on growth or yield reduction of specific plant parts such as seeds, roots, fruits, or leaves (Maas, 1990). However, for ornamental plants aesthetic value is more important than growth rate (Bernstein et al., 1972). Therefore, visual quality, although it can be highly subjective in its determination, is an important parameter for assessing salt tolerance of ornamental and landscape plants (Devitt et al., 2005; Fox et al., 2005; Niu and Rodriguez, 2006a, b; Niu et al., 2007a, b; Zollinger et al., 2007). Dry matter accumulation and/or relative growth rate are commonly used to evaluate salt tolerance, but this parameter is not available until the plants are harvested destructively. The timing when significant differences in foliar damage symptoms appear is another parameter used for the assessment of plant salt tolerance. The earlier the salt injury is exhibited in a plant, the less its tolerance to salinity.

In addition to these visible changes reflected in growth, size and foliar damage, plants also respond to salinity internally, expressed in changes in physiological and metabolic processes. To understand the mechanisms of salt tolerance and stress physiology, photosynthesis, leaf stomatal conductance, chlorophyll fluorescence, chlorophyll content, plant water status (i.e. water potential, relative water content) and ion uptake and accumulation in tissues are typically measured (Cabrera, 2003; Niu and Rodriguez, 2006a; Wu et al., 2001; Zollinger et al., 2007; Zribi et al., 2009).

General Responses of Ornamental Plants to Salinity

Plant salt tolerance is typically defined as the ability to withstand the effects of high or increasing salinity without significant adverse effects such as growth or yield reduction, or foliar salt damage. Salinity reduces the ability of plants to take up water, and this quickly causes reductions in growth rate, along with a suite of metabolic changes very similar to those caused by water stress (Marschner, 1995; Munns, 2002). For landscape plants, the typical symptoms of initial salt injury are stunted growth and foliar damage including marginal leaf burn or scorching, chlorosis, leaf necrosis and premature leaf drop (Bernstein et al., 1972). As salt stress becomes severe, older leaves show these symptoms more acutely, eventually extending to the whole laminae with less and less green leaves remaining in the plants. For flowering woody shrubs and trees, buds may fail to open or grow and branches may die. Salt injury symptoms are more obvious under hot, dry and windy conditions compared to cool, humid environments (Fox et al., 2005; Niu et al., 2007a; Wu et al., 2001). Salt tolerance of a plant varies largely with species or even cultivars within a species (Cabrera, 2009). The actual response of a plant to salinity is often affected by climate conditions, type of substrate or soil, irrigation management, and its growth or developmental stage. For each salinity level, it is possible to select a number of genotypes whose growth and/or aesthetic values are not affected or only slightly (Bernstein et al., 1972; Pasternak and Malach, 1994). Obviously, as salinity levels increase, the number of genotypes that can tolerate the stress becomes smaller.

High salinity induces a series of metabolic dysfunctions in landscape plants, including absorption of excessive minerals, nutrient imbalance, inhibition of photosynthesis, enzymatic activity, protein and nucleic metabolism and respiration (Azza Mazher et al., 2007;

Marschner, 1995; Munns and Tester, 2008). The degree of the negative effect of salinity on the above physiological processes depends on salinity levels and length of exposure.

MECHANISMS OF SALT TOLERANCE AND PHYSIOLOGICAL RESPONSES TO SALINITY

Ion uptake. Plant adaptations to salinity are of three distinct types: osmotic stress tolerance, Na^+ and/or Cl^- exclusion, and the tolerance of tissues to accumulated Na^+ and Cl^- (Munns and Tester, 2008). Salt tolerant genotypes usually have the ability to restrict Na^+ and Cl^- transport to shoots and/or tolerate high internal Na^+ and Cl^- concentrations. Ion uptake depends on species or genotype, salinity level, and the chemical composition of the soil solution.

Niu and Rodriguez (2006a) found out differences in overall salt tolerance and Na^+ or Cl^- tolerance among eight herbaceous species irrigated with solutions at four salinity levels. Four species died by the middle of the 12-week experiment. Among the surviving four species, the ion accumulation in shoots and roots differed (Figure 1). In all species both Na^+ and Cl^- in shoots and roots generally increased, albeit differentially, with irrigation water salinity. For example, *Gazania rigen* had the highest Na^+ concentration in shoots (39 $mg{\cdot}g^{-1}$ DW) but *Delosperma cooperi* had the highest Na^+ concentration in roots (22 $mg{\cdot}g^{-1}$ DW). Shoot Cl^- was highest in *D. cooperi* and *G. rigen*, and even in the control plants. The Cl^- concentration was 5 to 10 times higher in *D. cooperi* and *G. rigen* than the other two species. Regardless of the high Na^+ and Cl^- concentrations, no visual salt injury was observed in *G. rigen* and *D. cooperi*, indicating that these two species were tolerant to high internal concentrations of these ions. Compared to *G. rigen*, *D. cooperi* had a higher ability to exclude Na^+ from shoots, which is an important mechanism of salt tolerance.

In a 15-week salinity study, significant differences in ion accumulation were found between two azalea (*Rhododendron*) hybrids (Cabrera, 2003). The hybrid 'Delaware Valley White', which was more negatively affected by salinity stress (both in growth and quality), accumulated exceedingly higher levels of Cl^- and Na^+ (up to 3.33% and 5.65 $mg{\cdot}g^{-1}$ DW, respectively) than in the more salt-tolerant 'Red Hershey' (up to 1.31% and 0.46 $mg{\cdot}g^{-1}$ DW, respectively). Another 15-week salinity study with crape myrtles (*Lagerstroemia* spp.) also showed differences in Na^+ and Cl^- accumulation in leaf tissues, being lower in those cultivars that were deemed more salt tolerant on the basis of lesser foliar salt damage (Cabrera, 2009).

Salt composition also affects the response of landscape plants to salinity. Growth and ion uptake of four garden rose rootstocks responded to salinity and dominant salt type differently (Niu and Rodriguez, 2008). *Rosa ×fortuniana* was the most tolerant rootstock when irrigated with a saline solution dominated by sulfate, followed by *R.* 'Dr. Huey', *R. multiflora*, and *R. odorata* in descending order. However, *R. ×fortuniana* was the least tolerant rootstock when chloride was dominant in the saline solutions, *R. odorata* was the second least tolerant, and 'Dr. Huey' was the most tolerant. *Rosa ×fortuniana* and 'Dr. Huey' had a higher restriction ability for Na uptake than the other two rootstocks.

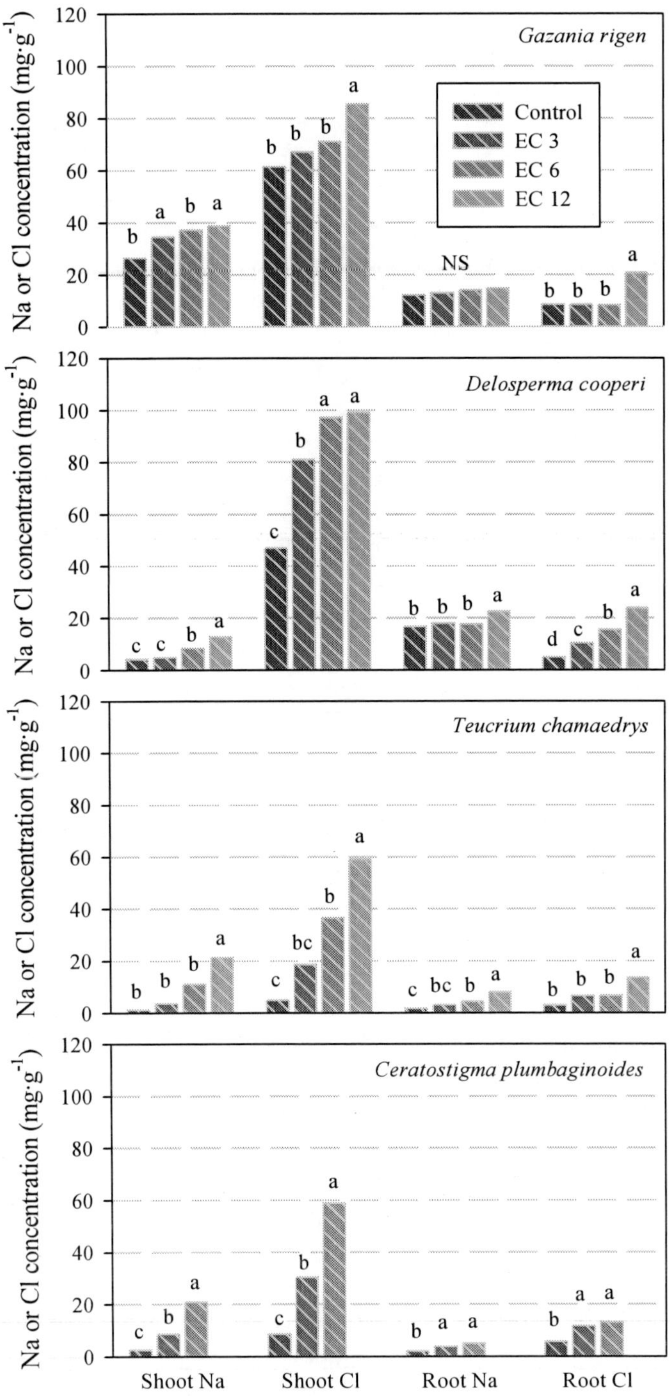

Figure 1. Shoot and root ion concentrations of *Gazania rigens*, *Delosperma cooperi*, *Teucrium chamaedrys* and *Ceratostigma plumbaginoides* as affected by irrigation water salinity (EC). Control, EC 3, EC 6, and EC 12 represent EC of 0.8, 3.2, 6.4 and 12 $dS \cdot m^{-1}$, respectively. Means (bars) followed by the same letters are not significantly different as tested by Student-Newman-Keuls multiple comparison at $P = 0.05$. Note: *Ceratostigma plumbaginoides* plants did not survive at EC 12 $dS \cdot m^{-1}$.

Photosynthesis and stomatal conductance. The most dramatic and readily measurable whole plant response to salinity is a decrease in stomatal aperture (Munns and Tester, 2008). Rates of photosynthesis per unit leaf area in salt-treated plants are often unchanged, even though stomatal conductance is reduced (James et al., 2002). Net photosynthetic rate of four rose rootstocks (*Rosa ×fortuniana*, *R. multiflora*, *R. odorata* and *R.* 'Dr. Huey'), which were irrigated with saline solutions up to 8.0 $dS{\cdot}m^{-1}$, were not affected (Niu, unpublished data), possibly because healthy, young leaves were selected for these measurements. The effect of salinity on photosynthesis and stomatal conductance depends on species and salinity level. For example, *Ziziphus nummularia* had greater reduction in photosynthesis and stomatal conductance than *Z. rotundifolia* (Meena et al., 2003). The effect of salinity on leaf photosynthesis possibly depends on leaf age or the time when the leaf was exposed to salinity and the level of salinity.

Chlorophyll fluorescence and chlorophyll content (SPAD reading). Chlorophyll fluorescence has been proved to be useful in the evaluation of salinity tolerance. The maximal photochemical efficiency of photosystem II (PSII), the ratio of variable fluorescence (F_v) to maximal fluorescence (F_m), F_v/F_m, is commonly measured to examine damage in the photosynthetic apparatus caused by salinity stress (Jimenez et al., 1997; Percival, 2005; Percival and Fraser, 2001). Percival (2005) concluded that measuring chlorophyll fluorescence (F_v/F_m) of excised, salt-treated leaves under laboratory conditions can be used to evaluate salt tolerance without destroying whole plants. In our studies, we found that F_v/F_m was reduced by elevated salinities at 3.0 and 6.0 $dS{\cdot}m^{-1}$ in three rose rootstocks, but the reduction was small (Niu et al., 2008). The advantages of using chlorophyll fluorescence include the fact that fluorescence measurements use a light portable piece of equipment, and measurements are non-destructive and non-invasive. The relative effectiveness of this physiological measurement depends on species and salinity levels. For example, no differences were found in F_v/F_m in tomato leaves treated with salinity up to 200 mM NaCl in dark-adapted leaves (Zribi et al., 2009). However, the PSII photochemistry was modified in salt-affected plants in light-adapted leaves.

Salt stress not only causes leaf scorching and necrosis, but chlorosis as well, as it has been shown that low to moderate salinity stimulates chlorophyll degradation, whereas higher salt concentrations more drastically affect chlorophyll synthesis (Santos, 2004). Leaf chlorophyll concentration could be determined routinely, but destructively, by chemical analyses. The development of non-destructive hand-held chlorophyll meters provides for the rapid measurement of a relative index of chlorophyll content, also called a leaf greenness index (Monje and Bugbee, 1992), and thus the degree of chlorosis. The most common chlorophyll meter, Minolta SPAD-502, provides a relative value of chlorophyll content, called a SPAD reading (SPAD, an acronym of Soil and Plant Analyzer Development), by measuring light transmission in wavelength regions in which the absorption of chlorophyll is different. These readings may be used as another parameter to screen differences in salt tolerance of multiple species or genotypes non-destructively. In a number of herbaceous and woody landscape plants, we found that less salt tolerant species had either reduced SPAD readings and/or these were significantly reduced with increasing salinity stress (Cabrera, 2003, 2009; Niu et al., 2007a).

Herbaceous Plants

Herbaceous plants do not have persistent woody stems above the ground. They can be annual, biennial and perennial. When they are used for landscapes, a significant growth reduction is often acceptable and sometime is desirable (Cameron et al., 2004). On the other hand, injury or loss of leaves or flowers due to salt stress is unacceptable. Since leaves and stems of herbaceous plants die back at the end of the growing season, salt accumulation in leaves and stems would not influence the growth in the next growing season, which is different from woody plants where accumulated ions would stay in the stem (Shannon et al., 1994).

Limited research results are found on the salt tolerance of herbaceous plants. Surprisingly, a wide range of salt tolerance exists in herbaceous plants. In a greenhouse container study with eight herbaceous species (Niu and Rodriguez, 2006a), plants were irrigated for 12 weeks with saline solutions at EC of 0.8, 3.2, 6.4, or 12 $dS{\cdot}m^{-1}$ created by adding NaCl, $MgSO_4$ and $CaCl_2$ at 87:8:5 (by weight). Three *Penstemon* species (*P. eatonii* A. Gray, *P. pseudospectabilis* M.E. Jones and *P. strictus* Benth.) and lavender (*Lavandula angustifolia* Mill.) were highly sensitive to salt stress and died in the middle of the 12-week experiment. Gazania (*Gazania rigen* (L.) Gaertn.) and purple ice plant (*Delosperma cooperi* (Hook.f.) L. Bolus) did not have any salt burn injuries when irrigated with a solution EC of 12.0 $dS{\cdot}m^{-1}$ (with EC_e up to 20 $dS{\cdot}m^{-1}$), although growth was reduced, especially in gazania. The salt tolerance of Germander (*Teucrium chamaedrys* L.) and hardy blue plumbago (*Ceratostigma plumbaginoides* Bunge) were moderate, between the sensitive and the tolerant groups. Plants resisted salt stress by limiting Na^+ accumulation in shoots or by tolerating high Na^+ and/or Cl^- uptake (Niu and Rodriguez, 2006a).

In another field study, ten herbaceous plants were grown in raised beds in a semi-arid environment (Niu et al., 2007a). The species included yarrow (*Achillea millefolium* L.), firewheel (*Gaillardia aristata* Pursh), lantana (*Lantana* x *hybrida* 'New Gold'), purple lantana (*Lantana montevidensis* (Spreng.) Brig.), honeysuckle (*Lonicera japonica* Thunb. 'Halliana'), rosemary (*Rosmarinus officinalis* L. 'Huntington Carpet'), black-eyed Susan (*Rudbeckia hirta* L.), purple verbena (*Glandularia canadensis* (L.) Nutt. 'Homestead Purple'), garden verbena (*Glandularia* x *hybrida* (Grönland and Rümpler) G. L. Nesom and Pruski), and spike verbena (*Verbena macdougalii* Heller). Based on their growth and visual quality, yarrow, firewheel, lantana, honeysuckle, and rosemary could be irrigated with EC up to 5.4 $dS{\cdot}m^{-1}$ with little reduction in growth and aesthetic appearance. Black-eyed Susan was the only species that was sensitive to salt stress and did not survive in the elevated salinity treatments. The rest of the species could tolerate EC up to 3.2 $dS{\cdot}m^{-1}$, provided that soils are well drained. Obviously, the salt tolerance of the above herbaceous plants varied considerably with genotype and species. Similarly, Zollinger et al. (2007) also found out variations in salt tolerance among eight herbaceous perennials native to intermountain western United States.

Floriculture crops are high values commodities and traditionally are irrigated with high quality water, but dwindling fresh water supplies have led growers to view saline wastewaters and ground waters are viable alternative irrigation sources. The salt tolerance of a number of floriculture crops, including cultivars of celosia, statice, stock and sunflower, was investigated at the U.S. Salinity Lab (Carter and Grieve, 2006) by exposing them to differing water ionic compositions and salinity levels. Salt tolerance of these crops was evaluated based on their stem length, which determines their marketability. Most were found to be marketable at moderate salinity levels, even though plant height tended to decline as salinity level increases. In Israel, where the supply of high quality water is extremely limited, saline wastewater is used for irrigating a variety of horticultural crops, including cut flowers (Shillo et al., 2002).

Table 1. Ten bedding plant species and cultivars used in two salinity experiments

Botanical name	Common name	Cultivar
Angelonia angustifolia	Angelonia	Lavender Pink
A. angustifolia	Angelonia	Purple
A. angustifolia	Angelonia	White
Capsicum annuum	Ornamental pepper	Black Pearl
C. annuum	Ornamental pepper	Calico
C. annuum	Ornamental pepper	Purple Flash
Catharanthus roseus	Vinca	Titan
Helenium amarum	Helenium	Dakota Gold
Helichrysum petiolatum	Licorice Plant	Silver Mist
Plumbago auriculata	Plumbago	Escapade Blue

Bedding plants are extensively used in urban landscapes, and the use of alternative waters in these settings justifies studies on the salt tolerance of these species. Two experiments were conducted, one in a 25% light exclusion shade house in the summer (Expt. 1) and the other in a greenhouse in the winter (Expt. 2) (Niu et al., 2010). Ten bedding plants and cultivars that performed from acceptable to excellent based on their growth and visual quality under different irrigation levels in a semi-arid environment (Niu and Rodriguez, 2007a) were selected (Table 1). Plants were irrigated with saline solutions created by adding NaCl, $MgSO_4$, and $CaCl_2$ to tap water to simulate the composition of local reclaimed water, and resulting in EC_w of 0.8, 2.8, 4.0, 5.1 or 7.4 $dS \cdot m^{-1}$. Shoot dry weight of all species and cultivars at the end of the experiments was reduced at 7.4 $dS \cdot m^{-1}$ compared to the control (Figure 2). However, the magnitude of reduction varied with species and cultivars. The salinity thresholds of irrigation water in which growth reduction occurred were 4.0 $dS \cdot m^{-1}$ for angelonia (*Angelonia angustifolia*) cultivars and ornamental pepper (*Capsicum annuum* 'Calico') and 4.0 to 5.1 $dS \cdot m^{-1}$ for helenium (*Helenium amarum*), licorice plant (*Helichrysum petiolatum*) and plumbago (*Plumbago auriculata*). All plants survived in Expt. 1 and did not show visual salt injuries, regardless of treatment, except for the ornamental pepper 'Purple Flash'. Leaf Na^+ and Cl^- concentrations varied with species and treatments (Figure 3). The ornamental pepper 'Black Pearl' had the highest leaf Cl^- concentrations at higher salinities compared to the other species and cultivars. Leaf Na^+ concentrations in licorice plant and plumbago were in the range of 10 to 30 $mg \cdot g^{-1}$ DW, higher than those in the other species. In Expt. 2, shoot DW was reduced by salinity treatments in the ornamental pepper 'Black Pearl', plumbago and angelonia but not in other species. The three ornamental peppers 'Black Pearl', 'Calico' and 'Purple Flash' exhibited slight foliar injuries on some plants in Expt. 2 due to high salinity in the rootzone in the highest salinity treatment. In general, the bedding plants tested in this study are moderately tolerant to salt stress and may be irrigated with saline water up to 4.0 dS·m^{-1} with little reduction on aesthetical appearance. The ornamental pepper 'Black Pearl' may become considerably smaller at moderate salinity stress.

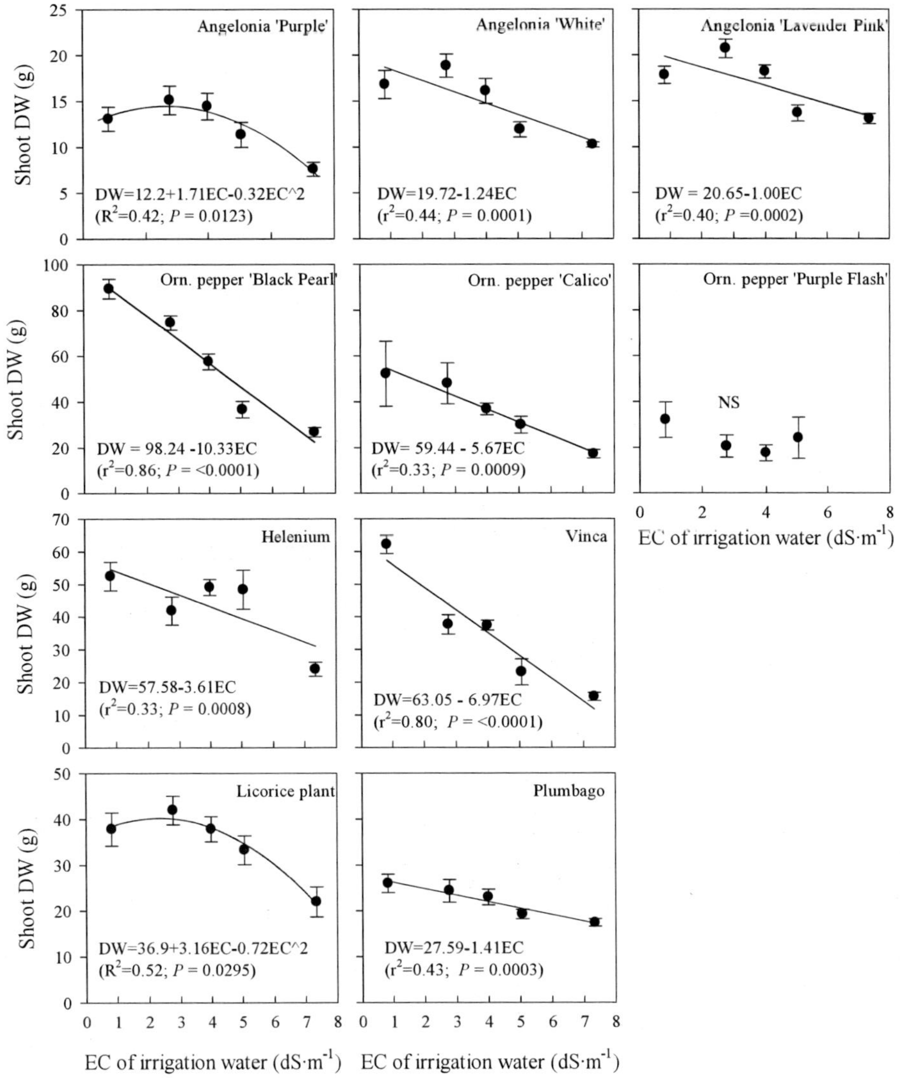

Figure 2. Shoot dry weight (DW) of angelonia (*Angelonia angustifolia*) 'Purple', 'White', and 'Lavender Pink', ornamental pepper (*Capsicum annuum*) 'Black Pearl', 'Calico', and 'Purple Flash', helenium (*Helenium amarum*), vinca (*Catharanthus roseus*), licorice plant (*Helichrysum petiolatum*) and plumbago (*Plumbago auriculata*) irrigated with water at EC of 0.8, 2.8, 4.0, 5.1, or 7.4 dS·m^{-1} (Expt. 1). Vertical bars represent standard errors. NS indicates nonsignificance.

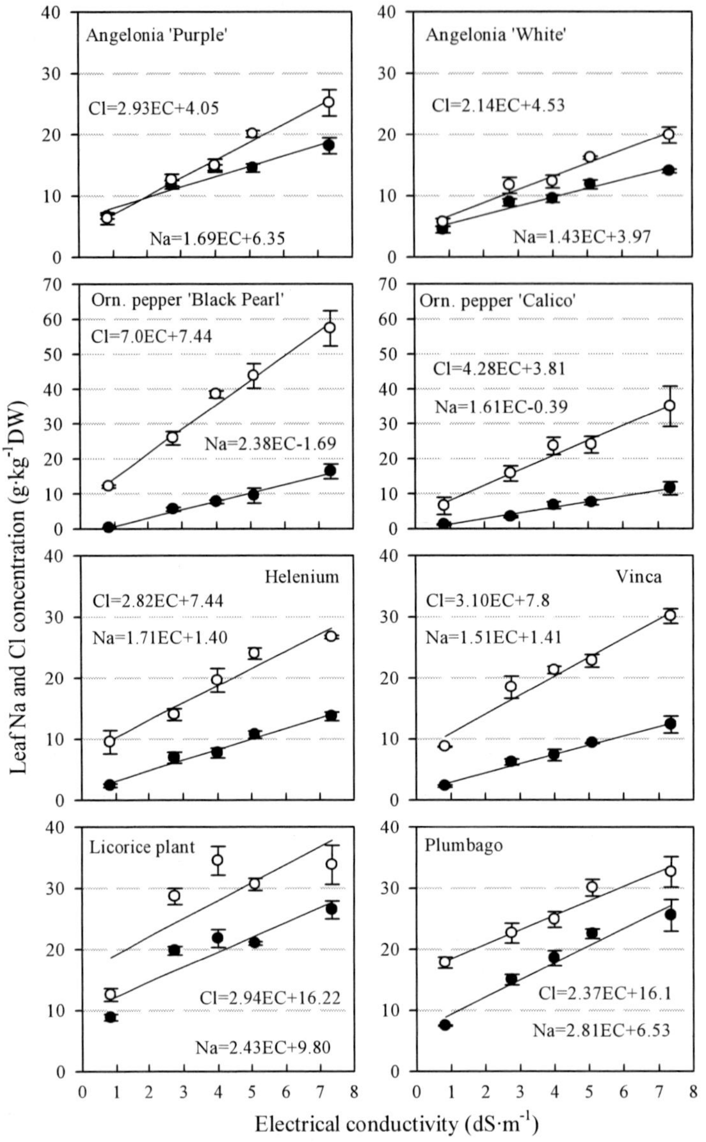

Figure 3. Shoot Na^+ (solid circle) and Cl^- (open circle) concentrations of angelonia (*Angelonia angustifolia*) 'Purple', 'White', and 'Lavender Pink', ornamental pepper (*Capsicum annuum*) 'Black Pearl', 'Calico', and 'Purple Flash', helenium (*Helenium amarum*), vinca (*Catharanthus roseus*), licorice plant (*Helichrysum petiolatum*) and plumbago (*Plumbago auriculata*) irrigated with water at EC of 0.8, 2.8, 4.0, 5.1, or 7.4 $dS{\cdot}m^{-1}$ (Expt. 1). Vertical bars represent standard errors. NS indicates nonsignificance.

Greenhouse Cut Roses and Garden Roses

The rose genus (*Rosa* spp.) is one of the most economically important ornamental crops in the world. Roses have been traditionally categorized as salt sensitive species with growth reductions and tissue salt injuries reported within a range of 0.5 to 3 $dS{\cdot}m^{-1}$ EC, depending on species, cultural medium, leaching fraction and environmental conditions (Cabrera and Perdomo, 2003; Urban, 2003). However, some studies reported that yield and quality of roses did not decrease when irrigated with drainage recycled water at EC of 3.5 $dS{\cdot}m^{-1}$, provided that an appropriate rootstock and well drained growing medium were used (Raviv *et al.*, 1998; Cabrera, 2003).

Container-grown 'Bridal Pink' roses (on *R. manetti* rootstock) in a peat moss-based growing medium were irrigated with complete nutrient solutions supplemented with NaCl at 0, 5, 10, 15 or 30 mM (Cabrera and Perdomo, 2003). No significant differences in cut-flower yield and quality were observed among the salt treatments. However, visual symptoms of salt injury were observed during the last three flowering cycles on older leaves. Given the economic importance of cut roses, many studies have been done on their salt tolerance, producing varying results, which are possibly due to the use of different cultivars, rootstocks and experimental conditions (Bernstein et al., 2006; Cabrera and Perdomo, 2003; de Vries, 2003; Fernandez Falcon et al., 1986; Hughes and Hanan, 1978; Wahome et al., 2001). Relatively little attention, however, has been paid to the salt tolerance of garden roses.

Most greenhouse roses used for cut-flower production (in greenhouses) and traditional (hybrid tea) garden roses are typically produced by clonal propagation, including the use of budding and grafting techniques (Pemberton, 2003). This stems from the observation that cultivar performance and flower production in grafted rose plants is higher than in non-grafted (own-rooted) plants (Cabrera, 2002; Cabrera et al., 2009). Different rootstocks are used in various areas of the world in accordance with climatic and soil conditions, and as expected, scion vigor and flower yields, with or without salt stress are significantly affected by rootstock selection (Pivetta *et al.*, 2004; Cabrera, 2002; Cabrera et al., 2009). In our previous studies, the garden rose rootstocks 'Dr. Huey', *R.* × *fortuniana*, *R. multiflora* and *R. odorata*, responded to salinity and dominant salts differently (Niu and Rodriguez, 2008). After 16 weeks of treatment, elevated salinities reduced growth of all rootstocks but the magnitude varied with rootstock and dominant salt type. At both moderate (3.9 $dS{\cdot}m^{-1}$ electrical conductivity, EC) and high salinities (7.9 to 8.2 $dS{\cdot}m^{-1}$), chloride dominated salinity led to a greater growth reduction in *R.* × *fortuniana,* followed by *R. odorata* and *R. multiflora.*

At high salinity dominated by sulfate, *R. odorata* had a greater growth reduction, followed by *R. multiflora*, 'Dr. Huey' and *R.* × *fortuniana*. Cut flower and dry biomass yields in 'Bridal White' and 'Red France' rose greenhouse cultivars exposed to increasing NaCl or mixed NaCl + $CaCl_2$ salinity stress were significantly and negatively affected more when grafted in the newer greenhouse rose rootstock 'Natal Briar' compared to the long-standing rootstock 'Manetti' (Cabrera et al., 2009).

Woody Shrubs and Trees

Almost all landscapes have woody shrubs and trees. Therefore, the salt tolerance of these ornamental woody landscape plants have been studied more extensively compared to other landscape plants. The relative salt tolerance of the following trees and shrubs were recently investigated: green ash (*Fraxinus pennsylvanica*), black cherry (*Prunus serotina*), Chinese elm (*Ulmus parvifolia*), desert willow (*Chilopsis linearis*), Russian olive (*Elaeagnus angustifolia*), sand cherry (*Prunus besseyi*) and sand plum (*Prunus angustifolia*) (Niu and

Rodriguez, 2007b). Bare-root seedlings were potted in 10-L containers filled with a peat-based commercial substrate in early January. Seven months later, the plants were irrigated with saline solution at 0.8, 2.0, or 4.0 $dS \cdot m^{-1}$ over a four-month period under outdoor conditions in a semi-arid environment. Daily maximum air temperatures often exceeded 40 °C during the experiment, and relative humidity ranged from 12% to 35%. Chinese elm did not have any visual injuries, regardless of salinity treatment. Desert willow and Russian olive had leaf burn and defoliation when irrigated with saline solution at 4 $dS \cdot m^{-1}$, and although these species were native to the region, they were not as tolerant to salinity as Chinese elm, which is native to China. Overall, Chinese elm was the most salt-tolerant species; green ash, desert willow, and Russian olive were moderately-tolerant (2 $dS \cdot m^{-1}$ of irrigation water, rootzone salinity was not quantified) and the rest were salt-sensitive. Black cherry and sand plum may perform better under partial shade in a semi-arid environment.

Several agencies in California began to investigate salt tolerance of landscape plants more than three decades ago. For example, Bernstein et al. (1972) determined the salt tolerance of 25 shrubs and ground-cover species. Tolerant species like bougainvillea (*Bouganvillea spectabilis*), Natal plum (*Carissa grandiflora*), and rosemary (*Rosmarinus lockwoodi*) were affected little at EC_e (soil saturation extracts) of 8.0 $dS \cdot m^{-1}$ whereas sensitive species like star jasmine (*Trachelospermum jasminoides*), pineapple guava (*Feijoa sellowiana*), holly (*Ilex cornuta* 'Burford'), and rose (*R.* 'Grenoble' on *R.* 'Dr. Huey') were severely damaged or killed at EC_e of 4 $dS \cdot m^{-1}$. Wu and Dodge (2005) compiled salt tolerance information in tables for over 200 species of trees and palms, shrubs, and ground covers. Grattan and Grieve (1999) reviewed comprehensively on the relations between salinity and mineral nutrition of horticultural crops and addressed the nutrients individually. The US Salinity laboratory has an updated version of comprehensive bibliography on response of plants to salinity (www.ars.usda.gov/services/docs.htm?docid=8908), which is open access to the public according to Grieve et al. (2008). The Southern California Salinity Coalition and National Water Research Institute maintains a website (www.salinitymanagement.org) with information related to salinity management guidelines for landscapes irrigated with recycled water (reclaimed water) (Tanji et al., 2008).

CONCLUSIONS

Salt tolerance of horticultural plants varies widely with genotype, environmental conditions and soil or substrate type. Ornamental horticultural plants and crops, most of which are glycophytes, have similar mechanisms of salt tolerance to agricultural crops. Salinity stress reduces plant size and growth, causes foliar injury (reducing aesthetic value) and salt accumulation in the root zone. Salt tolerant genotypes should be selected when alternative, poor-quality irrigation water sources are used. Due to the potential for significant salt accumulation, good drainage is the key to prevent salinity from rising to hazardous levels. Applying sufficient volumes of water, to allow for adequate salt leaching, and blending with less saline waters can alleviate salt stress and accumulation, and sustain the prolonged use of alternative water sources for the irrigation of ornamental horticultural crops.

References

Azza Mazher, A.M., E.M. Fatma El-Questni, and M.M. Farahat. 2007. Responses of ornamental plants and woody trees to salinity. *World J. Ag. Sci.* 3:386-395.

Bernstein, L., L.E. Francois and R.A. Clark. 1972. Salt tolerance of ornamental shrubs and ground covers. *J. Amer. Soc. Hort. Sci.* 97:550-556.

Bernstein, N., B.T. Asher, F. Haya, S. Pini, R. Ilona, C. Amram, and I. Marina. 2006. Application of treated wastewater for cultivation of roses (*Rosa hybrida*) in soil-less culture. *Scientia Hort.* 108:185-193.

Boland, A.M. 2008. Management of saline and/or recycled water for irrigated horticulture. *Acta Hort.* 792: 123-134.

Cabrera, R.I. 1998. Monitoring chemical properties of container growing media with small soil solution samplers. *Scientia Hort.* 75: 113-119.

Cabrera, R.I. 2002. Rose yield, dry matter partitioning and nutrient status responses to rootstock selection. *Scientia Hort.* 95: 75-83.

Cabrera, R.I. 2003. Growth, quality and nutrient responses of azalea hybrids to salinity. *Acta Hort.* 609:241-245.

Cabrera, R.I. 2009. Revisiting the salt tolerance of crape myrtles (*Lagerstroemia* spp.). *Arboric. Urban Fores.* 35(3): 129-134.

Cabrera, R.I., A.R. Solís-Pérez and J.J. Sloan. 2009. Greenhouse rose yield and ion accumulation responses to salt stress as modulated by rootstock selection. *HortScience* 44(7):2000-2008.

Cabrera, R.I. and P. Perdomo. 2003. Reassessing the salinity tolerance of greenhouse roses under soilless production conditions. *HortScience* 38:533-536.

Cameron, R.W.F., S. Wilkinson, W.J. Davies, R.S. Harrison-Murray, D. Dunstan, and C. Burgess. 2004. Regulation of plant growth in container-grown ornamentals through the use of controlled irrigation. *Acta Hort.* 630:305–312.

Carter, C.T. and C.M. Grieve. 2006. Salt tolerance of floriculture crops, p. 279-287. In: M.A. Khan and D.J. Weber (eds.). Ecophysiology of high salinity tolerant plants. Springer, Netherlands.

Cassaniti, C., C. Leonardi, and T.J. Flowers. 2009. The effects of sodium chloride on ornamental shrubs. *Scientia Hort.* 122:586-593.

Devitt, D.A., R.L. Morris, and L.K. Fenstermaker. 2005. Foliar damage, spectral reflectance, and tissue ion concentrations of trees sprinkle irrigated with waters of similar salinity but different chemical composition. *HortScience* 40:819-826.

de Vries, D.P. 2003. Rootstock, p. 2:633-638. In: A.V. Robert, T. Debener, and S. Gudin (eds.). Encyclopedia of Rose Science. Elsevier Academic Press, San Diego.

Duncan, R.R., R.N. Carrow and M. T. Huck. 2009. Turfgrass and landscape irrigation water quality: Assessment and management. CRC Press, Boca Raton, FL.

Dobrowolski, J., M. O'Neill, L. Duriancik, and J. Throwe (eds.). 2008. Opportunities and challenges in agricultural water reuse: Final report. USDA-CSREES, 89 p.

Eom, S.H., T.L. Setter, A. DiTommaso, and L.A. Weston. Differential growth response to salt stress among selected ornamentals. *J. Plant Nutri.* 30:1109-1126.

Fostad, O. and P.A. Pedersen. 2000. Container-grown tree seedling responses to sodium chloride applications in different substrates. *Environmental Pollution* 109:203-210.

Fernández Falcón, M., C.E. Álvarez, V. Garcia, and K. Báez. 1986. The effect of chloride and bicarbonate levels in irrigation water on nutrition content, production and quality of cut roses 'Mercedes'. *Scientia Hort.* 29:373-385.

Farnham, D.S., R.F. Hasek and J.L. Paul. 1985. Water quality: Its effects on ornamental plants. University of California Cooperative Extension Leaflet No. 2995, 15 p.

Fox, L.J., J. N. Grose, B.L. Appleton, and S.J. Donohue. 2005. Evaluation of treated effluent as an irrigation source for landscape plants. *J. Env. Hort.* 23:174-178.

Grattan, S.R. and C.M. Grieve. 1999. Salinity-mineral nutrient relations in horticultural crops. *Scientia Hort.* 78:127-157.

Grieve, C., L. Wu, L. Rollins, and A. Harivandi. 2008. Tolerance by landscape plants of salinity and of specific ions (Chapter V). In: A comprehensive literature review on salt management guide for landscape irrigation with recycled water in coastal southern California. www.salinitymanagement.org. 320 p.

Hughes, H. and J. Hanan. 1978. Effect of salinity in water supplies on greenhouse rose production. *J. Am. Soc. Hort.* 103:694-699.

Jimenez, M.S., A.M. Gonzalez-Rodriguez, D. Morales, M.C. Cid, A.R. Socorro, and M. Caballero. 1997. Evaluation of chlorophyll fluorescence as a tool for salt stress detection in roses. *Photosynthetica* 33(2):291-301.

James, R.A, A.R. Rivelli, R. Munns, and S. von Caemmerer. 2002. Factors affecting CO_2 assimilation, leaf injury and growth in salt-stressed durum wheat. *Funct. Plant Biol.* 29:1393–403.

Kjelgren, R., L. Rupp, and D. Kilgren. 2000. Water conservation in urban landscapes. *HortScience* 35:1037-1040.

Lang, H.J. 1996. Growing media testing and interpretation. In: D.W. Reed (ed.), Water, media and nutrition for greenhouse crops. Ball Publishing, Batavia, Illinois, p. 123-140.

Maas, E.V. 1990. Crop salt tolerance, p. 262-304. In: K.K. Tanji (ed.), Agricultural salinity assessment and management. Amer. Soc. of Civil Engineers, ASCE Manuals and reports on engineering practice No. 71. New York.

Marosz, A. 2004. Effects of soil salinity on nutrient uptake, growth, and decorative value of four ground cover shrubs. *J. Plant Nutr*. 27:977-989.

Marschner, H. 1995. Mineral nutrition of higher plants, 2nd Edition. Academic Press. San Diego, CA.

Meena, S.K, N.K. Gupta, S. Gupta, S.K. Khandelwal, and E.V.D. Sastry. 2003. Effect of sodium chloride on the growth and gas exchange of young Ziziphus seedling rootstocks. *J. Hort. Sci. Biotech.* 78:454-457.

Monje, O.A. and B. Bugbee. 1992. Inherent limitations of nondestructive chlorophyll meter: A comparison of two types of meters. *HortScience* 271(1): 69-71.

Munns, R. 2002. Comparative physiology of salt and water stress. *Plant Cell Env.* 25:239-250.

Munns, R. and M. Tester. 2008. Mechanisms of salinity tolerance. *Annu. Rev. Plant Biol.* 59:651-681.

Niu, G. and D.S. Rodriguez. 2006a. Relative salt tolerance of five herbaceous perennials. *HortScience* 41:1493-1497.

Niu, G. and D.S. Rodriguez. 2006b. Relative salt tolerance of selected herbaceous perennials and groundcovers. *Scientia Hort.* 110:352-358.

Niu, G. and D.S. Rodriguez. 2007a. Heat and drought tolerance of selected bedding plants. *Proceedings of Southern Nursery Association* 52:84–88.

Niu, G. and D.S. Rodriguez. 2007b. Salinity tolerance, water use and crop coefficients of selected shrubs and trees. *Proceedings of Southern Nursery Association* 52:133-137.

Niu, G. and D.S. Rodriguez. 2008. Responses of growth and ion uptake of four rose rootstocks to chloride- or sulfate-dominated salinity. *J. Amer. Soc. Hort. Sci.* 133(5):663–669.

Niu, G., D.S. Rodriguez, and L. Aguiniga. 2007a. Growth and landscape performance of ten herbaceous species in response to saline water irrigation. *J. Env. Hort.* 25:204–210.

Niu, G., D.S. Rodriguez, and T. Starman. 2010. Response of bedding plants to saline water irrigation. *HortScience* (in press).

Niu, G., D.S. Rodriguez, and Y.T. Wang. 2007b. Salinity and growing medium regulate growth, morphology and ion uptake of *Gaillardia aristata*. *J. Env. Hort.* 25(2):89-94.

Niu, G., D.S. Rodriguez, and L. Aguiniga. 2008. Effect of saline water irrigation on growth and physiological responses of three rose rootstocks. *HortScience* 43(5):1479–1484.

Pasternak, D. and Y.D. Malach. 1994. Crop irrigation with saline water. In: Pessarakli Mo (ed.), Handbook of Plant and Crop Stress. Marcel Dekker, New York, p. 599-622.

Pemberton, H.B. 2003. Overview of roses and culture, p. 2:570-573. In: A.V. Robert, T. Debener and S. Gudin. (eds.). Encyclopedia of Rose Science. Elsevier Academic Press, San Diego.

Percival, G.C. 2005. Identification of foliar salt tolerance of woody perennials using chlorophyll fluorescence. *HortScience* 40:1892-1897.

Percival, G.C. and G.A. Fraser. 2001. Measurement of the salinity and freezing tolerance of Crataegus genotypes using chlorophyll fluorescence. *J. Arbor.* 27(5):233-245.

Pivetta, K.F.L., P.U.C. Pizetta, and D.R. Pedrinho. 2004. Morphologic characterization and evaluation of the productivity of nine rootstocks of rose bush (*Rosa* spp.). *Acta Hort.* 630:213-217.

Raviv, M., A. Krasnovsky, S. Medina, and R. Reuveni. 1998. Assessment of various control strategies for recirculation of greenhouse effluents under semi-arid conditions. *J. Hort. Sci. and Biotec.* 73:485-491.

Santos, C.V. 2004. Regulation of chlorophyll biosynthesis and degradation by salt stress in sunflower leaves. *Sci. Hort.* 103: 93-99.

Schuch, U. Effect of reclaimed water and drought on salt-sensitive perennials. 2005. *HortScience* 40:1095 (Abstract).

Shannon, M.C., C.M. Grieve, and L.F. Francois. 1994. Whole-plant response to salinity. In: Wilkinson (ed.). Plant Environment Interaction. Marcel Dekker, New York, p. 199-244.

Shillo, R., M. Ding, D. Pasternak and M. Zaccai. 2002. Cultivation of cut flower and bulb species with saline water. *Scientia Hort.* 92:41-54.

Tanji, K.K. 1990. Nature and extent of agricultural salinity. p. 1-17. In: K.K. Tanji (ed.), Agricultural salinity assessment and management. ASCE Manual No. 71. *Amer. Soc. Civil Engrs.*, New York.

Tanji, K.K. 2002. Salinity in the soil environment. p. 21-51. In: A. Läuchli and U. Lüttge (eds.), Salinity: Environment - Plants – Molecules. Kluwer Academic Publishers, Netherlands.

Tanji, K., S. Grattan, C. Grieve, A. Harivandi, L. Rollins, D. Shaw, B. Sheikh, and L. Wu. 2008. A comprehensive literature review on salt management guide for landscape irrigation with recycled water in coastal southern California. www.salinitymanagement.org. 320 p.

Tanwar, B.S. Saline Water Management for Irrigation (3rd revised draft), Work Team on Use of Poor Quality Water for Irrigation (WT-PQW), International Commission on Irrigation and Drainage (ICID), New Delhi, India.

Wahome, P.K. 2003. Mechanisms of salt (NaCl) stress tolerance in horticultural crops – a mini review. *Acta Hort.* 609:127-131.

Wahome, P.K., Jesch, H.H., Grittner, I. 2001. Mechanisms of salt stress tolerance in two rose rootstocks: *Rosa chinensis* 'Major' and *R. rubiginosa*. *Scientia Hort.* 87: 207-216.

Wu, L. and L. Dodge. 2005. A Special Report for the Elvenia J. Slosson Endowment Fund: Landscape Salt Tolerance Selection Guide for Recycled Water Irrigation.

Wu, L., X. Guo, and A. Harivandi. 2001. Salt tolerance and salt accumulation of landscape plants irrigated by sprinkler and drip irrigation systems. *J. Plant Nutr.* 24:1473-1490.

Zollinger, N., R. Koenig, T. Cerny-Koenig, and R. Kjelgren. 2007. Relative salinity tolerance of intermountain western United States native herbaceous perennials. *HortScience* 42:529-534.

Zribi, L., G. Fatma, R. Fatma, R. Salwa, N. Hassan, and R. Mohamed Ne′ jib. 2009. Application of chlorophyll fluorescence for the diagnosis of salt stress in tomato (*Solanum lycopersicum*, variety Rio Grande). *Scientia Hort.* 120:367-372.

In: Horticulture in the 21st Century
Editors: Abigail N. Sampson, pp. 161-177

ISBN: 978-1-61668-582-9

Chapter 6

DEFICIT IRRIGATION STRATEGIES: USE OF STRESS PHYSIOLOGY KNOWLEDGE TO INCREASE WATER USE EFFICENCY IN TOMATO AND POTATO

Radmila Stikić[1], Sladjana Savić[2], Zorica Jovanović[1], Sven-Erik Jacobsen[3], Fulai Liu[3] and Christian R. Jensen[3]
[1] University of Belgrade, Faculty of Agriculture, Serbia
[2] Megatrend University, Faculty of Biofarming, Serbia
[3] Faculty of Life Sciences, University of Copenhagen, Tåstrup, Denmark.

ABSTRACT

Many vegetables, including tomato and potato, have high water requirements, so that supplemental irrigation is necessary for successful production. The predictions are that the demand for irrigation will increase considerably in years to come to alleviate the consequences of climate change and more frequent and severe droughts. Water is expected to become the main limiting factor in agricultural production. However, as a consequence of global climate changes and environmental pollution, water available for agriculture is often reduced. Emphasis therefore must be placed on crop physiology and crop management under dry conditions in order to make plants more efficient water users. Recent results have demonstrated that regulated deficit irrigation (RDI) and alternate partial root drying (ARD) are new deficit irrigation strategies which may decrease demand for agricultural use of water. The application of these techniques to crops such as tomato and potato has demonstrated benefits in terms of improved water-use efficiency and a stable yield, in addition to an increased yield quality. Both methods are developed on what is known on plant responses to drought stress and to the mechanism of root-to-shoot drought stress signalling. In this paper we review the research on the responses of tomato and potato to ARD and RDI irrigation strategies, with special emphasis on drought stress signalling mechanisms. Comparison of the beneficial effects of ARD and RDI were done with the aim to suggest which of these strategies may be used to get maximum tomato and potato water productivity and economic returns in dry environments.

INTRODUCTION

Tomato (*Lycopersicon esculentum* L.) and potato (*Solanum tuberosum* L.), both from the Solanaceae family, are the most widely grown vegetables in the world. Present world production of tomato is about 100 million t fresh fruit produced on 3.7 million ha. In the European Union, 17 million t fresh tomatoes are produced on 300,000 ha, and 2.3 million ha are planted with potato with an average yield of 30.3 t/ha (FAO, 2006). Water supply is essential for successful production of these crops, but water requirements depend on environmental conditions, soil type, developmental stage etc. Water requirement in tomato is 400-600 and in potato 300-700 mm ha^{-1} (Hanson and May, 2004; Wright and Strak, 1990).

In many countries, as a consequence of global climate changes and environmental pollution, water use in agriculture is reduced. Water supplies are also under pressure from other uses than agricultural, and saving of water resources and increasing agricultural productivity per unit of water ("more crop per drop") are becoming of strategic importance. Considerable emphasis is placed on crop physiology and crop management for dry conditions with the aim to increase crop water use efficiency (WUE) (Chaves et al., 2002; Liu et al, 2006a; Morrison et al., 2008; Wilkinson and Hartung, 2009). Use of deficit irrigation strategies is one of the actions that were suggested to increase WUE of agricultural crops (FAO, 2002). This review summarizes the information about deficit irrigation strategy application to tomato and potato plants with the emphasis on drought stress chemical signalling.

Drought Stress Long Distance Signalling-Hydraulic and Chemical

Drought induces a restriction of water supply which results in a reduction of tissue water content, stomatal conductance and growth. Plants have a mechanism for sensing the availability of water in the soil to regulate stomatal conductance and leaf growth accordingly. Jones (1980) was the first to suggest that this might involve the transfer of chemical information from the roots to shoots via the xylem. Such control has been termed non-hydraulic or chemical signalling. This distinguishes it from hydraulic signalling, which represents the transmission of reduced soil water via changes in the xylem sap tension (Dodd et al., 1996). Recent results suggest that chemical signals most probably dominate during early stages of stress, although hydraulic signals are produced and become more important under severe drought when leaf water potential declines (Bahrun et al., 2002; Schachtman and Goodger, 2008).

Abscisic Acid as a Signal

The plant hormone abscisic acid (ABA) plays a key role in chemical signalling. Substantial evidence obtained from a large number of experiments has demonstrated that reduced stomatal conductance, reduced leaf area and increased root biomass, are three of the major adaptive processes regulated by ABA. This can potentially be manipulated to improve crop water use efficiency (Wilkinson, 2004; Jacobsen et al. 2009). Cytokinins, a precursor of ethylene, pH and different ions also play roles in the root to shoot signaling under drought (Schachtman and Goodger, 2008).

Originally it was thought that ABA was synthesized in the leaf mesophyll cells and moved to the guard cells and to the meristem (Mansfield and Davies, 1981). Zhang and Davies (1990) provided evidence that increased xylem ABA concentration was root-sourced and quantitatively sufficient to account for the physiological changes observed, that is reduction of stomatal conductance and leaf growth rate. Similar evidence has come from numerous field and climate chamber experiments. The compartmental redistribution and long-distance transport of ABA influenced by environmental changes in the rhizosphere has been formulated in a biomathematical model (Slovic et al., 1995). However, the source of ABA (root or shoot) is still under debate. Experiments with drafted plants (mainly tomato) challenge the view that root-source chemical signals can be responsible for soil-drying-induced stomatal closure or reductions in leaf growth in the shoot (Dodd et al., 2009). Soil-drying-induced ABA signals arising in the shoot, upon receipt of other signals from the root, are responsible for the regulation of stomatal aperture (Wilkinson and Hartung, 2009). Recent research has also been centered on the mechanisms of ABA signal perception and transcription, and ABA-regulated gene expression (Novikova et al., 2009; Wilkinson and Hartung, 2009). Signalling of ABA could also be regulated by different factors including pH, root growth promoting rhizobacteria, ions and apoplastic β-glucosidases in the leaves (Wilkinson 1999; Bahrun et al., 2002; Davies et al., 2005; Jiang and Hartung, 2007).

ABA may conjugate to an ABA-glucose ester (ABA-GE) which has been referred not only to an ABA transported form, but also as an additional root-to-shoot stress signal. ABA-GE conjugated mainly with β-glucosidic linkages, were detected in the xylem of well-watered plants and under drought. Their importance for drought stress signalling is not yet well understood (Schachtman and Goodger, 2008).

pH as a Signal

Xylem sap and apoplastic pH is highly dependent upon water availability to the roots. Increases in xylem sap pH have been recorded in many plant species, including tomato, grown in drying soils, and may act as a root signal (Wilkinson, 1999).

The synergistic interaction between pH and ABA is under investigation. Wilkinson and Davies (2008) have demonstrated that spraying tomato plants with an alkaline buffer may mimic drought effects, and induce stomatal closure and leaf growth reduction via an effect on ABA signalling. However, the reaction of stomata on mutual effects of ABA and pH seems to be species or genotype dependent (Prokic et al., 2006). Sharp and Davies (2009) demonstrated that the sap alkalization response to soil drying is far from universal in higher plant species. They concluded that the species with alkalized sap to control internal water status were the most isohydric species, whereas none of the species exhibiting anisohydric responses alkalized xylem sap under drought stress. They also demonstrated that plants have conserved the ability to respond to changes in xylem pH and redistribute ABA, even if they do not currently utilize the mechanism when exposed to drought stress. To understand the role of pH in drought stress signalling it is important to understand the mechanistic basis of the change in xylem pH under drought and the relationship between xylem pH and the pH of the apoplast of the leaves.

Cytokinins and Ethylene as Signals

Other hormones, especially cytokinins and ethylene, are important for signalling. Cytokinins are considered as negative root-to-shoot stress signals whose production and transport decrease as the soil dries. They are supplied by turgid roots antagonistically to ABA, and they promote stomatal opening and shoot growth (Incoll et al., 1989). Stoll et al. (2000) showed a significant reduction of zeatin (Z) and zeatinriboside (ZR) in the shoots and stems of alternate root drying (ARD) irrigated grapevine. However, there is a lot of uncertainty of the intensity of the cytokinin signal in the xylem and universality of stomatal reaction to their decreased concentration in xylem in different plant species (Dodd, 2003; Vystskaya et al., 2009).

There is a lot of evidence in the literature that ethylene is a regulator of plant growth under drought. Several studies have pointed out the importance of the interaction between ABA and ethylene in plant development. Sharp (2002) proposed the following role of ABA on ethylene: (a) normal level of endogenous ABA are required to maintain shoot development, particularly leaf expansion, in well-watered tomato and Arabidopsis plants independently of effects on plant water balance, and (b) the impairment of leaf and shoot growth caused by ABA deficiency is at least partly attributable to increased ethylene production. The results of Sobeih et al. (2004) with wild-type and transgenic tomato (ACO1AS- with low stress-induced ethylene production) showed no inhibition of leaf growth under ARD, indicating that increased ethylene production under ARD is responsible for leaf growth inhibition.

Davies et al. (2005) reviewed the central role played by ABA in several signaling cascades and the possible interactions between ABA and long-distance signalling by other hormones and chemical components in the xylem or the apoplast.

The Practical Exploitation of Plant Stress Chemical Signalling Through the Use of Deficit Irrigation Strategies

In the past decade, a large number of field studies have validated the existence of drought signalling in the field, supporting the studies under controlled conditions. In recent years, significant effort was done on the practical exploitation of signalling mechanisms by deficit irrigation strategies, such as regulated deficit irrigation (RDI) and alternate partial root drying (PRD or ARD). These deficit irrigation methods have been studied in grapevine, vegetables, fruit trees, ornamentals and cereals (Costa et al., 2007; Kang and Zhang, 2004).

Regulated Deficit Irrigation (RDI)

In regulated deficit irrigation (RDI) the entire root zone is irrigated with an amount of water less than the potential evapotranspiration during specific periods of the crop cycle (English and Raja, 1996). The method is based on the observation that photosynthesis is less sensitive to drought stress than transpiration, so that consequently WUE is increased.

However, deficit irrigation may reduce yield and for optimal application of RDI plant and soil water status must be monitored in order to maintain a plant water regime within a certain degree of drought stress that could not limit yield. This is the major difficulty when RDI is applied under field conditions. If RDI is managed carefully, the negative impact on yield could be avoided, and assimilate partitioning to fruits sustained. Optimal RDI in water-limited situations might increase WUE, and maintain or even increase farmers' profits.

Alternate Partial Root Drying (PRD or ARD)

Alternate partial root drying (ARD) is a further development of RDI. Under ARD only half of the root zone is irrigated while the other half is allowed to dry out. The treatment is then reversed, allowing the previously well-watered side of the root system to dry while irrigating the previously dry side. The principle behind ARD is that irrigating part of the root system keep the leaves hydrated. When exposing the remaining part of the roots to soil drying, synthesis and transport of chemical signals (particularly ABA) from roots to the shoot via the xylem is triggered (Loveys et al., 2000). The frequency of the switch depends on soil type, genotype, rainfall and temperature. Most often the ARD cycle is 10-15 days (Davies et al., 2000; Stoll et al. 2000). Compared to RDI, implementing the ARD technique is simpler, requiring only the adaptation of irrigation systems to allow alternate wetting and drying of part of the rootzone (Loveys et al., 2000).

Effects of ARD on plant physiology are different from RDI because wet roots under ARD sustain shoot and fruit turgor important for growth (Mingo et al., 2003). Triggering partial stomatal closure under ARD irrigation may prevent excessive water loss and lead to better water balance of the plants, and this also may prevent the metabolic inhibition of CO_2 assimilation that otherwise would occur if drought stress was allowed to develop extensively (Chaves et al., 2002).

The ARD method was first tested in a vineyard of South Australia by Loveys (1991). These results showed that ARD reduced vigor, and increased yield and fruit quality. Excessive plant vigor is a major problem for many fruit crops, since the use of assimilates in leaf growth restricts fruit set and development, and may cause shading and more fungal diseases (Morison et al., 2008). The main benefit of this method is an increased WUE with a minor negative effect on yield, and in some cases with an improved quality (Davies et al., 2000; dos Santos et al., 2003; Zegbe-Domínguez et al., 2003, 2004; Liu et al., 2006a,b; Tahi et al., 2007). ARD was applied in the field as controlled alternate drip or subsurface drip irrigation, controlled alternate border irrigation, and controlled alternate furrow irrigation. However, a recent analysis challenges the beneficial effect of ARD technique and water saving in field conditions (Sadras, 2009).

Tomato and Potato Under Deficit Irrigation

Drought is a major limiting factor of tomato and potato productivity and therefore their successful production in many areas of the world requires irrigation. Results in recent years showed that deficit irrigation could be an efficient strategy to operate under conditions of water scarcity for these vegetables as for other crops. Many studies have demonstrated that RDI and ARD have significant effects on tomato and potato growth, yield and yield quality.

Tomato

Almost all studies confirmed significant increase in water use efficiency (WUE) in tomato under deficit irrigation strategies, and a significant saving of water for irrigation that could be more than 50%.

In consistency with the theory of shoot-to-root signalling effects on plant growth, most of the results also showed that the effect of ARD is smaller in fruits compared to vegetative parts of ARD treated plants (Davies et al., 2000). For explaining these different effects of ARD on shoot and fruit growth Davies et al. (2000) suggested that xylem area reduction, which occurred during tomato fruit development, might restrict the free movement of ABA from a shoot to fruit. Therefore, the ABA was not accumulated in fruit epidermis as in the leaves and, consequently, shoot growth was more reduced than the fruits. This so-called hydraulic hypothesis of shoot-to-fruit signalling is supported with anatomical observations and hydraulic resistance measurements of the tomato pedicel (Van Ieperen et al., 2003). Results of Rančić et al. (2008) confirmed the theory of hydraulic isolation of tomato fruits because they showed that ARD treatment affected the area of functional xylem in the tomato fruit pedicel, especially in the zone near the fruit. However, Davies et al. (2002) suggested that the extent of the hydraulic connection between fruit and the vegetative plant may depend on how the plant is grown. Mingo et al. (2003) demonstrated that restrictions in tomato fruit growth rate in plants growing in a drying soil can occur in the absence of any changes in fruit cellular turgor. These results suggest that signals born within the xylem can travel from root-to-shoot and shoot-to-fruit to elicit a powerful regulatory effect on fruit cell expansion.

Results from several experiments demonstrated that tomato fruit growth was more reduced by RDI than by ARD treatments (Davies et al., 2000; Mingo et al., 2003; Topcu et al., 2007). Different effect of ARD and RDI on fruit growth might be explained by differences in the activities of cell wall enzymes in the exocarp of fruits. During maturation or cessation of tomato fruit growth many results showed an increase in the activity of cell wall associated enzyme peroxidase or its different isoforms (Thompson et al., 1998; Andrews et al., 2001).

Our results (Savić et al., 2008) showed different roles of tomato fruit exocarp cell wall peroxidase in RDI and ARD treated plants (Fig.1). These results showed that in the fruits of RDI plants, peroxidase activity began to increase from the 15th day and this phase coincided with the phase when fruit growth started to decline. The peak of enzyme activity was reached in the phase of mature green fruits (36th day) when fruit growth rate was minimal. Contrarily, the results for ARD treated plants showed that a rapid increase in peroxidase activity occurred after 36th day. The peak of enzyme activity was at 43rd day at the end of fruit ripening in ARD treated plants

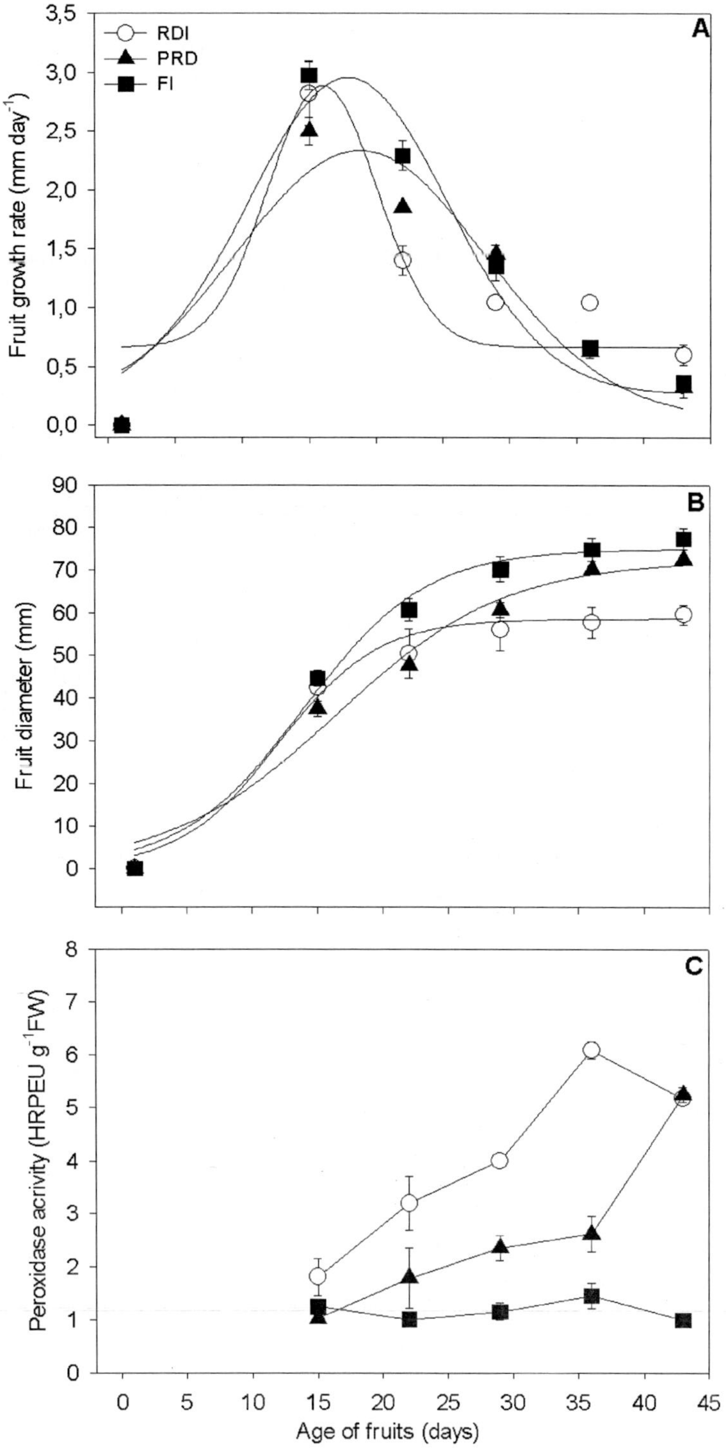

Fig 1. The effects of different irrigation treatments (FI-■, PRD-▲ and RDI-○) on fruit growth rate (A), fruit diameter (B) and cell wall peroxidase activity in exocarp (C) during tomato fruit development. (Reprinted with permissions of Scientia Horticulturae)

. These results indicate that increase in enzyme activity of RDI treated plants peroxidase may play a role in restricting fruit growth rate, although the increase in enzyme activity during ripening of ARD treated fruit indicated that peroxidase may also control fruit maturation by inducing more rapid process. This and other hypotheses concerning fruit growth under ARD and RDI need to be further investigated with fruits of other tomato genotypes or other crops.

An important mechanism of plant response to ARD, in addition to an increased WUE, may be an increased root growth. Mingo et al. (2004) have shown a significant increase in root growth of ARD-grown tomatoes associated with the re-watering of the previously dry compartment. The benefits of increased root growth might be significant under field conditions. Enhanced root growth will increase the plants' ability to explore a greater soil volume potentially increasing soil water and nutrient acquisition (Kang and Zhang, 2004)

Although almost all studies have confirmed that both ARD and RDI might be successfully applied as water saving strategy, there is a discrepancy in published results concerning the feasibility of these techniques to maintain yield. It has been demonstrated that under ARD yield of tomato was maintained and improved in both processing (Zegbe-Dominguez et al., 2003, 2004; Zegbe et al., 2006a,b) and fresh tomatoes (Kirda et al., 2004). On the contrary, there are results showing a decreased yield under ARD and RDI (Topku et al., 2007; Tahi et al., 2007).

Scheduling of RDI and ARD must be done by taking into account the phenological stage in which the vegetables are most vulnerable to soil water deficits. Reproductive tomato growth, especially flower and fruit set phases, are more sensitive to water deficit than the vegetative growth (Srinivasa et al., 2000). Results of Pulupol et al. (1996) showed that drought during these phases may induce abortion of flowers and as a result a lower number of fruits was produced, and lower yield was obtained. Results of Zegbe et al. (2006b) confirmed that the effect of ARD also depends on tomato phenological phases and that knowledge of the tomato response to drought is essential to achieve a minimal reduction of yield. When the plants were under ARD during the vegetative phase the yield was sustained although in the phase of fruit set yield reduction was significant.

Several studies have demonstrated that ARD or RDI can significantly contribute to tomato fruit quality. Pulupol et al. (1996) showed higher concentrations of soluble sugars and higher fruit color intensity in deficit irrigated tomato. However, this improvement in fruit quality was not enough to compensate for the reduction in yield with DI irrigation. Kirda et al. (2004) showed that under ARD an important fruit quality parameter, water soluble dry matter was increased. It consists of sugars, mainly glucose and fructose, and organic acids, mostly citric and malic acids, which have large influence on overall flavor of tomato fruits. Zegbe et al. (2006a) showed for processing tomato that a lower percentage of dry mass was partitioned into the ARD fruit, which had lower water content and higher total soluble solids concentration than the FI fruit. Blossom-end rot incidence was higher in ARD than in FI fruit, although their calcium concentration was the same. Our earlier results with ARD grown tomato also demonstrated an increase in fruit quality mainly due to an increased sugar content (Stikic et al., 2003). According to Zegbe et al. (2006b) the desirable quality attributes for processing tomatoes are high total soluble solids concentration, high dry matter concentration of the cortex, and low fruit water content because less energy would be needed to dry the fruit by the processing industry.

Our recent experiments with fresh tomato cv. *Abbelus* grown under ARD and FI in commercial polytunnel conditions are presented on Table 1. These results show that the number of harvested fruits per plant was reduced under ARD irrigation compared to FI, although fruit diameter, fresh and dry weight were similar. Due to the smaller number of fruits, yield of ARD plants was ca. 20% lower than FI. Similarly, Davies et al. (2000) reported that whole plant yield in ARD treated tomato variety *Ailsa Craig* was reduced by 15% and WUE increased, while Kirda et al. (2004) found 17-22% yield reduction with 50% deficit ARD irrigation of greenhouse tomato. Similarly to other authors (Davies et al., 2000; Kirda et al., 2004; Lei et al., 2009) our results also showed a significant increase in WUE. In addition to an increase in WUE, tomato fruit produced under ARD treatment had a higher total soluble solids, higher titrable acidity and lycopene content as well as antioxidant activity than FI fruits. Fruit quality improved in ARD treated tomato crops, especially due to increased antioxidative activity and lycopene content, which are very important nutrient characteristics (Middleton et al., 2000). This could compensate for the reduction in yield that was recorded in our tomato ARD experiments (Table 1).

Table 1. Yield and yield quality and irrigation water use efficiency (IWUE) of tomato cultivar Abbelus grown under the full irrigation (FI) and partial root zone (PRD) regime.

Traits	FI	PRD
No of fruit per plant	33.30	27.00
Average fruit diameter (mm)	65.21	61.24
Average fruit FW (g)	122.90	123.09
Average fruit DW (g)	6.43	6.59
Yield (t ha^{-1})	61.39a	49.85b
IWUE (kg m^{-3} H_2O)	45.47a	58.30b
Total soluble solids (°Brix)	5.50	5.80
Titrable acidity (citric acid μmol g^{-1} FW)	26.00	28.80
Lycopene (mg kg^{-1} FW)	15.40	14.10

Different letters show significant differences at $P<0.05$ between irrigation treatments (a and b)

Recently, Tahi et al. (2007) and Lei et al. (2009) showed that ARD-treated tomato plants accumulated more soluble and cell-wall bound phenolic compounds and changed the activity of several oxidative enzymes. They explained these reactions as defence mechanisms against oxidative stress that could be induced by ARD treatment.

Potato

RDI and ARD techniques are already being used for irrigation of potato. For most potato cultivars even short periods of drought stress can cause significant reductions in tuber yield and quality, and therefore irrigation is required for high yield (Onder et al., 2005). Drought stress imposed during early development (flowering and tuber set) might cause the greatest essential yield reduction (Yuan et al., 2003). RDI is difficult to manage in potato because the shallow and sparse root system of the potato plant makes it vulnerable to water deficits (Opena and Porter, 1999). Lynch et al. (1995) demonstrated that RDI irrigation is difficult to apply to potato because even short period of stress during tuber set phase might have a significant reducing effect on yield. However, Fabeiro et al. (2001) showed that RDI applied during growth and tuber bulking and resulted in the yield that was not significantly different from fully irrigated potatoes.

ARD or PRD irrigation has been successfully implemented in potato. Shahnazari et al. (2007; 2008) demonstrated a significant saving of water for irrigation and increase in water use efficiency when PRD was applied after tuber initiation until final harvest. In this period 30% of the irrigation water could be saved as compared to the fully irrigated treatment (FI) resulting in yield maintenance and significant increase in irrigation WUE (see treatment PRD2 in Table 2). Also the marketable yield was increased due to better size distribution as 15% more of the potatoes were within the marketable size 4-6 cm diameter in the PRD2 treatment (Shahnazari et al., 2007). PRD and DI treatments improved soil nitrogen availability late in the growing season maintaining top 'greeness' to a greater extent (Shahnazari et al., 2008).

Table 2. Yield and irrigation water use efficiency (IWUE) in 2005 and 2006 in potatoes grown under temperate conditions under automated rain out shelter (Modified after Shanazari et al., 2007; 2008).

Year	2005		2006			
Irrigation strategy	FI	PRD2	FI	DI	PRD1	PRD2
Irrigation, mm	267	206	285	214	208.1	240
Yield, FW (t ha^{-1})	60	66	42.4 ^{A}	36.8 ^{B}	33.3 ^{B}	41.7^{A}
IWUE (kg ha^{-1} mm-$^{1)}$	224 ^{B}	320 ^{A}	148.8^{B}	172.0^{A}	160.5^{B}	74.2^{A}

Treatments: the PRD1 and DI treatments received 70% of the FI treatment during whole growing season. The PRD2 treatment received 70% of the FI treatment from after tuber initiation until final harvest.

The efficiency of dry matter accumulation by tubers depends on the rate of photosynthesis and efficiency of assimilate partitioning to the tubers (bulking rate) and maturity period (Tekalign and Hammes, 2005). Initiation of tubers has been established to lead to a preferential partitioning of assimilates to tubers (Kooman and Rabbinge, 1996). Assimilate shifting from leaves and stems to tubers is important for tuber growth and could be regarded as the cause of leaf growth reduction and root growth cessation (Ewing and Struik, 1992). Our recent results (Jovanovic et al., 2008, 2010) demonstrate a tendency in ARD irrigated plants to allocate dry matter from leaves and stems towards tubers, and also increase N-use efficiency. Similar partitioning tendency was also found in other potato ARD experiments (Liu et al., 2006b; Shahnazari et al., 2007). In addition to increased WUE, nutrient use efficiency might be also increased. Shahnazari et al. (2008) found lower residual N in ARD than in FI treatments confirming that ARD treatment may improve soil nitrogen availability during the late phases of potato growing season indicating a higher N mineralization (Wang et al., 2009; Jovanovic et al., 2008; 2010).

We conducted potato ARD field trials with cultivar *Liseta* with the aim to compare static ARD management approach with a dynamic system when amounts of water irrigated in ARD were changed according to the plant growth phases (Jovanovic et al., 2010). In the 2007 season ARD plants received 70% of fully irrigated (FI), and in 2008 70% of ARD was decreased to 50% in the last 3 weeks of the irrigation period to further save water.

Comparison of the effects of ARD and FI irrigation technologies did not show significant differences in yield and marketable yield (Table 3). However, differences in IWUE between ARD and FI were larger in 2008 seasons when dynamic ARD was applied than in 2007 years when a static approach of ARD irrigation was used (by 14%). This study was also done with the aim to evaluate the effect of dynamic ARD criteria on yield quality. Quality assessment data did not show any significant differences in reducing sugars content and titrable acidity of ARD and FI irrigated crops. In both seasons a significant increase in antioxidant activity (by 3 μmol TE per 100g FW) and in starch content (from *ca.* 13 to 15 % FW) was found in ARD in comparison to FI tubers. These results similarly to the results for tomato fruits (Table 1) confirmed that ARD irrigation could be beneficial from the aspects of health-promoting value of potato tubers.

Table 3. Yield and yield quality and irrigation water use efficiency (IWUE) of potato grown under the PRD regime during 2007 and 2008 seasons.

Traits	$FI_{2007\ season}$	$PRD_{-2007\ season}$	$FI_{-2008\ season}$	$PRD_{-2008\ season}$
Yield (t ha^{-1})	45.31 ab	41.78 a	53.19 c	50.46 bc
Marketable yield (t ha^{-1})	43.10	40.00	42.52	41.28
IWUE (kg ha^{-1} mm^{-1})	241.00 a	334.27 b	236.40 a	380.14 c
Reducing sugars (g kg^{-1}FW)	3.38 a	3.11 a	3.10a	2.66b
Titrable acidity (%)	0.036 a	0.039 a	0.020 b	0.020b
Starch (% FW)	13.72 a	15.02 bc	13.45 ab	15.76 c
Antioxidant activity (μmol TE 100 g^{-1} FW)	19.92 a	22.63 b	19.13 a	22.81 b

Different letters show significant differences at $P<0.05$ between irrigation treatments and seasons (Modified from Jovanovic et al., 2010).

Greater savings of water for irrigation and increase in IWUE when 70% ARD was replaced by 50% ARD during drought tolerant growth stages indicate that dynamic approach is a promising strategy for saving water for potato irrigation. Further research of ARD dynamic approach should be done to a wider variety of field-grown crops and on a larger scale.

ADVANTAGES AND DISADVANTAGES OF RDI AND ARD

The main risk in using RDI irrigation is a reduction of tomato and especially potato yield. By using RDI the grower must monitor plant and soil moisture status and precisely regulate and direct water during irrigation. The timing of RDI irrigation is important in order to maintain yield, and it should be applied during growth phases with tolerance to drought stress.

In most studies ARD has been superior to RDI as a strategy for irrigation of tomato and potato. In addition to an increased WUE the benefits of ARD technology are in an increase in quality of fruits or tuber, and an increased root growth. Increased yield quality in many different crops could minimize the negative effects of ARD on yield (Kang and Zhang, 2004). According to Davies and Hartung (2004) possible limitations to the use of ARD could be nutritional and salinity problems associated with the limited soil volume occupied by plant roots in soil that is periodically dried. They suggest that for overcoming this problem fertigation should be applied via drip lines, and soil profiles could be flushed periodically with excess water. This will provide the possibility of optimising nutrient use efficiency and using nutrient treatments to perturb the plant's signalling system to the benefit of the grower.

ARD irrigation must be switched regularly from one side of the root to the other to keep roots in dry soil alive and sustain the supply of root signals. The time of switching is a major issue of ARD irrigation. This is one of the main reasons that Sadras (2009) challenges the beneficial effects of ARD technique. He concludes that substantial improvement in water use efficiency can be achieved by closely monitoring deficit irrigation, without the complexity and additional cost of ARD. However, two decades ago Zhang and Davies (1990) suggested that the early wilting of older leaves may indicate the right time for irrigation. Usually in the most applied ARD systems the switching is based on soil water depletion or stomatal reactions. A precise indicator for this method is still missing.

Liu et al. (2008) developed a model for prediction of when to shift side based on accumulation of xylem ABA in potatoes grown in pots. Their model simulation indicated that irrigation should be shifted between the two sides when Ψsoil of ARD dry side decreased to −80 kPa, and xylem ABA of ARD plants reach a peak of 150 nM. However, the xylem ABA value predicted by the model was significantly lower than the measured value and, therefore, further verification of the proposed model should be done especially with plants grown under field conditions.

To quantify the level of DI below full irrigation, the full crop ET requirements might be calculated using the Penman–Monteith equation times crop coefficient (Allen et al., 1998). Fereres and Soriano (2007) concluded that it is possible to perform deficit irrigation in water-limited situations when sufficient knowledge exists to manage RDI optimally and hereby increase farmers' profits while reducing irrigation water use. They also concluded that substantial improvement in water use efficiency can be achieved by closely monitored deficit irrigation, without the complexity and additional cost of the partial root-zone drying. However, the technical future development may ease the use of the ARD technology and hereby its implementation may be less costly in the future.

Future Perspectives and Conclusions

It is clear that successful implementation of RDI and ARD strategies will provide the potential to improve the production of tomato and potato under water scarcity. It could be expected that the potential improvements in WUE, fruit and tuber quality will be beneficial economic output and provide a significant gain in on-farm profitability, particularly during periods of limited water supply. The techniques may also be used as fertilization strategies. However, the application of RDI and ARD will be easier to use once we understand better the perception and transduction of root-to-shoot signals in different soils, climatic conditions and different genotypes. This will allow a sustainable production of tomato and potato and other vegetables under water scarcity in the future.

Acknowledgments

This study was supported by EU Commission (FP6 projects WATERWEB, SAFIR and CROPWAT) and Serbian Ministry of Science (TR 20025).

References

Allen R.G., Pereira L.S., Raes D., Smith M. (1998) Crop evapotranspiration. Guidelines for computing crop water requirements - *FAO Irrigation and drainage paper 56.*

Andrews, J., Adams, S.R., Burton, K.S., Edmondson, R.N. 2001. Partial purification of tomato fruit peroxidase and its effect on the mechanical properties of tomato skin. *J. Exp. Bot. 53,* 2393-2399.

Bahrun A., Jensen C.R., Asch, F., and Mogensen, V.O. 2002. Drought-induced changes in xylem pH, ionic composition and ABA concentration act as early signals in field grown maize (Zea mays L.). *Journal of Experimental Botany, 53,* 251-63.

Chaves, M.M., Pereira, J.S., Maroco, J.P., Rodrigues, M.L., Ricardo, C.P.P., Osorio, M.L., Carvalho, I., Faria, T., Pinheiro, C. 2002. How plants cope with water stress in the field: Photosynthesis and growth. *Ann. Bot. 89,* 1-10.

Costa, J.M., Ortuno, M.F., Chaves, M.M. 2007. Deficit Irrigation as a strategy to save water: Physiology and potential application to horticulture. *J. Integ. Plant Biol. 49,* 1421–1434.

Davies, W.J., Bacon, M.A., Thompson, D.S., Sobeigh, W., Rodriguez, L.G. 2000. Regulation of leaf and fruit growth in plants in drying soil: exploitation of the plant's chemical signalling system and hydraulic architecture to increase the efficiency of water use in agriculture. *J. Exp. Bot. 51,* 1617-1626.

Davies W.J., Wilkinson, S., Loveys, B. 2002. Stomatal control by chemical signalling and the exploitation of this mechanism to increase water use efficiency in agriculture. *New Phytol. 153,* 449-460.

Davies, W.J., Hartung, W. 2004. Has extrapolation from biochemistry to crop functioning worked to sustain plant production under water scarcity? In: *Proceeding of the 4th International Crop Science Congress,* 26 Sept.-1 Oct. 2004, Brisbane, Australia.

Davies, W.J., Kudoyarova, G., Hartung, W. 2005. Long distance ABA signalling and its relations to other signalling pathways in the detection of soil drying and the mediation of the plant's response to drought. *J. Plant Growth Regul. 24,* 285-295.

Dodd, I.C. 2003. Hormonal interactions and stomatal responses. *J. Plant Growth Regul. 22,* 32–46.

Dodd, I.C., Stikić, R., Davies, W.J. 1996. Chemical regulation of gas exchange and growth of plants in drying soil in the field. *J. Exp. Bot. 47,* 1475-1490.

Dodd, I.C., Theobald, J.C., Bacon, M.A., Davies, W.J. 2006. Alternation of wet and dry sides during partial rootzone drying irrigation alters root-to-shoot signalling of abscisic acid. *Funct. Plant Biol. 33,* 1081-1089.

Dodd, I.C., Theobald, J.C., Sarah, K., Richer, S.K., Davies, W.J. 2009. Partial phenotypic reversion of ABA-deficient flacca tomato (*Solanum lycopersicum*) scions by a wild-type rootstock: normalizing shoot ethylene relations promotes leaf area but does not diminish whole plant transpiration rate. *J. Exp. Bot. 60*, 4029-4039.

English, M.J., Raja, S.N. 1996. Perspectives on deficit irrigation. *Agric. Water Manage. 32,* 1-14.

Ewing, E.E., Struik, P.C., 1992. Tuber formation in potato. Induction, initiation and growth. *Hort. Rev. 14,* 89-198.

FAO 2006. FAO statistical databases. *Agriculture Data Collection* http://faostat.fao.org/. FAO, Rome, Italy.

FAO 2002. Deficit Irrigation Practices. *Water Reports 22,* Rome, Italy.

Fabeiro, C., Olalla, F.M.D., de Juan, J.A. 2001. Yield and size of deficit irrigated potatoes. Agric. *Water Manage. 48:*255-266.

Fereres, E., Soriano, M.A. 2007. Deficit irrigation for reducing agricultural water use. *J. Exp. Bot. 58,* 147-159.

Hanson, B., May, M. 2004. Effect of subsurface drip irrigation on processing tomato yield, water table depth, soil salinity, and profitability. *Agric.Water Manage. 68,* 1–17.

Incoll, L., Ray, J.P., Jewer, P.C. 1989. Do cytokinins act as root to shoot signals? In: Importance of root to shoot communication in the responses to environmental stress. *Brit. Soc. Plant Growth Reg., Monograph No. 21,* pp185-200.

Jacobsen, S.-E., F. Liu & C.R. Jensen. 2009. Does root-sourced ABA play a role for regulation of stomata under drought in quinoa (*Chenopodium quinoa Willd.*). *Scientia Horticulturae 122,* 281–287

Jiang, F., Hartung, W. 2007. Long-distance signalling of abscisic acid (ABA): the factors regulating the intensity of the ABA signal. *J. Exp. Bot.,* Special Issue, June, 1-7.

Jones, H.G. 1980. Interaction and integration of adaptive responses to water stress: the implication of an unpredictable environment. In: *Adaptation of plants to water and high temperature stress.* Turner, N.C., P.J. Kramer, Ed. Wiley, New York, 353-365.

Jovanovic, Z., Brocic, Z., Stikic, R., 2008. Effects of Partial Root Drying on Nitrogen Distribution in Potato. Ital. *J. Agron. 3, Suppl 3,* 337-338, X European Congress for Agronomy, Bologna, Italy.

Jovanovic, Z., Stikic, R., Vucelic Radovic, B., Paukovic, M., Brocic, Z., Matovic, G., Rovcanin, S., Mojevic, M. 2010. Partial root zone drying increase WUE, N and antioxidant content in field potatoes. *Eur. J. Agron.* (in press).

Kang, S.Z., Zhang, J.H. 2004. Controlled alternate partial root-zone irrigation: its physiological consequences and impact on water use efficiency. *J. Exp. Bot. 55*, 2437-2446.

Kirda, C., Cetin, M., Dasgan, Y., Topcu, S., Kaman, H., Ekici, B., Derici, M.R., Ozguven, A.I. 2004. Yield response of greenhouse-grown tomato to partial root drying and conventional deficit irrigation. *Agric. Water Manage. 69,* 191-201.

Kooman, P.L., Rabbinge, R. 1996. An analyses of the relation between dry matter allocation to tuber and earliness of potato crop. *Ann. Bot. 77,* 235-242.

Lei, S., Yunzhou Q., Fengchao, J., Changhai S., Chao, Y., Yuxin, L., Mengyu L., Baodi, D. 2009. Physiological mechanism contributing to efficient use of water in field tomato under different irrigation. *Plant Soil Environ. 55,* 128–133.

Liu, F., Shahnazari, A., Andersen, M.N., Jacobsen, S.E., Jensen, C.R., 2006a. Physiological responses of potato (*Solanum tuberosum* L.) to partial root-zone drying: ABA signalling, leaf gas exchange, and water use efficiency. *J Exp. Bot. 57,* 3727-3735.

Liu, F., Shahnazari, A., Andersen,M.N., Jacobsen, S.E., Jensen, C.R., 2006b. Effects of deficit irrigation (DI) and partial root drying (PRD) on gas exchange, biomass partitioning, and water use efficiency in potato. *Sci. Horti. 109*, 113-117.

Liu, F., Song, R., Zhang, X., Shahnazari, A., Andersen, M.N., Plauborg, F., Jacobsen, S., Jensen, C.R. 2008. Measurement and modelling of ABA signalling in potato (*Solanum tuberosum* L.) during partial root-zone drying. *Environ. Exp. Bot. 63,* 385-391.

Loveys, B.R. 1991. Water use is knowledge of ABA physiology for crop improvement. In: *Abscisic acid, Oxford*: Bios Scientific Publishers (W.J. Davies., H.G. Jones, eds.), pp. 245-259.

Loveys, B.R., Dry, P.R., Stoll, M., McCarthy, M.G. 2000. Using plant physiology to improve the water use efficiency of horticultural crops. *Acta Horti. 537,* 187-199.

Lynch, D.R., Foroud, N., Kozub, G.C., Farries, B.C., 1995. The effect of moisture stress at three growth stages on the yield components of yield and processing quality of eight potato cultivars. *Am. Potato J.* 72, 375-386.

Mansfield, T.A., Davies, W.J. 1981. Stomata and stomata] mechanisms. In: Paleg LG, Aspinall D, eds. *The physiology and biochemistry of drought resistance.* Sydney: Academic Press, 315-46.

Middleton, E., Kandaswami, C., Theoharides, T.C. 2000. The effects of plant flavonoids on mammalian cells: implications for inflammation, heart disease, and cancer. *Pharmacol. Rev. 52,* 673-751.

Mingo, D., Bacon, M.A., Davies, W.J. 2003. Non-hydraulic regulation of fruit growth in tomato plants (*Lycopersicon esculentum* cv. Solairo) growing in drying soil. *J. Exp. Bot. 54,* 1205–1212.

Mingo, D.M., Theobald, J.C., Bacon, M.A., Davies, W.J., Dodd, I.C. (2004). Biomass allocation in tomato (*Lycopersicum esculentum*) plants grown under partial rootzone drying: Enhancement of root growth. *Funct. Plant Biol. 31,* 971–978.

Morison, J.I.L., Baker, N.R., Mullineaux, P.M., Davies, W.J. 2008. Improving water use in crop production. *Phil. Trans. R. Soc. B. 363,* 639-658.

Novikova, G.V., Stepanchenko, A.V., Nosov, N.S., Moshkov, I.E. 2009. At the Beginning of the Route: ABA Perception and Signal Transduction in Plants. Russ. *J. Plant Physiol. 56,* 727-741.

Onder, S., Caliskan, M.E., Onder, D., Caliskan, S., 2005. Different irrigation methods and water stress effects on potato yield and yield components. *Agric. Water Manage.* 73, 73–86.

Opena, G.B., Porter, G.A., 1999. Soil management and supplemental irrigation effects on potato. I. Root growth. *Agron. J. 91*, 426-431.

Passioura, J.B. 1988. Root signals control leaf expansion in wheat seedlings growing in drying soil. Aust. J. *Plant Physiol. 15*, 687-693.

Prokic, Lj, Jovanovic, Z., McAinsh, M., Vucinic, Z., Stikic, R. 2006. Species-dependent changes in stomatal sensitivity to abscisic acid mediated by external pH. *J. Exp. Bot. 57,* 675–683.

Pulupol, L.U., Behboudian, M.H., Fisher KJ 1996. Growth, yield and postharvest attributes produced under deficit irrigation. *Hort. Sci. 31*, 926–929.

Rančić, D., Pekić Quarrie, S., Savić, S, Stikić, R. 2008. Comparison of light and fluorescence microscopy for xylem analysis in tomato pedicels during fruit development. *J. Microsc. 232,* 618-622.

Sadras, V.O. 2009. Does partial root-zone drying improve irrigation water productivity in the field? A meta-analysis. *Irrig. Sci., Volume 27, Number 3,* 183-190.

dos Santos, T.R., Lopes, C.M., Rodrigues M.L., de Souza, C.R., Maroco, J.P., Pereira, J.S., Silva, J.R., Chaves, M.M. 2003. Partial root zone drying: effects on growth and fruit quality of field-grown grape vines (*Vitis vinifera). Func. Plant Biol. 30*, 663-671.

Savić, S., Stikić, R., Vucelić Radović, B., Bogičević, B., Jovanović, Z., Hadži-Tašković Šukalović, V. 2008. Comparative effects of regulated deficit irrigation (RDI) and partial root-zone drying (PRD) on growth and cell wall peroxidase activity in tomato fruits. *Sci. Horti. 117/1,* 15-20.

Shahnazari, A., Liu, F., Andersen, M.N., Jacobsen, S.E., Jensen, C.R., 2007. Effects of partial root zone drying (PRD) on yield, tuber size and water use efficiency in potato under filed conditions. *Field Crop Res. 100*, 117-124.

Shahnazari, A., Ahmadi, S.H., Laerke, P.E., Liu, F., Plauborg, F., Jacobsen, S.E., Jensen, C.R., Andersen, M.N., 2008. Nitrogen dynamics in the soil – plant system under deficit and partial root–zone drying irrigation strategies in potatoes. *Eur. J. Agron. 28*, 65-73.

Sharp, R.E. 2002. Interaction with ethylene: changing views on the role of abscisic acid in root and shoot growth responces to water stress. *Plant Cell Environ.* 25, 211-222.

Sharp, R.E., Davies, W.J. 2009. Variability among species in the apoplastic pH signalling response to drying soils. *J Exp. Bot. 60*, 4363–4370.

Schachtman, D.P., Goodger, J.Q.D. 2008. Chemical root to shoot signaling under drought. *Trends Plant Sci. 13,* 281-287.

Slovic S., Daeter W, and Hartung W 1995. Compartmental redistribution and long-distance transport of abscisic acid (ABA) in plants as influenced by environmental changes in the rhizosphere- a biomathematical model. *J. Exp. Bot. 46*, 881-894.

Sobeih, W.Y., Dodd, I.C., Bacon, M.A., Grierson, D., Davies, W.J. 2004. Long-distance signals regulating stomatal conductance and leaf growth in tomato (*Lycopersicon esculentum* L.) plants subjected to partial root-zone drying. *J. Exp. Bot. 55*, 2353-2363.

Srinivasa, R.N.K., Bhatt, R.M., Sadashiva, A.T. 2000. Tolerance to water stress in tomato cultivars. *Photosynthetica 38*, 465–467.

Stikić, R., Popović, S., Srdić, M., Savić, D., Jovanović, Z., Prokić, Lj., Zdravković, J. 2003. Partial root drying (PRD): a new technique for growing plants that saves water and improves the quality of fruit. Bulg. J. *Plant Physiol.*, Special Issue, 164-171.

Stoll, M., Loveys, B., Dry, P. 2000. Hormonal changes induced by partial rootzone drying of irrigated grapevine. *J. Exp. Bot. 51,* 1627-1634.

Tahi, H., Wahbi, S., Wakrim, R., Aganchich, B., Serraj, R., Centritto, M. 2007. Water relations, photosynthesis, growth and water use efficiency in tomato plants subjected to partial rootzone drying and regulated deficit irrigation. *Plant Bios. 141*, 265-274.

Tekalign, T., Hammes, P.S., 2005. Growth and productivity of potato as influenced by cultivar and reproductive growth. I Stomatal conductance, rate of transpiration, net photosynthesis, and dry matter production and allocation. *Sci. Horti. 105*, 13-27.

Thompson, D.S., Davies, W.J., Ho, L.C. 1998. Tomato fruit growth regulation by epidermal cell wall enzymes. *Plant Cell Environ. 21*, 589-599.

Topcu, S., Kirda, C., Dasgan, Y., Kaman, H., Cetin, M., Yazici, A., Bacon, M.A. 2007. Yield response and N-fertiliser recovery of tomato grown under deficit irrigation. Eur. *J. Agron. 26*, 64-70.

Van Ieperen, W., Volkov, V.S., Van Meeteren, U. 2003. Distribution of xylem hydraulic resistance in fruiting truss of tomato influenced by water stress. *J. Exp. Bot. 54*, 317-324.

Vysotskaya, L.B., Korobova, A.V., Veselov, S.Y., Dodd, I.C., Kudoyarova, G.R. 2009. ABA mediation of shoot cytokinin oxidase activity: assessing its impacts on cytokinin status and biomass allocation of nutrient-deprived durum wheat. *Funct. Plant Biol. 36*, 66–72.

Wang, H., Liu, F., Andersen, M.N., Jensen, C.R., 2009. Comparative effects of partial root-zone drying and deficit irrigation on nitrogen uptake in potatoes (*Solanum tuberosum* L.*). Irri. Sci. 27*, 434-448.

Wilkinson, S. 1999. pH as a stress signal. *Plant Growth Regul. 29*, 87-99.

Wilkinson, S., Davies W.J. 1997. Xylem sap pH increase: A drought signal received at the apoplastic face of the guard cell which involves saturable ABA uptake by the epidermal symplast. *Plant Physiol. 113*, 559-573.

Wilkinson, S., Davies, W.J. 2002. ABA-based chemical signalling: the coordination of responses to stress in plants. *Plant Cell Environ. 25,* 195-210.

Wilkinson, S. 2004. Water use efficiency and chemical signalling. In: Bacon MA, ed. *Water use efficiency in plant biology*. Oxford, UK: Blackwell Publishing Ltd, 75–112.

Wilkinson, S., Davies, W.J. 2008. Manipulation of the apoplastic pH of intact plants mimics stomatal and growth responses to water availability and microclimatic variation. *J. Exp. Bot. 59,* 619–631.

Wilkinson, S., Hartung, W. 2009. Food production: reducing water consumption by manipulating long-distance chemical signalling in plants. *J. Exp. Bot. 60*, 1885–1891.

Wright, J.L., Stark, J.C. 1990. Potato. In: Stewart, B.A. and D.R. Nielson (eds.), *Irrigation of Agricultural Crops,* pp: 859–889. American Society of Agronomy, Crop Science Society of America, Soil Science Society of America, Madison, USA

Zegbe-Domínguez, J.A, Behboudian, M.H., Lang, A., Clothier, B.E. 2003. Deficit irrigation and partial rootzone drying maintain fruit dry mass and enhance fruit quality in "Petopride" processing tomato (*Lycopersicon esculentum* Mill.). *Sci. Horti. 98*, 505-510.

Zegbe-Domínguez, J.A., Behboudian, M.H., Clothier, B.E. 2004. Partial rootzone drying is a feasible option for irrigation processing tomatoes. *Agric. Water Manage., Vol. 68*, Issue 3, 195-206.

Zegbe, J.A., Behboudian, M.H., Clothier, B.E. 2006a. Yield and fruit quality in processing tomato under partial rootzone drying. *Eur. J. Hort. Sci. 71*, 252–258.,

Zegbe, J.A., Behboudian, M.H., Clothierm B.E. 2006b. Responses of "Petopride" processing tomato to partial rootzone drying at different phenological stages. *Irrig. Sci. 24*, 203-210.

Zhang, J., Davies, W.J. 1990. Changes in the concentration of ABA in the xylem sap as a function of changing soil water status can account for changes in leaf conductance and growth. *Plant Cell Environ. 13*, 277-285.

Yuan, B., Nishiyama, S., Kang, Y. 2003. Effects of different irrigation regimes on the growth and yield of drip irrigated potato. *Agri. Water Manage. 63*, 153-167.

In: Horticulture in the 21st Century
Editors: Abigail N. Sampson, pp. 179-192
ISBN: 978-1-61668-582-9

Chapter 7

HIGH TEMPERATURE STRESS TOLERANCE AND THE CHANGES IN ANTIOXIDATIVE ABILITY IN MYCORRHIZAL STRAWBERRY PLANTS

Yoichi Matsubara[1]
Faculty of Applied Biological Sciences, Gifu University, Gifu 501-1193, Japan

ABSTRACT

In strawberry cultivation, high temperature in summer often restricts rooting, root growth and induces browning of roots and shoots during nursery production resulting in withering and causing great losses of plants. Then, it is necessary to establish stable technique for nursery production in strawberry cultivation during the high temperature period in summer including the problems of global warming. Arbuscular mycorrhizal fungus (AMF) has the effect of promoting host plant growth mainly by enhancing phosphorus uptake through symbiosis. On the other hand, promotion of active oxygen synthesis in plants occurs under several stress conditions including temperature stress, and antioxidative ability has relation with the resolution. In this study, the changes in activities of antioxidative ability through symbiosis with AMF in strawberry plants under high temperature stress were investigated. The strawberry (*Fragaria×ananassa* Duch., cv. Nohime) runner plants were treated with split root system method. Ten weeks after AMF (*Glomus mosseae*) inoculation, plants were transferred under 35°C-constant condition using growth chamber. Three weeks after high temperature treatment, dry weight of shoots and roots became greater in AMF plants than in non-AMF, resulting that growth promotion appeared in mycorrhizal plants. The severity of browning in shoots and roots caused by high temperature stress was eased in AMF plants than in non-AMF. In this case, the alleviation of root browning appeared in both AMF- and AMF+ root system, so that induced tolerance to high temperature stress was recognized in mycorrhizal plants. AMF plots showed higher activities in SOD and APX than non-AMF; the effect differs with plant portions and investigated periods. In addition, some AMF plots showed higher values in followed items; chlorophyll content, protein content, polyphenol content, ascorbic acid content. These effects occurred especially after high temperature stress treatment. From these findings, tolerance to high temperature stress

1 E-mail: ymatsu@gifu-u.ac.jp.

and the increase in antioxidative ability occurred in mycorrhizal strawberry plants, so that the activation of antioxidative ability is supposed to be closely associated with the high temperature stress tolerance.

INTRODUCTION

In strawberry forcing cultivation, high temperature in summer often restricts rooting, root growth and induces browning of shoots and roots during nursery production resulting in withering and causing great losses of plants, especially in the warm climate regions in Japan (Saito et al., 2003). Then, it is necessary to establish stable technique for nursery production in strawberry cultivation during the high temperature period in summer including ever-bearing strawberry cultivation. However, characteristics of high temperature tolerance in strawberry plants remain unclear. Recently, Tsukiji et al. (2001, 2002) reported the evaluation method of high temperature tolerance in strawberry plants using leaf disc thermostability, and high temperature acclimation might be effective for affording high temperature tolerance in strawberry plants. On the other hand, cooling medium system by heat of vapourization in strawberry bench culture was tried (Takaichi, 2004), but some problems such as high cost are remained.

Arbuscular mycorrhizal fungus (AMF) has the effect of promoting host plant growth mainly by enhancing phosphorus uptake through symbiosis (Marschner and Dell, 1994). As for strawberry plants, growth enhancement by AMF inoculation was reported in several combinations of AMF species and strawberry cultivars (Williams et al., 1992; Niemi and Vestberg, 1992; Varma and Schuepp, 1994). In addition, AMF increased yields in strawberry plants cultivated in elevated bench culture method (Li et al., 2009). Schenck and Smith (1982) and Haugen and Smith (1992) also reported that AMF plants produced more dry weight of shoots and roots than non-inoculated ones under high soil temperature in soybean and mung bean, respectively. On the other hand, Matsubara and Harada (1996) reported that high and low temperature stress tolerance occurred in mycorrhizal asparagus plants, and supposed that temperature stress tolerance might be induced by AMF colonization. Actually, the author confirmed high temperature stress tolerance in mycorrhizal strawberry plants (Matsubara et al., 2004), and suggested that the high temperature stress tolerance could be biologically controlled by AMF. However, it has been unclear the mechanisms on high temperature stress tolerance in mycorrhizal strawberry plants.

On the other hand, promotion of active oxygen synthesis in plants occurs under environmental stresses, such as temperature, drought and plant disease, and the antioxidative abilities have relation with the resolution (Fridovich, 1986; Jahnke, et al., 1991; Feierabend et al., 1992; Kang and Saltveit, 2002; Gulen and Eris, 2003; Li et al., 2004; Moghaddam et al., 2006; Sahoo et al., 2007). As for mycorrhizal plants, disease tolerance and the increase in superoxide dismutase (SOD) activity occurred in tomato (Pozo et al., 2002) and pepper (Garmendia et al., 2006), and drought tolerance and SOD increase appeared in lettuce (Lozano et al., 1996). However, it remains unclear the relationship between high temperature stress tolerance and antioxidative abilities in mycorrhizal plants.

In this study, the influence of AMF colonization on tolerance to high temperature stress and antioxidative abilities in strawberry plants were investigated.

MATERIALS AND METHODS

Inoculation of AMF

The strawberry (*Fragaria×ananassa* Duch., cv. Nohime) runner plants were treated with split root system method. Plant root system was separated into halves and each was placed in a different pot (10.5 cm in diameter) with bedding soil (autoclaved at 1.2 kg□cm^{-2} and 121°C for 1 h), and inoculated with *Glomus mosseae* (supplied by Idemitsukosan Co. Ltd.), according to Matsubara et al. (2004). Two weeks after AMF inoculation, the inoculated plants and the non-inoculated control plants were administered by mixed fertilizer (Long total 70 type, Asahi Kasei Co. Ltd.; N : P : K=13 : 11 : 13, 1 g per pot). Thirty plants per plot with no replication, were irrigated by the capillary watering method as regularly and grown in a greenhouse. Ten weeks after inoculation, plants were transferred under high temperature stress (HS) condition using growth camber; 35±2 °C, RH60±5%, 12hr-daylength.

ESTIMATION OF BROWNING IN SHOOTS AND ROOTS

Two days, 1 and 3 weeks after HS treatment, the ratio of browning symptoms in shoots and roots of ten plants per plot were investigated.

Evaluation of AMF Colonization Level

Roots of AMF plants were harvested before HS treatment (10 weeks after AMF inoculation), 2 days, 1 and 3 weeks after HS. The roots were stained to observe the colonization according to Phillips and Hayman (1970). Then, the rate of AMF colonization in 1-cm segments of lateral roots (RFCSL) was calculated. Hence, RFCSL expresses the percentage of 1-cm AMF-colonized segments to the total 1-cm segments of all the lateral roots; the number of total segments was approx. 50 per a plant. The average colonization was calculated from the values of three plants.

Analysis of Antioxidative Abilities

Ten weeks after AMF inoculation, 2 days, 1 and 3 weeks after HS treatment, plants were sampled and partitioned into compound leaves, petioles, crowns and main roots characterized by no colonization and freezed in liquid nitrogen. Protein content was determined by Bradford (1976), and chlorophyll content in compound leaves was estimated by Porra et al. (1989). Analysis of superoxide dismutase (SOD) activity, ascorbic peroxidase (APX) activity, polyphenol content and ascorbic acid content were carried out according to the methods of Beauchamp and Fridovich (1971), Li et al. (2004), Folin-denis (1915), Roe et al. (1948), respectively.

RESULTS AND DISCUSSION

AMF plants had higher dry weight of shoots and roots than did control ones among all the plots before and after HS treatment (Figure 1 and 2), suggesting that growth promotion effect appeared in mycorrhizal strawberry plants in all the investigated periods including under HS. AMF colonization levels reached more than 40% in all the sampling periods, and little fluctuation appeared before and after HS (Figure 3). The severity of browning in shoots and roots caused by successive stress of high temperature became lower in AMF plants than in control plants 3 weeks after HS treatment; no symptom appeared in 2 days and 1 week after HS treatment (Figure 4). In this case, severity of symptoms in roots was alleviated in both AMF+ and AMF- plots than in control, thus, AMF- roots also showed HS tolerance. From these findings, tolerance to high temperature stress and induced tolerance to high temperature were recognized in mycorrhizal strawberry plants. Schenck and Smith (1982) reported that AMF plants produced more dry weight of shoots and roots than non-inoculated ones under high soil temperature in soybean, and the effect differed with fungal species. Shrestha et al. (1995) demonstrated that satsuma mandarin trees colonized with AMF showed greater growth and photosynthetic activity compared with non-inoculated ones under high temperature in August, though the effect differed with fungal species. Matsubara et al. (1996) reported that the temperature stress tolerance occurred in mycorrhizal asparagus plants. On the other hand, Pozo et al. (2002) reported that in tomato plants with split root system, tolerance to *Phytophthora parasitica* appeared in both non-AMF inoculated roots and inoculated roots in AMF plants, so that induced tolerance to plant disease was recognized. The results in this study resembled with their findings.

In this study, chlorophyll content became higher in AMF plots than in control among all of the investigated periods except 2 days after HS treatment (Figure 5). Protein content increased in some AMF plots before and after HS compared with control, which depends on plant portions; no consistent difference appeared (Figure 6). As for antioxidative abilities, SOD activity became higher in all the AMF plots 3 weeks after HS except main roots (AMF+, AMF-) and in AMF- roots before HS (Figure 7). SOD activities in compound leaves and petiole of control prominently decreased after HS, however, the reduction was eased in AMF plots. AMF plots showed higher values in APX activity among most of the compound leaves and main roots 1 and 3 weeks after HS, though the activity showed no difference among the pots before HS (Figure 8). Polyphenol content was greater in AMF plots of compound leaves 3 weeks after HS, petiole 2 days after HS, AMF+ roots before HS (Figure 9). Ascorbic acid showed no difference between AMF and control plots in all the plant portions before HS, however, AMF plots became higher content in main roots (AMF+, AMF-) 2 days after HS, and compound leaves and AMF+ roots 3 weeks after HS (Figure 10). From these findings, tolerance to high temperature stress and the increase in antioxidative abilities occurred in mycorrhizal strawberry plants, though the changes in antioxidative abilities varied depending on plant portions and before and after HS, so that the activation of antioxidative abilities supposed to be associated with the temperature stress tolerance. However, we could not clarify the reasons for the increase in antioxidative abilities in mycorrhizal strawberry plants, further investigations are needed on this point.

Figure 1. Growth enhancement in mycorrhizal strawberry plants 10 weeks after AMF inoculation. Left, control; right, AMF (*Glomus mosseae*)-inoculated.

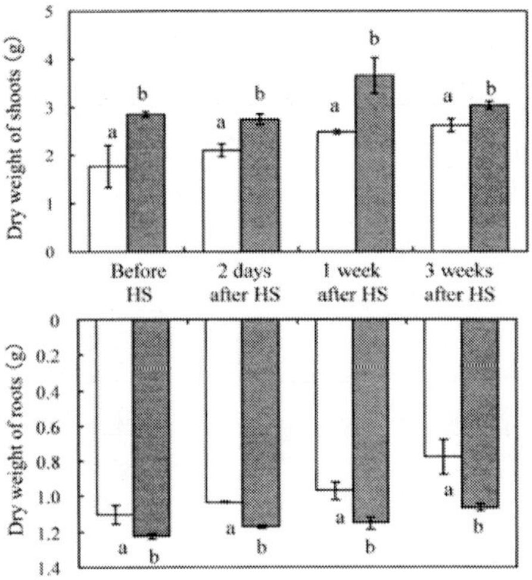

Figure 2. Influence of high temperature stress (HS) on dry weight of strawberry plants., control , AMF. Bars represent standard errors (n=5). Different letters represent significant difference among thetreatments according to Student's t-test (P<0.05).

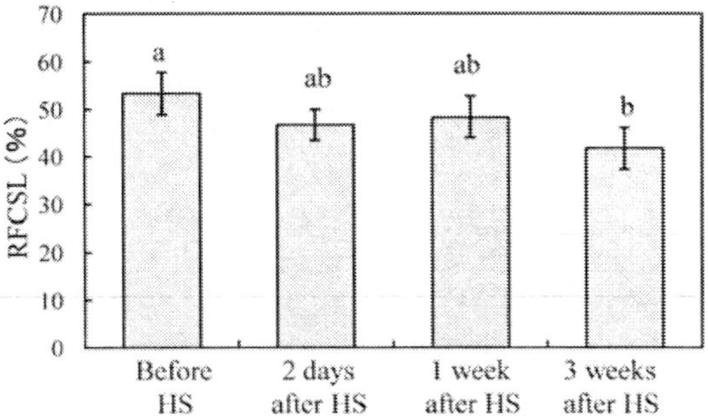

Figure 3. AMF colonization level (RFCSL) in strawberry plants under HS. Different letters represent significant difference amongthe treatments according to Tukey's test (P<0.05).

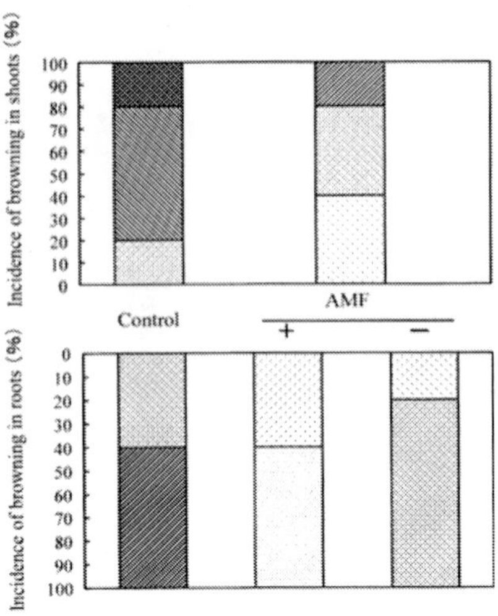

Figure 4. Incidence of browning in shoots and roots of strawberry plants 3 weeks after HS. Ratio of browning in shoots and roots: , □ 0~20; ▒ 20~40; ▨ 40~60; ▩ 60~80; ■ 80~100 (%).AMF+ : AMF-inoculated roots, AMF-: AMF-noninoculated roots.

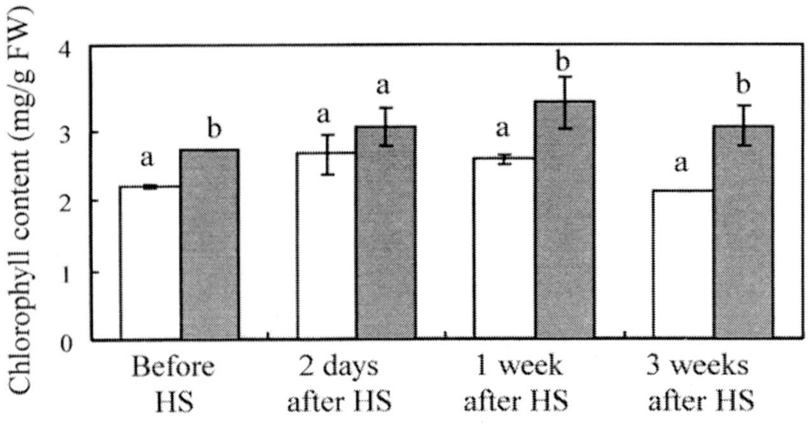

Figure .5. Chlorophyll content in compound leaves of strawberry plants under HS. □ Control ■ , AMF. Different letters represent significant difference among the treatments according to Tukey's test ($P<0.05$).

SOD play a primary role in defensive reactions and detoxify superoxide (O_2^-) among the antioxidative enzymes, thus, SOD activity is considered as the most important key enzyme in antioxidative abilities in plants (Fridovich, 1986). Garmendia et al. (2006) reported that tolerance to *Verticillium dahliae* and increase in SOD activity occurred in mycorrhizal pepper plants. The results in this experiment had similar patterns with the findings as the increase in SOD activity related to the stress tolerance, so that the antioxidative enzyme activity might be closely related with the tolerance to environmental stress (HS and plant disease, etc.) as a common factor in mycorrhizal plants. On the other hand, Pozo et al. (2002) demonstrated that tolerance to *Phytophthora parasitica* appeared in mycorrhizal tomato plants and SOD activity increased in both non-AMF inoculated and AMF-inoculated roots in split root system. In this study, non-AMF inoculated roots showed tolerance to HS, however, no increase in SOD activity occurred in AMF- roots. On the other hand, APX activity and ascorbic acid in AMF- roots increased compared with control, supposing that antioxidative abilities might be also some association with the induced tolerance to HS in mycorrhizal strawberry plants. Further detail investigation would be needed to clarify the relationship between induced tolerance to HS and antioxidative abilities in mycorrhizal plants.

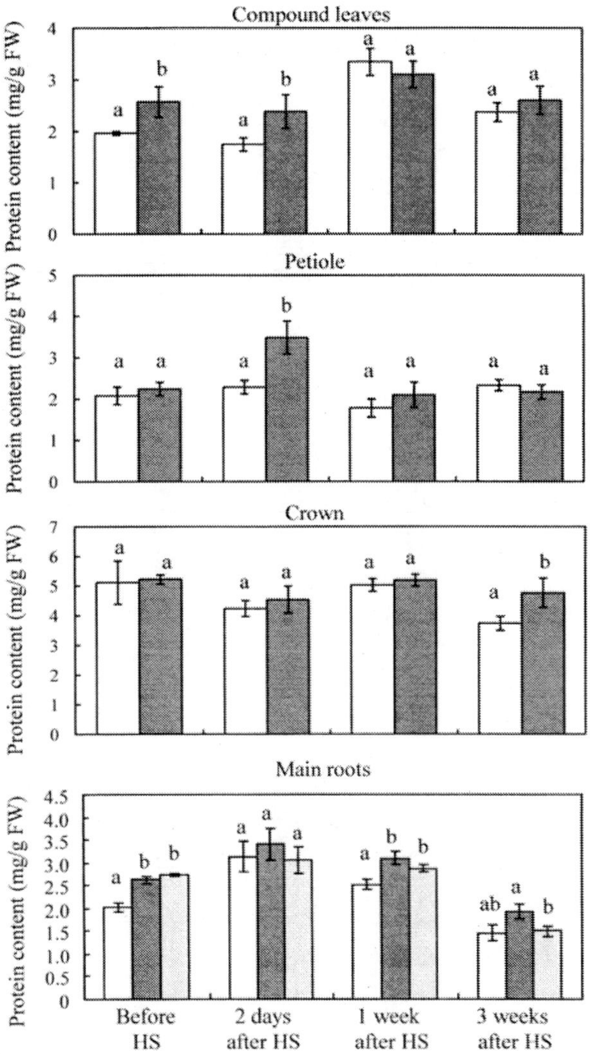

Figure 6. Protein content in strawberry plants under HS. □control , ■AMF+, □AMF-. Different letters represent significant difference among the treatments according to Student's t-test or Tukey's test (P<0.05).

In the present study, AMF symbiosis promoted the growth of strawberry plants, and both the severity of symptoms in HS were reduced by pre-inoculation with AMF. Schenck and Smith (1982) reported that AMF colonization level was not associated with the growth response to high soil temperature in soybean. In this study, no clear relationship between AMF colonization level and the response to high temperature was observed. Sutton (1973) demonstrated that the AMF colonization consisted of three phases: (1) a lag phase during which spore germination, germ tube growth, and initial penetration occur, (2) a rapid growth phase, coinciding with the development of external mycelium, and spread of the fungus within the roots and (3) a stable phase during which the proportion of colonized roots to non-colonized ones remains nearly constant. In this study, the RFCSL was investigated few times throughout the experiment, so that it is difficult to estimate the RFCSL showed the maximum values and the relationship between the response to HS and RFCSL.

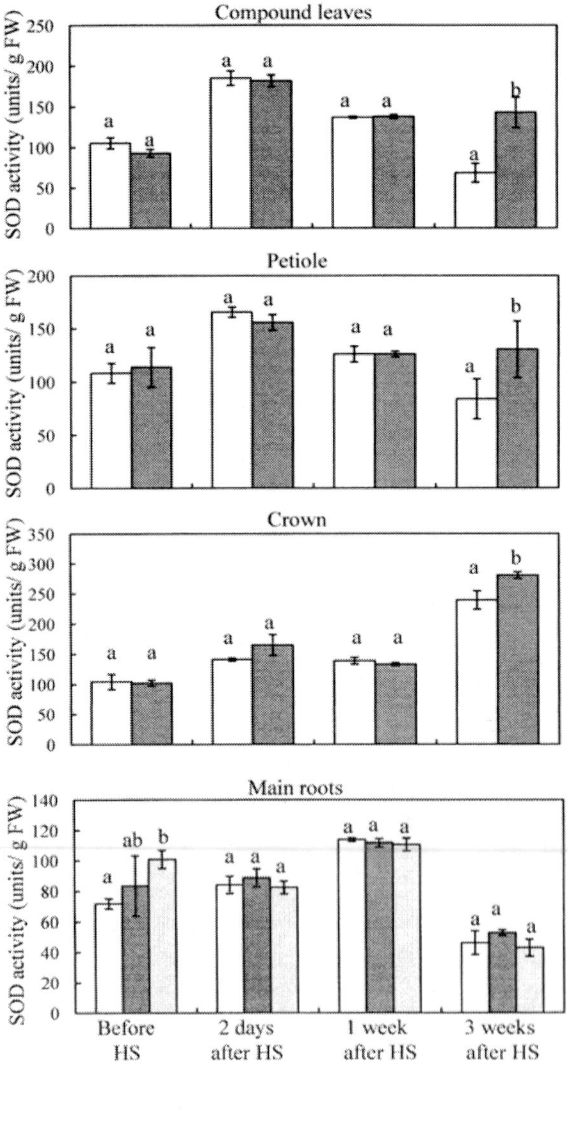

Figure 7. SOD activity in strawberry plants under HS. □control, ■AMF+, □AMF-.

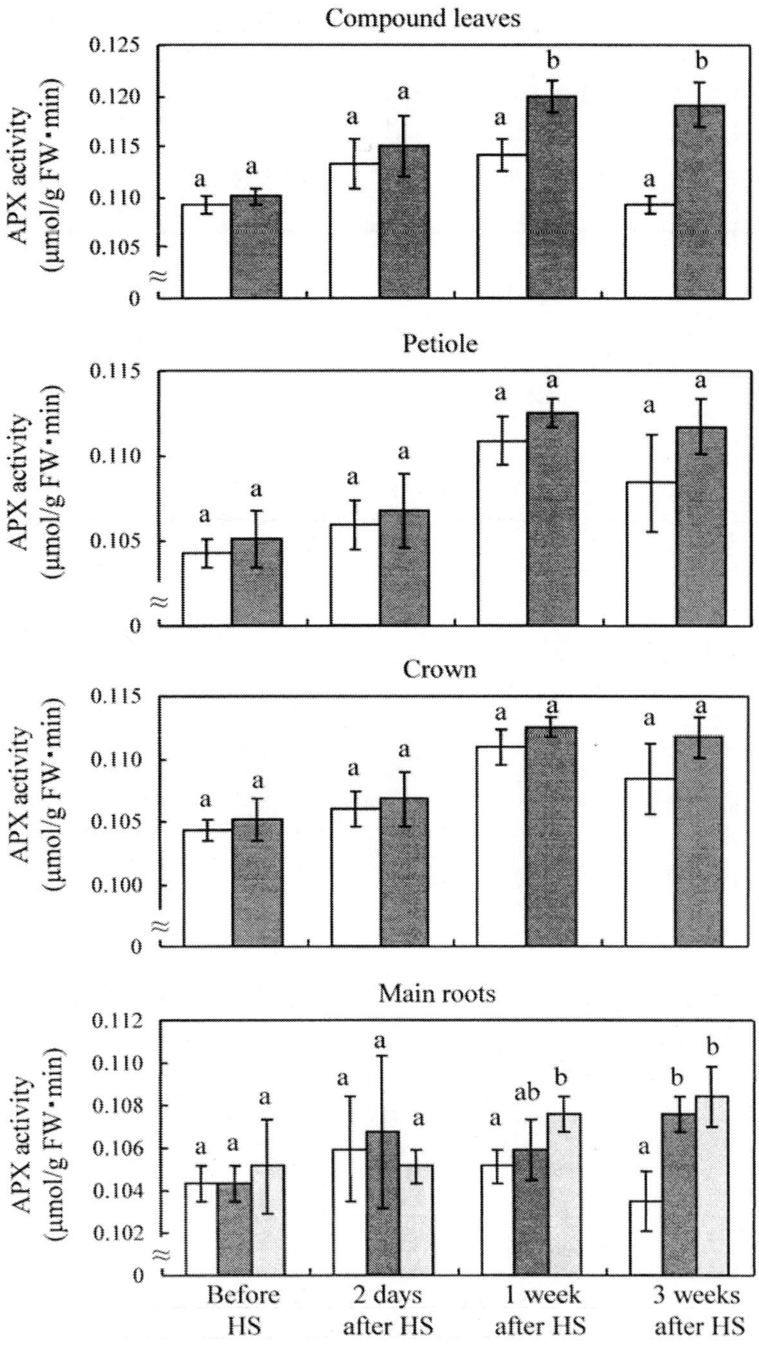

Figure 8. APX activity in strawberry plants under HS. , ☐ control, ■AMF+ , ☐AMF-.

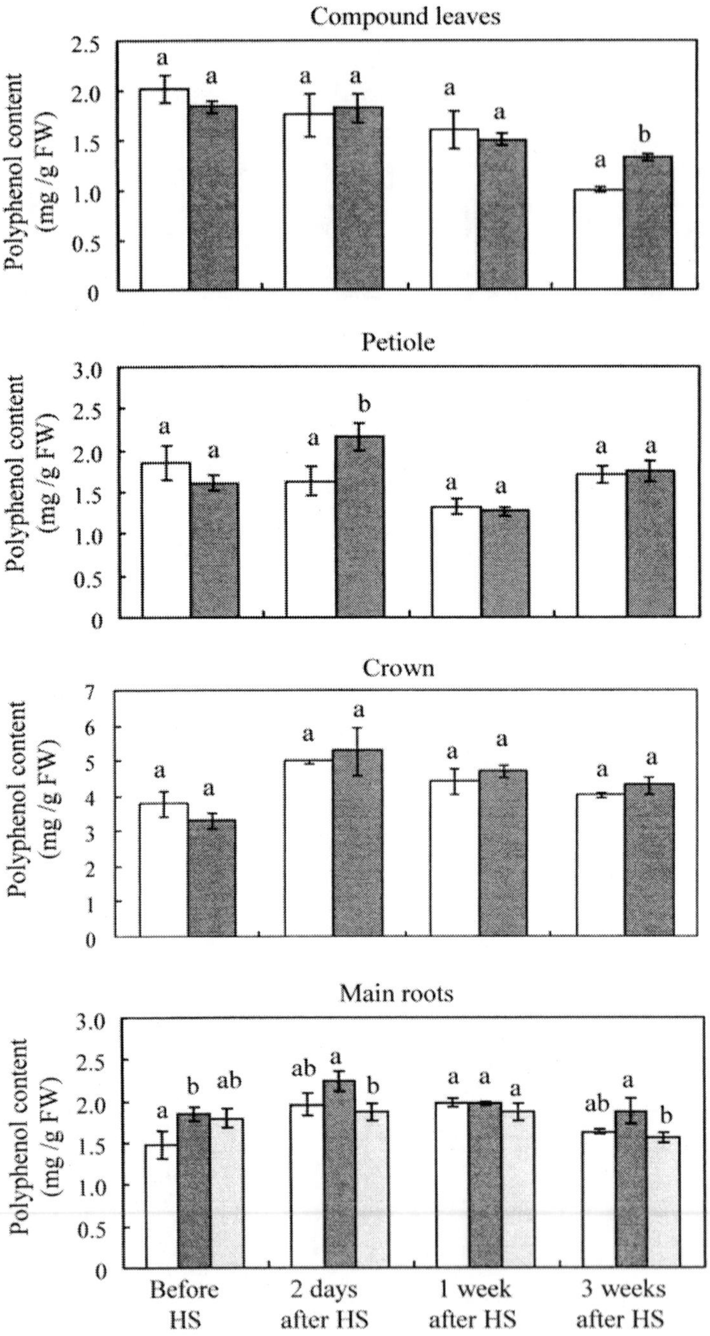

Figure 9. Polyphenol content in strawberry plants under HS. , ☐ control , ■AMF+,☐ AMF-.

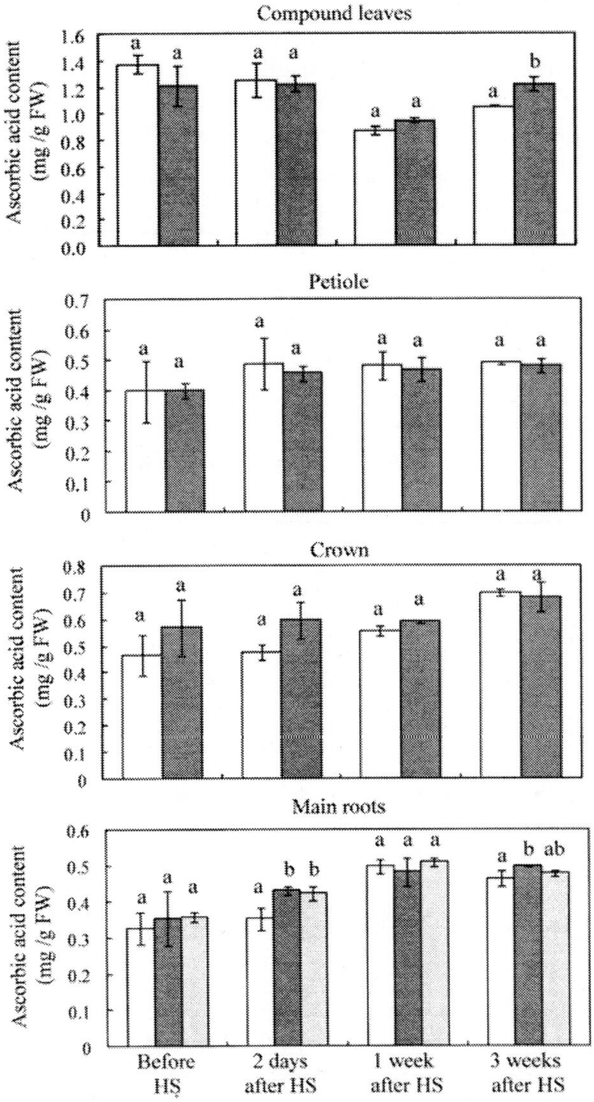

Figure 10. Ascorbic acid content in strawberry plants under HS. □ control, ■AMF+, □AMF-.

Ozgonen and Erkilic (2007) reported that growth promotion and tolerance to *Phytophthora capsici* had no correlation with mycorrhizal colonization level in pepper. Lozano et al. (1996) reported that the increase in SOD activity and drought tolerance showed no correlation with mycorrhizal colonization level in lettuce. In this study, no major characteristics between colonization level and the antioxidative abilities were observed. Thus, the colonization level might have less association with stress tolerance and antioxidative abilities. On the other hand, some reports described that mycorrhizal colonization itself induced temporal increase in antioxidative enzymes (APX, CAT), H_2O_2, flavonoid content, suggesting that colonization might become temporal stress for host plants (Volpin et al., 1995; Salzer et al., 1999; Blilou et al., 2000). In this study, no characteristics concerning colonization were found in the changes in antioxidative abilities. However, most of the antioxidative factors became higher levels in mycorrhizal plants after HS rather than those before HS. Hence, the increase in antioxidative abilities might be induced especially under the stress conditions in mycorrhizal strawberry plants, resulted in the tolerance to HS in this study.

The results in this study reveal that the symbiotic relationship with AMF provides vigorous shoot and root growth under uncertainly fluctuating temperatures, resulting in fast growing, vigorous plants with a high yield potential in strawberry.

ACKNOWLEDGMENTS

This work was supported partially by a Grant-in-Aid for Scientific Research (no.21580029) from Japan Society for the Promotion of Science.

REFERENCES

Beauchamp, C. and Fridovich, I. (1971). Superoxide dismutase: Improved assays and an assay applicable to acrylamide gels. *Anal. Biochem.,* 44, 276-287.

Blilou, I., Bueno, P., Ocampo, J. A. and Garrido, J. M. (2000). Induction of catalase and ascorbate peroxidase activities tobacco roots inoculated with the arbuscular mycorrhizal *Glomus mosseae*.., 104, 722-725.

Bradford, M. M. (1976). A *Mycol. Res* rapid and sensitive method for the quantitation of microgram quantities of protein utilizing the principle of protein-dye binding. Anal. *Biochem.,* 72, 248-254.

Feierabend, J., Schaan, C. and Hertwig. B. (1992). Photoactivation of catalase occurs under both high and low temperature stress conditions and accompanies photoinhibition of photosynthesis. *Plant Physiol.,* 100, 1554-1561□

Folin, O. and Denis, W. (1915). A colorimetric method for the determination of phenols (and phenol derivatives) in urine. *J. Biol. Chem.* 22, 305-308.

Fridovich, I. (1986). Biological effects of the superoxide radicals. *Arch. Biochem. Biophys.,* 247, 1-11.

Garmendia, I., Aguirreolea, J. and Goicoechea, N. (2006). Defence-related enzymes in pepper roots during interactions with arbuscular mycorrhizal fungi and/or *Verticillium dahliae*. Biocont., 51, 293-310.

Gulen, H. and Eris, A. (2003). Effect of heat stress on peroxidase activity and total protein content in strawberry plants. *Plant Sci.,* 166, 739-744.

Haugen, L. M. and Smith, S. E. (1992). The effect of high temperature and fallow period on infection of mung bean and cashew roots by the vesicular-arbuscular mycorrhizal fungus *Glomus intraradices*. *Plant Soil,* 145, 71-80.

Jahnke, L. S., Hull, M. R. and Long, S. P. (1991). Chilling stress and oxygen metabolizing enzymes in *Zea mays* and *Zea diploperennis*. *Plant Cell Environ.,* 14, 97-104.

Kang, M. and Saltveit, M. E. (2002). Effect of chilling on antioxidant enzymes and DPPH-radical scavenging activity of high- and low-vigour cucumber seedling radicles. *Plant Cell Environ.*, 25, 1233-1238.

Li, Y., Imai, K., Ohno, H. and Matsui, S. (2004). Effects of acclimation temperatures on antioxidant enzyme activities in mericlones of a cattleya hybrid. *J. Japan. Soc. Hort. Sci.*, 73, 386-392 (In Japanese with English abstract).

Li, Y., Sassa, D., Matsubara, Y. and Koshikawa, K. (2009). Yield in mycorrhizal strawberry plants by elevated bench culture using fabric pots. *Hort. Res.*, 8, 215-219 (In Japansese with English abstract).

Lozano, J. M., Azcon, R. and Palma, J. M. (1996). Superoxide dismutase activity in arbuscular mycorrhizal *Lactuca sativa* plants subjected to drought stress. *New Phytol.*, 134, 327-333.

Marschner, H. and Dell, B. (1994). Nutrient uptake in mycorrhizal symbiosis. *Plant Soil*, 159, 89-102.

Matsubara, Y. and Harada, T. (1996). Effect of constant and diurnally fluctuating temperatures on arbuscular mycorrhizal fungus infection and the growth of infected asparagus (*Asparagus officinalis* L.) seedlings. *J. Japan. Soc. Hort. Sci.*, 65, 565-570.

Matsubara, Y., Hirano, I., Sassa, D. and Koshikawa, K. (2004). Alleviation of high temperature stress in strawberry plants infected with arbuscular mycorrhizal fungi. Environ. *Control Biol.*, 42, 105-111.

Moghaddam, B., Charles, M. T., Carisse, O. and Khanizadeh, S. (2006). Superoxide dismutase responses of strawberry cultivars to infection by *Mycosphaerrella fragariae*. *J. Plant Physiol.*, 163, 147-153.

Niemi, M. and Vestberg, M. (1992). Inoculation of commercially grown strawberry with VA mycorrhizal fungi. *Plant Soil*, 144, 133-142.

Ozgonen, H. and Erkilic, A. (2007). Growth enhancement and Phytophthora blight (*Phytophthora capsici* Leonian) control by arbuscular mycorrhizal fungal inoculation in pepper. *Crop Protection*, 26, 1682-1688.

Phillips, J. M. and Hayman, D. S. (1970). Improved procedures for clearing roots and staining parasitic and vesicular-arbuscular mycorrhizal fungi for rapid assessment of infection.Trans. *Br. Mycol. Soc.*, 55, 158-163.

Porra, R. J., Thompson, W. A. and Kriedemann, P. E. (1989). Determination of accurate extinction coefficients and simultaneous equations for assaying chlorophylls a and b extracted with four different solvents: verification of the concentration of chlorophyll standards by atomic absorption spectroscopy. *Biochem. Biophys. Acta*, 975, 384-394.

Pozo, M. J., Cordier, C., Gaudot, E. D., Barea, J. M. and Aguilar, C. A. (2002). Localized versus systemic effect of arbuscular mycorrhizal fungi on defence responses to *Phytophthora* infection in tomato plants. *J. Exp. Bot.*, 53, 525-534.

Roe, J. H, Mary, B. M., Oesterking, M. J. and Charlotle, M. D. (1948). The determination of diketo-L-gluconic acid, dehydro-L-ascorbic acid, and L-ascorbic acid in the same tissues by 2,4-dinitrophenyl hydrazine method. *J. Biol. Chem.*, 174, 201-208.

Sahoo, M. R., Dasgupta, A. M., Paresh, A., Kole, C., Jayant, A., Bhat, S. and Mukherjee, A. (2007). Antioxidative enzymes and isozymes analysis of taro genotypes and their implications in Phytophthora blight disease resistance. *Mycopathol.*, 163, 241-248.

Saito, Y., Nagasaka, Y. and Imagawa, M. (2003). Repression of rooting in high temperature and effect of the pretreatment of short time low temperature storage in strawberry runner cutting. J. Japan. Soc. Hort. Sci., 72 (suppl. 1), 244 (In Japanese).

Salzer, P., Corbiere, H. and Boller, T. (1999). Hydrogwn peroxidase accumulation in *Medicago truncatula* roots colonized by the arbuscular mycorrhiza-forming fungus *Glomus intraradices*. Planta, 208, 319-325.

Schenck, N. C. and Smith, G. S. (1982). Responses of six species of vesicular-arbuscular mycorrhizal fungi and their effects on soybean at four soil temperatures. *New Phytol.,* 92, 193-201.

Shrestha, Y. Ishii, T. and Kadoya, K. (1995). Effect of vesicular-arbuscular mycorrhizal fungi on the growth, photosynthesis, transpiration and the distribution of photosynthates of bearing satsuma mandarin trees. *J. Japan. Soc. Hort. Sci.,* 64, 517-525.

Sutton, J. C. (1973). Development of vesicular-arbuscular mycorrhizae in crop plants. *Can. J. Bot.,* 51, 2487-2493.

Takaichi, M. (2004). Medium cooling system by heat of vapourization in strawberry bench culture. *Agr. Hort.,* 59, 55-57 (In Japanese).

Tsukiji, K., Imada, S. and Dan, K. (2001). Evaluation of a technique for assessing high temperature tolerance in strawberry by leaf discs thermostability. *J. Japan. Soc. Hort. Sci.*, 70 (suppl. 2), 170 (In Japanese).

Tsukiji, K., Imada, S. and Dan, K. (2002). Evaluation of high temperature acclimation in strawberry by leaf discs thermostability. *J. Japan. Soc. Hort. Sci.,* 71 (suppl. 2), 373 (In Japanese).

Varma, A. and Schuepp, H. (1994). Infectivity and effectiveness of *Glomus intraradices* on micropropagated plants. *Mycorrhiza,* 5, 29-37.

Volpin, H., Phillips, D. A., Okon, Y. and Kapulnik, Y. (1995). Suppression of an isoflavonoid phytoalexin defense response on mycorrhizal alfalfa roots. *Plant Physiol.*, 108, 1449-1454.

Williams, S., Vestberg, M., Uosukainen, M., Dodd, J. C. and Jeffries, P. (1992). Effects of fertilizers and arbuscular mycorrhizal fungi on the post-vitro growth of micropropagated strawberry. *Agronomie,* 12, 851-857.

In: Horticulture in the 21st Century
Editor: Abigail N. Sampson, pp. 193-201
ISBN 978-1-61668-582-9

Chapter 8

IN VITRO PROPAGATION OF THE BISHOP'S HAT, *EPIMEDIUM ALPINUM* L.

Snježana Mihaljević[1]
Ruđer Bošković Institute, P.O. Box 180, 10002 Zagreb, Croatia

ABSTRACT

The genus *Epimedium* (Bishop's Hat, Barrenwort) in the family *Berberidaceae* comprises about 50 species of herbaceous perennials. A number of species, hybrids and cultivars are grown as ornamentals in woodland gardens and naturalized areas. Certain species are also in use in traditional Asian medicine. The plant can be propagated by division of its long rhizomes or by seeds. The difficulties associated with gathering the seeds and with slow growth make bishop's hat a suitable species for propagation by tissue culture. *In vitro* propagation of *Epimedium alpinum* L. was carried out using organogenic callus induced from immature seeds transiently exposed (48 hours) to 20 µM CPPU or 80 µM TDZ followed by culture on hormone-free woody-plant medium (WPM). To promote adventitious shoot induction, the effects of different concentrations of 6-benzyladenine (BA) in combination with 2,4-dichlorophenoxyacetic acid (2,4-D) and gibberellic acid (GA_3) were investigated. WPM containing 2.2 µM BA and 1.1 µM 2,4-D was optimal for callus proliferation while shoot induction and elongation was enhanced on media with 44 µM BA and 0.55 µM 2,4-D. Histological examination of the cultures revealed that the regenerated plants were derived from *de novo* developed shoots. Shoots were successfully rooted on hormone-free WPM and well-developed plantlets were transferred to soil. Somatic chromosome number of the regenerated plants was found to be the same as the source plants.

1 Author for correspondence: Snježana Mihaljević, Ruđer Bošković Institute, P. O. Box 180, 10002 Zagreb, Croatia, fax: ++385 1 4561 177, e-mail: mihaljev@irb.hr.

Introduction

Epimediums are shade-loving herbaceous perennials. Beautiful heart-shaped leaves on wiry stems make attractive low-growing ground cover in woodland gardens (Figure 1). They survive in dry shade beneath the trees and by that keep down weeds, which shoot up in mild weather between autumn and spring. In spring, attractive red marked new foliage and delicate four-petaled, bell shaped flowers perk up the ground. Due to the shape of their flowers, *Epimediums* are commonly called bishop's hat; another common name is barrenwort. In Europe and the United States, the genus has received recent attention from gardeners and nurseries with an interest in more unusual ornamental plants. According to W. T. Stearn (2002), there are 54 species and a considerable number of hybrids and cultivars. Most of them are native to China, where the genus reaches its peak of diversity. Certain species of *Epimedium* are in use in traditional Asian medicine and have proven therapeutic effects (Lee et al. 1995, Kim et al. 2001, Wang et al. 2004, Zhang et al. 2008).

Epimediums are commonly propagated vegetatively, by rhizome division or through dissemination. Vegetative buds grow from rhizomes, but lateral bud development is suppressed by apical dominance of the rhizomic apex (Lubell et al. 2005). Together with a slow growth, this makes vegetative propagation through the division of rhizomes time-consuming: only a few division propagules can be produced per plant within two growing seasons (Lubell et al. 2005). Controlled dissemination of this species is hard because of problems with collecting and germinating the seeds. When a ripe capsule of *Epimedium* bursts, it spreads seeds containing differentiated but underdeveloped embryos. These seeds should pass a period of morphophysiological dormancy before germination (Finch-Savage and Leubner-Metzger 2006). Also, a considerable portion of seeds happen to be dispersed by ants (Stearn 1938).

The difficulties associated with propagation by traditional techniques may be overcome using tissue culture methods. Micropropagation is an alternative method of vegetative propagation, which can be employed for large-scale propagation of disease-free clones, gene pool conservation and genetic improvement of ornamental plants (Rout et al. 2006). Micropropagation via shoot organogenesis has been reported for three *Epimedium* species: *E. grandifolium* (Al-Matar et al. 1999), *E. diphyllum* (Koga et al. 1994) and recently *E. alpinum* (Mihaljević and Vršek 2009). All authors characterized *Epimedium* sp. as slow-growing and recalcitrant to regeneration *in vitro*. Among various explant sources tested in these studies, only the very young rachis and immature seeds gave a regeneration response. In addition, regeneration potential and rooting efficiency varied among *Epimedium* species (Koga et al. 1994).

Epimedium alpinum L. is one of two representatives of this genus natively growing in Europe. It usually grows in thermophyle oak and beech forests. This species has dull red flowers (red petals and yellow nectaries) in 12–20 flowered racemes appearing in spring or early summer. New leaves in spring emerge with a pink tinge, mature to green and turn deep reddish bronze in fall. Decorative foliage is attractive not only to gardeners but also to florists. However, uncontrolled collecting of foliage may endanger the native population of this slow-growing species.

Figure 1. *Epimedium alpinu* L. grows well under trees and makes groundcover that can serve as an attractive edger for walkways.

In attempt to optimize an *in vitro* regeneration procedure, we examined the effects of 2,4-D, BA and GA_3 on organogenic callus propagation and adventitious bud elongation. Since the application of PGR may affect chromosomal variations and behavior in plant tissue culture (D'Amato 1985), the karyotype of *in vitro* regenerated plantlets was analyzed at the level of chromosome number to study genetic stability of regenerants. Development of a genetically stabile *in vitro* propagation system for *Epimedium alpinum* L. would raise the propagation success rate, shorten the cultivation cycle, and enable breeders to introduce new variations.

Material and Methods

In Vitro Culture and Acclimatization

In vitro propagation of *Epimedium alpinum* L. was carried out using organogenic callus lines induced from immature seeds exposed to short treatments of thidiazuron (TDZ) and forchlorfenuron (CPPU), as described by Mihaljević and Vršek (2009). In short, the capsules (15 mm in size, green in color) were surface sterilized, immature seeds (4 mm in size, green in color) were isolated from pericarps and submitted to the two-day treatment with 80 μM TDZ or 20 μM CPPU in liquid woody plant medium (WPM, McCown and Lloyd 1981). After the treatment, explants were transferred to solid hormone-free WPM medium (pH 5.8) supplemented with 2% sucrose and 2% Phytagel until the callus developed.

To promote shoot bud induction, organogenic calli were transferred from hormone-free WPM to WPM supplemented with 2,4-D (0.56 or 1.1 μM) in combination with BA (2.2, 22 or 44 μM). One set of explants was grown in the presence of 2.2 μM BA and 1.15 μM GA_3. The tissue was subcultured to fresh medium three times in four week-intervals before callus growth and shoot regeneration were estimated. The initial fresh weight of the explants was about 200 mg. For elongation and rooting, single shoots (1.5-2.0 cm long) were transferred to solid hormone-free WPM. All cultures were subcultured to fresh medium at four-week intervals to avoid darkening of the culture media. Organogenic cultures were maintained in a culture room at 23±1°C and a 16-h photoperiod (irradiance 30 to 35 μmol $m^{-2}s^{-1}$, provided by cool-white fluorescent tubes).

Rooted regenerants were potted in a mixture of vermiculite and humus (3:1) sterilized by autoclaving. Plants were grown covered with plastic bags to maintain high humidity. After two weeks, the cover was removed and the acclimatization was extended under conditions of moderate humidity, in covered clear plastic containers, until new leaves appeared.

Light Microscopic Studies

Standard procedures (Ruzin 1999) were followed to carry out histological studies. Organogenic callus explants were collected, fixed in FAA (50% ethanol, 10% formalin, 5% glacial acetic acid) for 16 hours at 4°C, and dehydrated in a series of ethanol and xylene. Specimens were embedded in Paraplast (Fluka), and then sectioned with a rotary microtome (Leica RM 2255). The 7 μm thick sections were deparaffinized, rehydrated, stained with Hematoxylin and counterstained with Orange G. The sections were observed under a light microscope (Olympus BX51).

For karyotype analysis, specimens of root tip cells of acclimatized plants were pretreated with 0.05% colchicines at room temperature for 3.5 - 4.0 hours, and then fixed in ethanol-acetic acid (3:1) for at least 24 hours. Samples were stained in a few drops of 1% acetocarmin, the cells were gently macerated and squashed in 45% acetic acid. The coverslips were removed by dry-ice method. Samples were examined under light microscope Opton Axioplan (Zeiss).

Statistical Analysis

The effects of the growth regulators in the culture medium on callus growth and bud induction were monitored during four subcultures. Each treatment consisted of at least three replicates. Organogenic callus proliferation was expressed as the arithmetic mean of the fresh weight increase. Regeneration efficiency was expressed as the mean number of shoots produced per explant. Data were compared by analysis of variance (ANOVA), using STATISTICA 7.1 (StatSoft, Inc., USA) software package. Differences among means were considered as statistically significant at $P < 0.05$.

Results and Discussion

Shoot Regeneration

To promote callus proliferation and shoot bud induction, organogenic callus lines were cultured onto WPM supplemented with a range of concentrations of BA in combination with 2,4-D or GA_3. Fresh weight increase and the frequency of bud induction were measured for five organogenic lines. All experimental treatments tested supported callus tissue proliferation with an average fresh weight increase from 240 to 330 mg per month (Figure 2A). There were no significant differences in fresh weight increase regardless of the hormone levels in the media. However, there were differences for the callus growth regarding to callus lines (Figure 3A). Callus lines were induced from seeds of different plants and thus, represent material with different potential for regeneration.

Organogenic callus grown on shoot regeneration medium was compact with numerous nodules (Figure 4A). Light green adventitious buds often formed in clusters. Light microscopic observation of *Epimedium alpinum* organogenic cultures grown on WPM supplemented with different concentrations of 2,4-D and BA revealed meristematic zones in the peripheral layer and in the inner tissue of organogenic callus (Figure 4B). The meristematic zones were comprised of cells which were small in size with dense cytoplasm. Fully developed and elongated adventitious buds with heart-shaped leaf blades were visible after about 30 days of culture.

Among the different concentrations of BA and 2,4-D used for shoot bud induction, 44 μM BA in combination with 0.56 μM 2,4-D exhibited the best response (2.5 ± 2 shoots per callus explant) (Figure 2B). Slightly less effective was the combination of 22 μM BA and 1.1 M 2,4-D, while combinations of 2.2 μM BA and 1.1 M 2,4-D, or 2.2 μM BA and 1.15 μM GA_3 induced the lowest number of shoots per explant. In general, the number of shoots increased in response to increasing the concentration of BA. Statistical analysis confirmed the significance of the differences in the number of shoots depending on the concentration of BA used. Results were consistent with general knowledge that cytokinins stimulate cell division and shoot initiation/bud formation in tissue culture (Davies 2004). Although the addition of GA_3 to regeneration medium with 2.2 μM BA did not improved shoot bud induction, shoots produced on this medium had bigger leaf surface (data not shown). This could be explained by the stimulatory effect of gibberellins on cell division and elongation (Davies 2004). Koga et al. (1994) reported that the MS medium containing 2.2 μM BA and 1.45 μM GA_3 was suitable for shoot regeneration of *Epimedium diphyllum*.

The number of shoot buds for each combination of PGR varied regarding the regeneration potential of callus lines tested (Figure 3B). This was expected due to a possible difference in genetic background of seeds used as primary explants.

Shoot Rooting and Acclimatization

For root induction, well developed shoots, 1 to 2 cm in length, were separated from the maternal tissue and transferred to hormone-free WPM. Four weeks after the transfer, the first rooted shoots could be observed (Figure 4D). After two subcultures, 75-85 % of shoots had rooted successfully. In addition, more than 80% of the *in vitro*-regenerated plants survived in substrate. The acclimatized plants showed normal growth and a morphology corresponding to that of outdoor-grown plants (Figure 4E).

The analysis of mitosis in root-tip of *in vitro*-regenerants showed normal chromosome segregation. As the result of regular mitosis, chromosome counting revealed diploid chromosome complement, 2n= 12, consisted of six metacentric and six submetacentric pairs (Figure 4C). This number corresponds to the number of chromosomes determined for *E. alpinum* and 12 other *Epimedium* species determined by Sheng and Chen (2007).

A

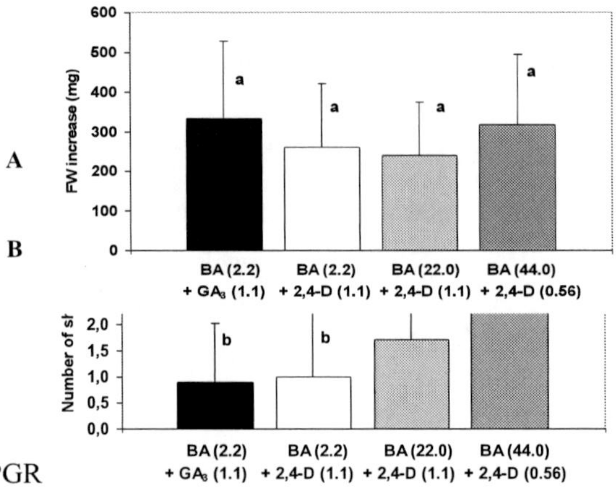

Figure 2. Effects of PGR present in the WPM medium on shoot organogenesis in *E. alpinum*. (A) Fresh weight (FW) increase of callus tissue; (B) Shoot bud induction (number of shots per callus explant after 30 days of culture). Results are the mean of callus growth of all tissue lines cultured on each combination of PGR. Bars represent ± STDEV. Different letters indicate significant difference at $P<0.05$.

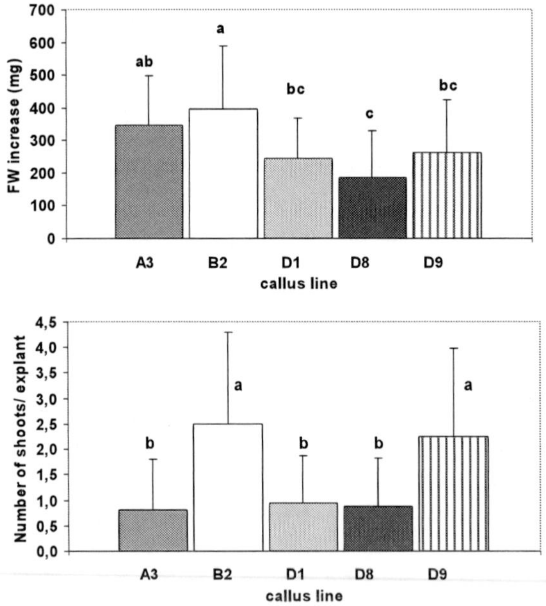

Figure 3. Effects of the callus tissue lines on shoot organogenesis in *E. alpinum*. (A) Fresh weight increase as influenced by the tissue lines; (B) Shoot bud induction (number of shots per callus explant after 30 days of culture) as influenced by the tissue lines. Results are the mean of the tissue line growth on all tested combination of PGR. Bars represent ± STDEV. Different letters indicate significant difference at $P<0.05$.

Figure 4. *In vitro* propagation of *Epimedium alpinum*. (A) Organogenic callus induced in immature seeds after the two-day treatment with 80 μM TDZ followed by 6 months of culture on hormone-free WPM. (B) *De novo* shoots induction at the surface of organogenic callus line B2 grown on WPM supplemented with 2.2 μM BA and 1.3 μM 2,4-D, *bar* = 500 μm (arrows point to shoot induction). (C) Rooting of regenerated shoot after one months of culture on hormone-free WPM, (D) Metaphase chromosomes in a root-tip meristematic cell of an *in vitro* regenerated plant (2n=12). (E) *In vitro* regenerated plants of *Epimedium alpinum* growing in the culture room.

Thus, the 2-year-old organogenic culture of *Epimedium alpinum* presents a stable culture at the cytogenetic level with the ability for production of true-to-type plantlets. However, we could not exclude the fact that the mutation potential could be increased with the aging of the tissue culture. The numerous studies of chromosomal variations and behavior in plant tissue culture are mostly affected by growth regulators, especially auxins (D'Amato 1985).

Conclusion

In this chapter we described the successful establishment of an *in vitro* regeneration system for *Epimedium alpinum* L. As reported previously (Mihaljević and Vršek 2009), the organogenic callus induction in immature seeds treated with TDZ or CCPU was not as effective, and the improvement of the procedure is possible. However, once established, the described system for organogenic callus proliferation and plant regeneration makes possible the production of a large number of plantlets. The number of shoots was significantly affected by the BA/2,4-D ratio. The most efficient medium for organogenic callus proliferation and shoot regeneration in *E. alpinum* was WPM supplemented with 44.0 μM BA and 1.1 μM 2,4-D. The rooting on hormone-free WPM was highly efficient and the acclimatized plants showed normal growth and morphology. Karyotype analysis showed that 2-year-old organogenic culture of *Epimedium alpinum* presents a stable culture at the cytogenetic level with the ability for production of true-to-type plantlets. We believe that the procedure reported in this chapter may be useful for commercial growers, enabling them to propagate this and related species more efficiently and, at the same time, it may offer a permanent source of plant material for further genetic improvement.

Acknowledgments

This research was funded by the Ministry of Science, Education and Sport, Republic of Croatia, Projects No. 0098903 and 098-0982913-2829. The author of this chapter thanks the Group for ichtiopathology (Ruđer Bošković Institute, Zagreb, Croatia) who made their laboratory facilities available for histological studies, and to Prof. Višnja Besendorfer (Faculty of Sciences, Uni Zagreb, Croatia) for help with karyotype analysis.

References

Al Matar, SM; Al-Khayri, JM; Brar, MS; Klingaman, GL. *In vitro* regeneration of long spur barrenwort (*Epimedium grandiflorum* Morr.) from rachis explants. *In Vitro Cell Dev Biol Plant*, 1999 35, 245-248.

D'Amato, F. Cytogenetics of plant cell and tissues cultures and their regenerants. *Crit Rew Plant Sci*, 1985 3, 73-112.

Davies, PJ. *Plant Hormones: Biosynthesis, Signal Transduction, Action!* Netherlands: Kluwer Academic Publishers; 2004.

Finch-Savage, WE; Leubner-Metzger, G. Seed dormancy and the control of germination. *New Phytol*, 2006 171, 501-523.

Kim, J-H; Mun, Y-J; Im, S-J; Han, J-H; Lee, H-S; Woo, W-H. Effects of the aqueous extract of Epimedii Herba on the antibody responses in mice. *Int Immunopharmacol*, 2001 1, 935-944.

Koga, S; Shoyama, Y; Nishioka, I. *In Vitro* Propagation of *Epimedium* species. *Nat Med*, 1994 48, 49-52.

Lee, MK; Choi, YJ; Sung, SH; Shin, DI; Kim, JW; Kim, YC. Antihepatotoxic activity of icariin, a major constituent of *Epimedium koreanum*. *Planta Med*, 1995 61, 523-526.

Lubell, JD; Thompson, DM; Brand, MH. Foliar sprays of benzyladenine increase bud and propagule production in *Epimedium* x *rubrum* Morren and *Helleborus* x *hybridus* L. *Propag Ornam Plants*, 2005 5, 19-22.

McCown, BH; Lloyd, G. Woody Plant Medium (WPM) - a mineral nutrient formulation for microculture for woody plant species. *HortScience*, 1981 16, 453.

Mihaljević, S; Viršek, I. *In vitro* shoot regeneration from immature seeds of *Epimedium alpinum* induced by thidiazuron and CPPU. *SciHorti*, 2009, 129: 406-410

Rout, GR; Mohapatra, A; Jaim SM. Tissue culture of ornamental pot plants: A critical review on present scenario and future prospects. *Biotechnol Adv*, 2006 24, 531-560.

Ruzin, SE. *Plant microtechnique and microscopy.* USA: Oxford Uni. Press; 1999.

Stearn, WT. *Epimedium* and *Vancouveria* (Berberidaceae), a monograph. *J Proc Linn Soc Bot*, 1938 51, 409–555.

Stearn, WT. *The genus Epimedium and other herbaceous Berberidaceae*. Portland, Oregon: Timber Press; 2002.

Wang, S; Zheng, Z; Weng, Y; Yu, Y; Zhang, D; Fan, W; Dai, R; Hu, Z. Angiogenesis and anti-angiogenesis activity of Chinese medicinal herbal extracts. *Life Sci*, 2004 74, 2467-2478.

Zhang, DW; Cheng, Y; Wang, NL; Zhang, JC; Yang, MS; Yao, XS. Effects of total flavonoids and flavonol glycosides from Epimedium koreanum Nakai on the proliferation and differentiation of primary osteoblasts. *Phytomedicine*, 2008 15, 55-61.

In: Horticulture in the 21st Century
Editors: Abigail N. Sampson, pp. 203-211
ISBN 978-1-61668-582-9

Chapter 9

DEVELOPMENT OF SAFE NEW CULTIVATION SYSTEM FOR APPLES BASED ON *S*-RNASE AND SSR INFORMATION

***Shogo Matsumoto*[1]**
Graduate School of Bioagricultural Sciences, Nagoya University, Chikusa, Nagoya 464-8601, Japan

1. ABSTRACT

Apple production without agrochemicals is not easy, and consumers demand safe edible fruit. Reducing the amount and frequency of spraying with agrochemicals is important not only for safe fruit production, but also for the protection of the environment. Until now, as the apple has gametophytic self- and cross-incompatibility, growers have planted several genetically distinct cultivars in an orchard for stable fruit production. In such cases, agrochemicals sprayed on cultivars for harvest at late season may well drift onto the cultivars at harvest time. To avoid this, we developed a new apple cultivation system consisting of one cultivar with suitable pollinizers. We selected 'Maypole' and 'Dolgo' as pollinizers for 'Fuji' cultivar by *S*-RNase allele analyses of more than 430 cultivars and species. We also determined the arrangement of the pollinizers and the cultivar with the pollinator, *Osmia cornifrons*, based on *S*-RNase and SSR analyses of seeds in fruits pollinated mainly by *O. cornifrons* and pollen loaves collected by *O. cornifrons*.

2. INTRODUCTION

Apples (*Malus x domestica* Borkh.) are susceptible to infectious diseases caused by pathogenic fungi, bacteria, and viruses. Disease control is a major annual expense for the apple grower. For instance, the management of apple scab, bitter rot, Japanese apple canker and fry speck in Japan, respectively require nine, five, seven and nine protective fungicide

1 Corresponding author. Tel./fax : +81 52 789 4025. E-mail address: shogo@agr.nagoya-u.ac.jp (S. Matsumoto).Graduate School of Bioagricultural Sciences, Nagoya University, Chikusa, Nagoya 464-8601, Japan.

applications annually. The volume and number of pesticide applications in apple orchards are high, but apple production without agrochemicals is difficult (Figure 1). Today, consumers want apples grown without pesticides, which can cause adverse human health effects. Reducing the amount and frequency of spraying with agrochemicals is important not only for safe fruit production, but also for the protection of the environment. The spraying of pesticides on mature fruit must particularly be avoided.

Since the apple has gametophytic self- and cross-incompatibility, growers have until now planted several genetically distinct cultivars in an orchard for stable fruit production. In such cases, agrochemicals sprayed on cultivars for harvest at late- or mid-season may well drift onto the cultivars at harvest time. To avoid this, we developed a new apple cultivation system consisting of one cultivar with suitable pollinizers and pollinators (Figure 2). In this paper, I introduce this new cultivation system in detail.

Figure 1. Apple production without an agrochemical. A, B: 'Jonathan' fruit (28/8/2007) without agrochemical (A), with agrochemical (B), C, D: 'Fuji' fruit (28/8/2007) without agrochemical (C), with agrochemical (D).

Figure 2. Example of a block of an apple orchard consisting of a pollinizer 'Maypole' (red petal color) and commercial cultivar 'Fuji' (white petal color) at the Nagano Fruit Tree Experiment Station in Nagano, Japan.

Table 1. S-RNase genotype of crab-and wild-apple pollinizers

Pollinizer	I.D. number	*S*-RNase genotype
Atrosanguiner	4522	$S_{26}S_X$[z)]
Dolgo	3032	S_xS_x
Makamik	3104	$S_{1,20,24\text{-}like}S_X$
Maypole		$S_{10}S_{16b}$
M. baccata	79091	$S_{16a}S_{20}$
Red Bud	2019	$S_{25}S_X$
Sentinel	3108	$S_{26}S_X$
Snowdrift	3110	$S_{25}S_X$

z) $S_x \neq S_1, S_2, S_3, S_4, S_5, S_7, S_9, S_{10}, S_{16}, S_{20}, S_{24}, S_{25}, S_{26}, S_{28}$

3. Selection and Arrangement of Pollinizers

As mentioned above, apple orchards consisting of single cultivars are increasingly preferred in Japan in light of the hazard of agrochemical spray drift from late- or mid-season cultivars to early-season cultivars at harvest time. For stable fruit production, we have usually selected suitable pollinizers such as flowering crab-apples and wild species from the perspective of cross-compatibility with commercial cultivars (Matsumoto et al., 2003, 2007, 2009a). In the apple, a pistil-specific glycoprotein with ribonuclease activity (S-RNase) is involved in gametophytic self-incompatibility response, and *S*-genotypes of the apple have been identified by cross-pollination methods followed by stylar protein (S-RNase) analyses, and by *S*-RNase allele-specific PCR amplification and digestion analyses (Broothaerts et al., 1995; Kobel et al., 1939; Sassa et al., 1994). Using S-RNase allele-specific PCR amplification methods developed by us and other research groups, we have selected seven crab-apples and one wild species as pollinizers (Table 1). 'Dolgo' did not contain any known *S*-RNase allele, and 'Atrosanguiner,' 'Makamik,' 'Red bud,' 'Sentinel' and 'Snowdrift' each contained one unidentified allele (Table 1). 'Dolgo' seems to be a useful pollinizer for all cultivars in Japan, and the remainder also seems to be useful pollinizers of commercial cultivars, except for those having an identical *S*-RNase allele. 'Maypole' and *Malus baccata* 79091 may also be useful pollinizers for commercial cultivars since the alleles S_{16a} and S_{16b} are infrequent in domestic Japanese cultivars (Table 1). However, 'Maypole' and *M. baccata* 79091 are not useful against commercial cultivars having S_{10} [ex. 'McIntosh' ($S_{10}S_{25}$) and 'Shinano Red' (S_3S_{10})] and S_{20} [ex. 'American Summer Pearmain' (S_1S_{20}), 'Indo' (S_7S_{20}) and 'Kitanosachi' (S_7S_{20})], respectively, because they exhibit semi-compatibility. These selected pollinizers have other desirable characteristics, such as profuse blossoms, high pollen viability, annual rather than biennial flowering, distinctive fruit at harvest (Dennis, 2003). In addition, their flowering times are synchronized with or overlap that of the commercial cultivars being pollinated. *M. baccata* 79091, 'Dolgo,' and 'Maypole' flower a few days earlier than does the typical commercial cultivar 'Fuji,' and 'Sentinel' and 'Makamik' flower one day earlier than does 'Fuji.' On the other hand, 'Atrosanguinar' flowers on the same day as 'Fuji.' 'Red Bud' and 'Snowdrift' flower two days later than does 'Fuji,' but the flowering periods of 'Red Bud' and 'Snowdrift' are two days longer than that of 'Fuji.' In apple orchards consisting of pollinizers and a single commercial cultivar, at least two different pollinizers with early flowering times, such as 'Maypole' and 'Dolgo,' and late flowering times, such as 'Red Bud' and 'Snowdrift,' should be planted to have overlapping flowering times with commercial cultivars.

We recommend at least two different types of pollinizers for an apple orchard consisting of a single cultivar. If the number of pollinizers is adequate, pollinators easily fertilize the commercial cultivars with the pollinizer's pollen. However, it may cause a reduction in the yield due to the increase of the planting area of the pollinizers. Growers would like to reduce the number of pollinizers to produce higher yields of the commercial cultivars. To learn the minimum number of required pollinizers, we investigated the rates of 'Fuji' fruit and seeds in 'Fuji' fruit produced by the pollen of the pollinizers 'Maypole' and 'Dolgo' (Matsumoto et al., 2008). We developed a method for determining the distance insects (mainly *Osmia cornifrons*) carry pollen from the pollinizers to 'Fuji' plants based on the leaf color of the progeny and the *S*-RNase allele of the 'Maypole,' and on the Simple Sequence Repeat (SSR) analyses of the 'Maypole' and 'Dolgo.' Although the fruit set of 'Fuji' apples was not reduced by increasing the distance between the 'Fuji' trees and the pollinizers (probably due to pollen flow from other commercial cultivars planted outside the area), the rate of 'Fuji' fruit produced by the pollen of the pollinizers decreased with the increasing distance, dropping below 50% when the distance between the pollinizers and the 'Fuji' was 10 m (Figs. 3, 4). In an apple orchard consisting of pollinizers and 'Fuji' trees, the pollinizers should be planted no more than 10 m apart to guarantee an adequate yield and the high quality of such factors as fruit shape since the number of seeds influences the shape of the fruit.

4. SELECTION AND ARRANGEMENT OF POLLINATORS

We have selected *Osmia cornifrons*, not honeybees (*Apis mellifera* Linnaeus, which are the most important natural carrier of pollen in an apple orchard), as the pollinator of apple orchards in Japan consisting of one kind of commercial cultivar with pollinizers. Since the working area of *O. cornifrons* is smaller than that of honeybees, *O. cornifrons* are useful in relatively small and densely planted apple orchards in Japan. Moerover, *O. cornifrons* seems to be superior to the honeybee as it does more pollinating, lacks stinger, its foraging activity coincides with the apple blooming period, and it flies and pollinates apples in cooler and damper weather (Matsumoto and Maejima, 2010; Sekita, 2001; Way, 1978).

We have investigated the foraging behaviour of *O. cornifrons* in apple orchards consisting of single cultivars with pollinizers, and clarified its characteristics as shown below (Matsumoto and Maejima, 2010; Matsumoto et al., 2009b).

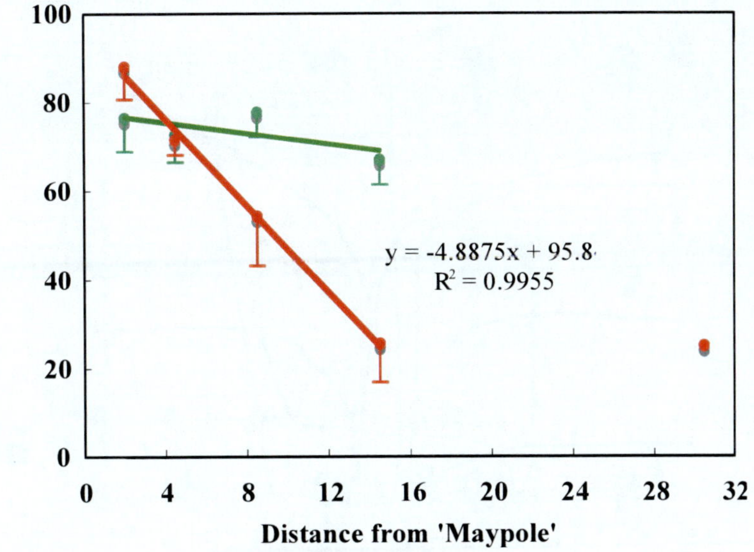

Figure 3. Rate of 'Fuji' fruit produced by 'Maypole' pollination at 23 block in Nagano (see Matsumoto et al., 2008). Red line: 28 to 30 'Fuji' fruits were collected over 3 years (2004-2006) at distances of 2, 4.5, 8.5 and 14.5 m from the 'Maypole'. Twelve fruits were collected in 2006 at 30.5 m from the 'Maypole'. The rate of 'Fuji' fruit produced by 'Maypole' pollination was estimated from the data of red-leaf trait, S-RNase allele, and SSR analyses. Data of each year from 8 to 10 fruits were averaged (mean±S.D.). Green line: 'Fuji' fruit set under open-pollination in block 23.

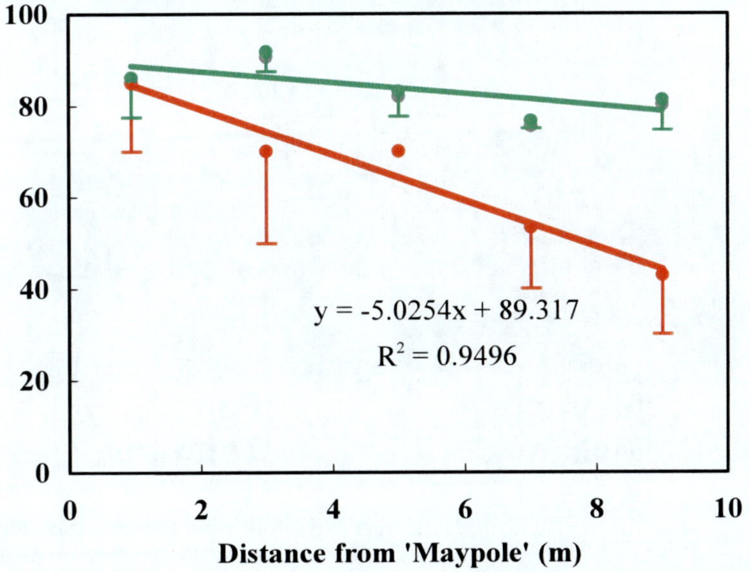

Figure 4. Rate of 'Fuji' fruit produced by 'Dolgo' pollination at 28 block in Nagano (see Matsumoto et al., 2008). Red line: 19 to 20 'Fuji' fruits were collected over 2 years (2005-2006) at distances of 0.95, 3, 5, 7 and 9 m from the 'Dolgo'. The rate of 'Fuji' fruit produced by 'Dolgo' pollination was estimated by the data of SSR analyses. Data of each year from 9 to 10 fruits were averaged (mean±S.D.). Green line: 'Fuji' fruit set under open-pollination in block 28.

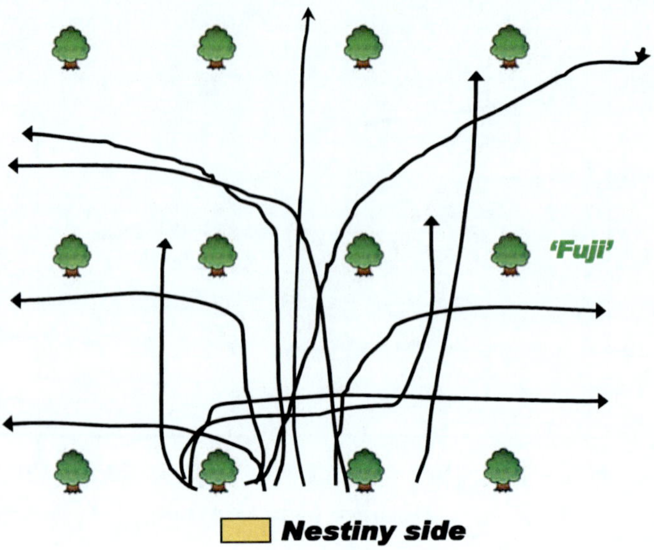

Figure 5 Flying direction of *O. cornifrons* from the nestiny side. Arrows indicate the tracks of *O. cornifrons*.

Figure 6 *O. Cornifrons* seems to visit different flowers at the following foraging trip. In that case, 'Maypole' pollen attached to *O. cornifrons* bodies might be used for the fertilization of 'Fuji'.

1. *O. cornifrons* flies in all directions, rather than simply moving along the rows like the honeybee (Figure 5).
2. *O. cornifrons* shows strong flower constancy for 4-8 min during one pollen-nectar foraging trip. However, the bees seem to forage for different types of flowers during their 16-22 pollen nectar foraging trips (Figure 6).

3. After one pollen-nectar foraging trip, the bees brushed off the pollen in the reed tube, but approximately 10% of the pollen was not brushed from their bodies. The remaining pollen on their body retained the ability to fertilize apples for 12 days (see Matsumoto et al., 2009b).
4. *O. cornifrons* acts as a good pollinator of apple trees between a commercial cultivar, 'Fuji' (white), and a pollinizer, 'Maypole' (red) (see Matsumoto et al., 2009b).
5. *O. cornifrons* is a useful pollinator for 'Delicious' (see Matsumoto et al., 2009b).

In addition, the fruit set rate of 'Seirin Spur' (sport of 'Fuji') by the pollen of the 'Maypole' carried mainly by *O. cornifrons* remains high (74.7-83.8%), even at 33 m from the nestiny side (Table 2).

Table 2. Fruit set of 'Serin Spur'fruit formed by 'Maypole' pollen preferentially carried by *Osmia cornifrons*

Distance from nestiny side (m)	Year	No. of flowers	No. of fruits	Rate of fruit set (mean±S.E.%)[z)]
3.7	2006-2008	1774	1519	81.3±5.4
7.6	2006-2008	1730	1286	74.7±4.9
13.4	2006-2008	2698	2198	79.3±4.7
33	2006-2008	2385	1962	83.8±1.8

[Z)] Data of each year were averaged (mean ±S.E.).

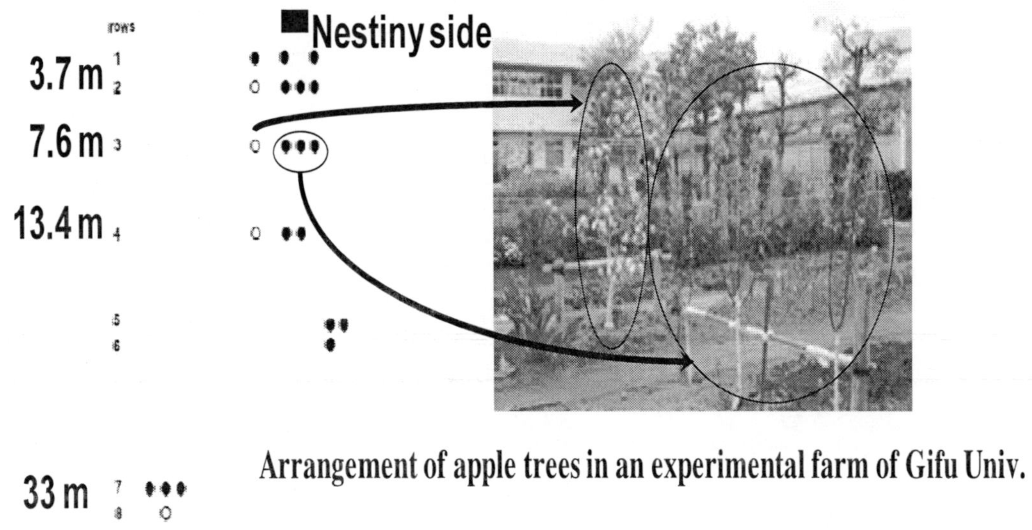

Arrangement of apple trees in an experimental farm of Gifu Univ.

5. Conclusion (Apple Orchard Design)

We have designed a block (approximately 800 m^2) of an apple orchard consisting of pollinizers, pollinator and single commercial cultivars (Figure 7). The block is composed of 10 rows of 11 trees each, with commercial cultivars and three pollinizers at the head, middle, and bottom of each row. The distance between rows was 4 m, and the distance between two adjacent trees within a row was 2 m. Pollinizers were planted every 10 m in each row. As in the case of *O. cornuta*, 530 females per hectare are enough for apple pollination (Vicens and Bosch, 2000), and we put one shelter about 100-200 females in front of the pollinizer's line (Figure 7).

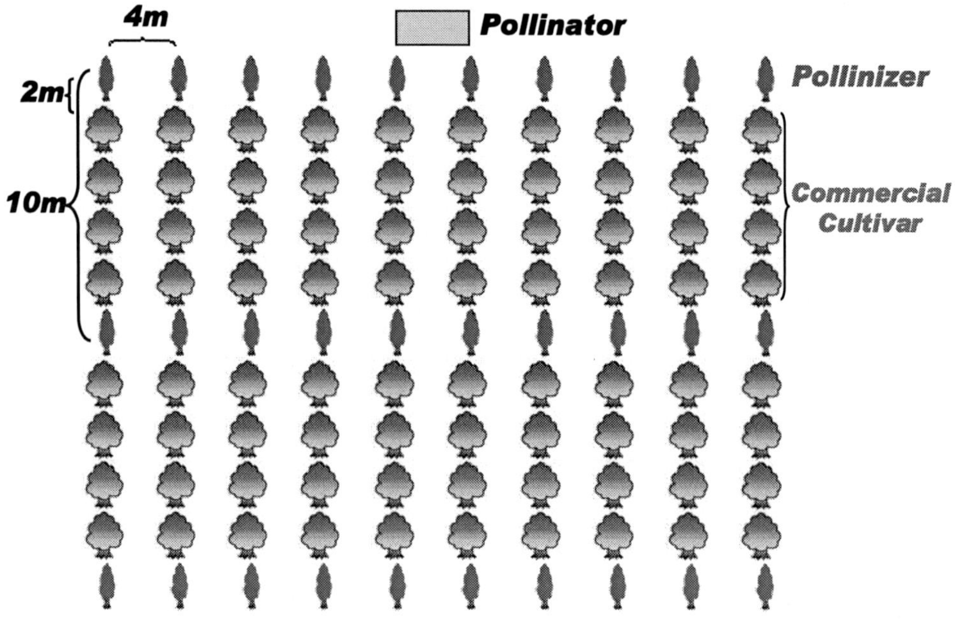

Figure 7 A block (about 800 m^2) of an apple orchard consisting of pollinizers, pollinator and one kind of commercial cultivar.

Acknowledgments

I wish to thank Dr. Hideo Bessho and Mr. Tsutom Maejima for their useful suggestions. This research was supported by the Research Project for Utilizing Advanced Technologies in Agriculture, Forestry and Fisheries, Japan

REFERENCES

Broothaerts, W., Janssens, G.A., Proost, P., Broekaert, W.F. 1995. cDNA cloning and molecular analysis of two self-incompatibility alleles from apple. *Plant Mol. Biol.* 27, 499-511.

Dennis, F., Jr. 2003. Flowering, pollination and fruit set and development. In: D.C. Ferree and I.J. Warrington (eds.). Apples, Botany, Production and Uses. CABI Publishing, Oxon, UK. pp. 153-166.

Kobel, F., Steinegger, P., Anliker, J. 1939. Weitere Untersuchungen uber die Befruchtungsverhaltnisse der Apfel- und Birnsorten. Landvirtschaftliches Jahrbuch der Schweiz 53, 160-191.

Matsumoto, S., Maejima, T. 2010. Several new aspects of the foraging behavior of *Osmia cornifrons* in an apple orchard. *Psyche: A Journal of Entomology* Article ID 384371, 6 pages.

Matsumoto, S., Furusawa, Y., Komatsu, H., Soejima, J. 2003. *S*-allele genotypes of apple pollinizers, cultivars and lineages including those resistant to scab. *J. Hortic. Sci. Biotechnol.* 78, 634-637.

Matsumoto, S., Eguchi, T., Bessho, H., Abe, K. 2007. Determination and confirmation of *S*-RNase genotypes of apple pollinators and cultivars. *J. Hortic. Sci. Biotechnol.* 82: 323-329.

Matsumoto, S., Eguchi, T., Maejima, T., Komatsu, H., 2008. Effect of distance from early flowering pollinizers 'Maypole' and 'Dolgo' on 'Fuji' fruit set. *Scientia Hortic.* 117, 151-159.

Matsumoto, S., Morita, J., Abe K., Bessho, H., Yamada, K., Shiratake, K., Fukui, H., 2009a. *S*-RNase genotypes of wild apples necessary for utilization as pollinizers. *Hort. Environ. Biotechnol.* 50, 1-4.

Matsumoto, S., Abe, A., Maejima, T., 2009b. Foraging behavior of *Osmia cornifrons* in an apple orchard. *Scientia Hortic.* 121, 73-79.

Sassa, H., Mase, N., Hirano, H., Ikehashi, H. 1994. Identification of self-incompatibility-related glycoproteins in styles of apple (*Malus* x *domestica*). *Theor. Appl. Genet.* 89, 201-205.

Sekita, N. 2001. Managing *Osmia cornifrons* to pollinate apples in Aomori Prefecture, Japan. *Acta Hort.* 561, 303-307.

Vicens, N., Bosch, J. 2000. Weather-dependent pollinator activity in an apple orchard, with special reference to *Osmia cornuta* and *Apis mellifera* (Hymenoptera: Megachilidae and Apidae). *Environmental Entomology* 29, 413-420.

Way, R.D., 1978. Pollination and fruit set of fruit crops. *New York Food and Life Sci. Bull.* 76, 1-9.

In: Horticulture in the 21st Century
Editors: Abigail N. Sampson, pp. 213-221
ISBN: 978-1-61668-582-9

Chapter 10

HERITABILITY AND CORRELATIONS AMONG ORNAMENTAL TRAITS OF PANSY (*VIOLA WITTROKIANA*)

Jian Wang[1,2] and Man Zhu Bao[2]
[1]College of Horticulture and Garden Sciences, Hainan University, Danzhou, China
[2]College of Horticulture and Forestry Sciences,
Huazhong Agricultural University, Wuhan, China

ABSTRACT

The heritabilities and the correlations of some ornamental traits of six pansy (*Viola wittrokiana*) inbred lines, such as size of flower (SF), number of flowers per plant (NF), length of flower stalk (LFS), number of branches per plant (NB), height of plant (HP), width of plant (WP), flower index (FI), value of HP/WP (H/W), length of leaf (LL), width of leaf (WL), leaf index (LI), number of nodes (NN) and area of leaf (AL), were calculated. 9 out of 13 traits were significantly different and therefore the broad sense heritability, genetic advances, relative genetic advances, coefficient of genetic variation and coefficient of correlations of these traits were calculated. The results showed that the broad sense heritability values of most traits were high (>80%) except for FI (0.73). The coefficients of genetic variation of SF, FI, WP, HP, LI and NN were not high (0.20-0.35). Genotypic correlations among these traits were generally high. Strong negative genotypic correlations among SF with NF, NB, HP, WP and NN could be observed, and NF was strongly positively correlated to NB, HP, WP and NN. Significant SF×FI positive correlation (r=0.911) and NF×FI negative correlation (r=-0.927) indicated that the flower size should be the more important trait than the number of flower for the breeding of pansy.

Abbreviations: size of flower (SF), number of flowers per plant (NF), length of flower stalk (LFS), number of branches per plant (NB), height of plant (HP), width of plant (WP), flower index (FI), value of HP/WP (H/W), length of leaf (LL), width of *leaf (WL), leaf index (LI), number of nodes (NN) and area of leaf (AL).*

INTRODUCTION

Pansy (*Viola wittrokiana*) is an important biennial flower in the world, and a lot of breeding work of the species has been carried out in the past decades. Novotna (1977, 1978/1979) studied the effect of different pollination methods, and proved that the open-pollinated progenies or artificial self-pollinated progeny plants could achieve similar fruit set, flower size and duration of flowering. The high yield of seed and large flower size were incompletely dominant traits (Novotna 1984, 1985). However, there are very few reports about the heritability and correlations of traits of pansy, except that a positive correlation between the width and the height of pansy flowers and the negative correlation between the flower size and number of flowers were estimated (Novotna 1981, Lee 1998). Thus, identifying the heritability and correlations of some ornamental traits of pansy is important to increase our understanding of the inheritances of pansy, which is critical to the selection of beautiful pansy.

Some quantitative analysis of heritability and correlations among flower traits were reported in other ornamental plants, e.g., *Dimorphotheca* (Hof et al.1999) and *Gerbera* (Drennan et al.1986, Harding 1981, 1990, Huang et al.1983, Huang and Harding 1990, Harding et al. 1998). Variance components analysis has been widely used in these researches. Our study presented here aimed to estimate the heritability and the coefficients of correlation among some important traits of pansy using variance and covariance components analysis to provide a reliable reference for the selection of pansy in breeding practice.

MATERIALS AND METHODS

Six inbred lines of pansy were sown in plugs in autum in the experimental field of Huazhong Agricultural University, Wuhan, China. The names and sources of the inbred lines were listed in Table 1.

Table 1.The inbred lines of pansy used in the experiment

No.	Name	Cultivar Source	No.	Name	Cultivar Source
1	1103	'Swiss giant Berna '	4	2011	'Swiss giant Red wing'
2	DZ01	Mixed cv.	5	2135	Mixed cv.
3	1104	'Swiss giant Berna'	6	1011	Mixed cv.

The seedlings were transplanted to pots at a density of 8 pots/m^2 in substrate of 1:1 mixture of soil and sphagnum when they had 2 leaves. Each inbred line involved 3-replicate randomized complete blocks design with 20 plants in each replicate. All the experiments were arranged in greenhouse to keep the temperature above 5℃ in winter (Nov. 20 to Feb. 30). Plants were evaluated for the following traits at full bloom: size of flower (SF), number of flowers per plant (NF), length of flower stalk (LFS), number of branches per plant (NB), height of plant (HP), width of plant (WP), flower index (FI), value of HP/WP (H/W), length of leaf (LL), width of leaf (WL), leaf index (LI), number of node (NN) and area of leaf (AL). Among these traits,

FI=NF $(SF/WP)^2$

LI=LL/WL

$$AL = \frac{1}{4}\pi(LL)(WL)$$

The heritability and correlation coefficients of these traits were calculated by the variance components analysis using the following model:

$Y_{ij}=\mu+g_i+r_j+e$

with Y_{ij}=observation of genotype i in replication j; μ= general mean; g_i= effect of genotype i; r_j= effect of replication j; e=effect of environment.

The genetic parameters were calculated using the following equations (Milligan 2003, Falconer 1989, Drennan et al.1986, Liu et al. 1984, Ma 1980) where:

$$\sigma_P^2 = \sigma_G^2 + \sigma_E^2, \quad COV_P = COV_G + COV_E, \quad h_B^2 = \frac{\sigma_G^2}{\sigma_P^2}, \quad r_{P12} = \frac{COV_{P12}}{\sqrt{\sigma_{P1}^2 \times \sigma_{P2}^2}},$$

$$r_{G12} = \frac{COV_{G12}}{\sqrt{\sigma_{G1}^2 \times \sigma_{G2}^2}}, \quad r_{E12} = \frac{COV_{E12}}{\sqrt{\sigma_{E1}^2 \times \sigma_{E2}^2}}$$

$$GCV = \frac{\sqrt{\sigma_G^2}}{\overline{\chi}}, \quad GA = k\sigma_G\sqrt{h_B^2}, \quad GA^* = kGCV\sqrt{h_B^2}$$

h_B^2 is broad sense heritability; σ^2_P, σ^2_G, σ^2_E are phenotypic variance, genotypic variance and environmental variance of each trait , respectively; COV_P, COV_G, COV_E are phenotypic covariance, genotypic covariance and environmental covariance of each trait respectively; r_P, r_G, r_E are phenotypic correlation coefficient, genotypic correlation coefficient and environmental correlation coefficient of each trait, respectively; σ_G is genotypic standard deviation; GCV is coefficient of genetic variation; GA is genetic advance; GA^* is relative genetic advance; $\overline{x}$ is the mean of each trait; k is the intensity of selection (the hypothetic k=2.06 in our experiment). The variance and covariance components analyses were presented in Table 2. The data was analyzed with SAS (SAS institute, 2000).

Table 2. The variance and covariance analysis of traits

Source	Degrees of freedom	Variance exception	Covariance exception
Replicates	$r-1$		
Lines	$n-1$	$\sigma_E^2+r\sigma_G^2$	COV_E+rCOV_G
Error	$(r-1)(n-1)$	σ_E^2	COV_E

r: the number of replicates
n: the number of inbred lines

RESULTS

Variance Analysis, the Broad Sense Heritability and Genetic Advance

Out of the 11 traits, 9 had significant differences among inbred lines for the means (Table 3). The LFS and H/W, without significant difference among inbred lines, were unnecessary for further genetic analysis. The columns of means of these 9 traits were showed in Fig.1 to visualize the differences among six inbred lines. Inbred line 3, having large leaf with few flowers and branches, had the maximal value of AL and the least value of NF and NB. Inbred line No.5 had the largest NF, NB, WP and NN, and the least SF, AL, LFS and FI. For the important ornamental traits of flower size, inbred line 2 had the largest value.

Three traits, NF, NB and AL, had high values of *GCVs* (over 5.0), which meant these traits have large genetic variations. Some important traits, such as SF, FI, HP and WP, however, had low value of *GCV* (0.20-0.35), indicating that the ranges of genetic variation of these traits were limited.

The broad sense heritability values of SF, NF, NB, HP, WP, LI, NN and AL were generally high (0.84-0.99), indicating that the selection for these traits would be effective even in the initial generations (Table 3).

Table 4 summarized the values of *GA* and GA^* of the 9 traits. According to the estimates of *GA*, the size of flower of the next generation could reach to 6.85cm (4.10+2.85cm) when the ratio of selection was 5%. NF, NB, and AL had very high relative genetic advance over 100%, indicating that the selections for these traits within the experimental population would be very effective.

PHENOTYPIC, GENOTYPIC AND ENVIRONMENTAL CORRELATION

Table 5 summarized the coefficients of phenotypic, genotypic and environment correlations of 9 traits. In general, the data of coefficients of phenotypic correlations were similar to the coefficients of genotypic correlations, indicating that it was reliable to predict the correlation between traits using the coefficients of phenotypic correlation to some extent. Strong negative genotypic correlations among SF with NF, NB, HP, WP and NN could be observed. NF was strongly positively correlated to NB, HP, WP and NN. SF had positive correlation with AL, although not significant, showing that the plants with large leaves had generally large flowers. The trait of FI, related to SF, NF and WP, had strong positive correlation with SF and strong negative correlations with NF and WP.

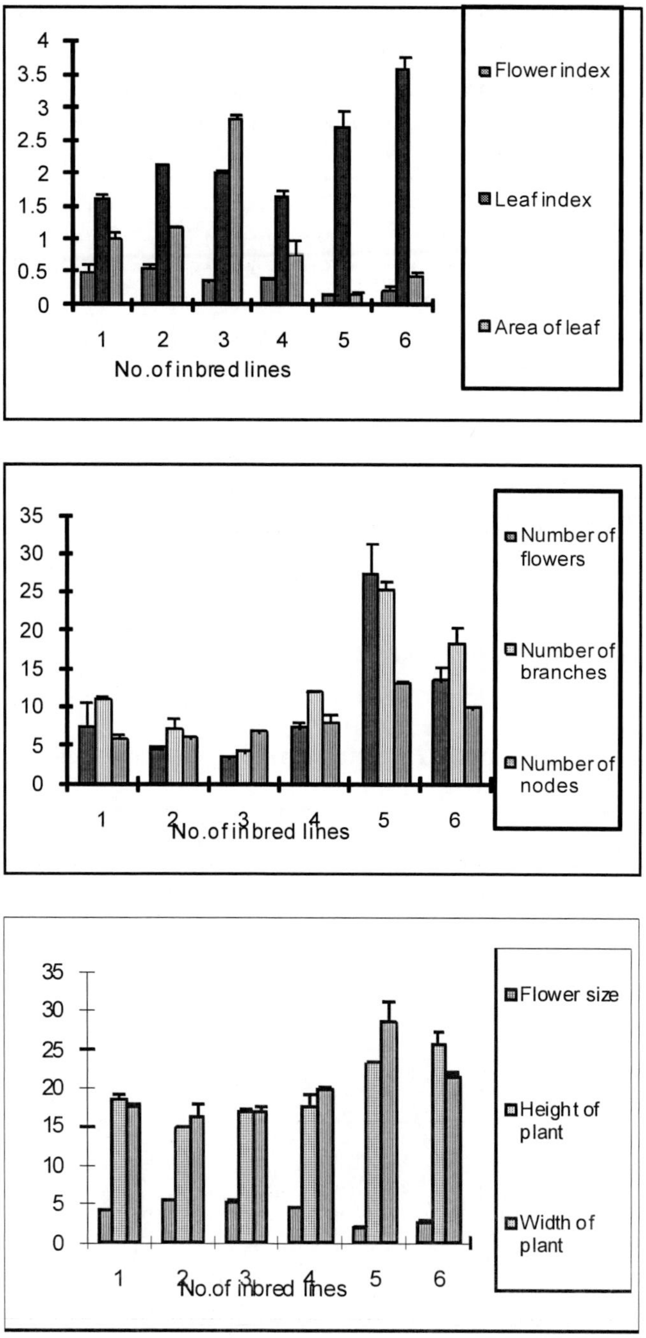

Figure 1. The columns with error bars of means of 9 traits with significant differences among six inbred lines. The traits depended on the fill patterns of the columns showed on the right of the every figure.

Table 3. Means and range along with estimates of phenotypic variance (□P2), genotypic variance (with standard errors), environmental variance (□E2 with standard errors), genetic coefficients of variations (GCV), and broad sense heritability (hB2 with standard errors)

Traits	Mean	Range[#]	σ_P^2	σ_G^2	σ_E^2	GCV	h_B^2
SF(cm)	4.10 **	2.11(5) -5.505(2)	1.95	1.93±1.04	0.02±0.01	0.34	0.99±0.01
NF	10.63 **	3.405(3)-27.365(5)	83.02	75.86±42.50	7.16±3.58	0.82	0.91±0.07
LFS(cm)	8.80 NS	4.245(5)-11.57(1)	/	/	/	/	/
FI	0.36 *	0.155(5)-0.545(2)	0.03	0.02±0.01	0.01±0.003	0.35	0.73±0.21
NB	13.00 **	4.195(3)-25.265(5)	59.70	57.49±31.33	2.21±1.11	0.58	0.96±0.03
HP(cm)	19.52 **	14.885(2)-25.565(6)	17.57	15.30±8.81	2.27±1.14	0.20	0.87±0.11
WP(cm)	20.14 **	16.415(2)-28.575(5)	22.75	19.20±11.25	3.55±1.77	0.22	0.84±0.13
H/W	0.99 NS	0.905(4)-1.2(6)	/	/	/	/	/
LI	2.27 **	1.605(1)-3.58(6)	0.58	0.56±0.30	0.03±0.01	0. 33	0.96±0.04
NN	8.28 **	5.865(1)-13.2(5)	8.31	7.83±4.32	0.48±0.24	0.34	0.94±0.05
AL(cm^2)	1.07 **	0.151(5)-2.831(3)	0.900	0.88±0.48	0.02±0.01	0.88	0.98±0.34

*, Significant at 5% level

**Significant at 1% level

#: Parenthesis indicate the lines (see Table 1).

NS: not significant

Table 4. Genetic advance (GA) and relative genetic advance (GA*) at k=2.06

Traits	GA	GA*
SF(cm)	2.845	69%
NF	17.151	161%
FI	0.215	60%
NB	15.328	118%
HP(cm)	7.519	39%
WP(cm)	8.293	41%
LI	1.495	66%
NN	5.048	61%
AL(cm^2)	1.362	128%

The trait of HP showed positive genetic correlations with NB, WP, LI and NN, however, the correlations were not significant except for NB and LI. AL only showed significantly negative correlation with NB, suggesting that plants with larger leaves had fewer branches.

Table 5. The coefficients of phenotypic, environmental and genotypic correlation

Trait		SF	NF	FI	NB	HP	WP	LI	NN
NF	P	-0.893*							
	E	0.118							
	G	-0.945**							
FI	P	0.789	-0.726						
	E	0.223	0.215						

	G	0.911**	-0.927**						
NB	P	-0.954**	0.927**	-0.731					
	E	0.174	-0.130	-0.026					
	G	-0.982**	0.996 **	-0.866*					
HP	P	-0.891*	0.731	-0.724	0.778				
	E	0.429	0.496	0.505	-0.564				
	G	-0.980**	0.761	-0.995**	0.893 *				
WP	P	-0.854*	0.966 **	-0.818 *	0.905 *	0.697			
	E	0.176	0.755	-0.422	0.076	0.148			
	G	-0.944 **	0.993 **	-0.930 **	0.998 **	0.788			
LI	P	-0.677	0.503	-0.633	0.596	0.778	0.470		
	E	-0.112	-0.630	0.172	0.571	-0.323	-0.577		
	G	-0.694	0.579	-0.778	0.597	0.879 *	0.576		
NN	P	-0.877 *	0.935 **	-0.850 *	0.888 *	0.777	0.949 **	0.606	
	E	-0.260	0.502	-0.032	-0.607	0.645	0.382	-0.298	
	G	-0.902 *	0.969 **	-0.993 **	0.961**	0.796	0.913 *	0.654	
AL	P	0.749	-0.685	0.350	-0.829*	-0.590	-0.644	-0.421	-0.619
	E	0.265	-0.102	-0.098	-0.003	-0.338	-0.022	-0.472	-0.681
	G	0.757	-0.719	0.421	-0.853 *	-0.621	-0.704	-0.420	-0.620

P, E, G: The coefficient of phenotypic, environmental and genotypic correlation, respectively.
*, **: Significant at 5% and 1% level, respectively.

DISCUSSION

The flower size, flower number and plant height are probably the most important traits in the annual or biennial ornamental plant breeding. Big and plentiful flowers and short plant can produce high ornamental value. In this study, the inbred line 2 with the biggest flower size should be the best parent for large flowers, and the inbred line 5 with the most number of flowers should be the best parent for plentiful flowers. However, the flower size and the number of flowers were negatively correlated (Table 5), indicating that it was very difficult to breed variety with both big and plentiful flowers. The result was similar to Lee (1998). Therefore, the flower index, an index of ornamental value, was positively correlated to flower size and negatively correlated to number of flowers, which suggested that the flower size should be the main trait to promote the ornamental value of pansy.

Certainly, the pansy with small and plentiful flowers was also an important type in the market (Bailey 1998). In this study, number of flowers was strongly positively correlated to NB, HP, WP and NN. The result could be explained that pansy had no strong apical dominance, so the individuals with more branches and more nodes, which caused greater HP and WP, could produce more flowers. These relationships suggested that the inbred lines with plentiful branches and nodes such as inbred line 5, should be the best parents for the multiflorous pansy.

The height of plant was also an important trait for the annuals or biennials. In this study, HP had high inheritability (0.87), indicating that selection for the plant height in early generation would be efficient. Therefore, HP show significantly positive correlation with LI

(Table 5), suggesting the plants with long and narrow leave were generally high. Furthermore, HP shows significantly positive correlation with NB, so the dwarf pansy would have few flowers.

The heritability of flower traits of pansy, SF and NF, were high (0.91 and 0.99), and the results were higher than the heritablity of the number of flowers of *Dimorphotheca pluvialis* (h^2=0.30-0.90) (Hof et al. 1999) and the heritablity of the inflorescence size of *gerbera hybrida* (hn^2=0.08) (Drennan et al. 1986). The high heritabilities seen for both traits would lend support for success in attaining better ornamental when choosing these parents. However, this study utilized only "blocks" as the environmental variable. Further evaluations incorporating different locations and years would be required to accurately estimate genotype, environment and genotype× environment effects.

ACKNOWLEDGMENTS

This research was supported by a grant from the Ministry of Agriculture of China (No: 2003-Z36).

REFERENCES

Bailey, A. D., 1998: *Commercial pansy production. Horticulture Information Leaflet.* No. 521, 1-8. College of Agriculture & Life Sciences of North Carolina State University, USA.

Drennan, D., J. Harding, and T. G. Byrne, 1986: Heritability of inflorescence and floret traits in gerbera. *Euphytica*. 35, 319-330

Falconer, D.S., 1989: *Introduction to Quantitative Genetics*, 3rd ed. Longman Scientific & Technical Press, United Kingdom.

Harding, J., H. Huang, T. Byren, and N. Huang, 1990: Quantitative analysis of correlations among flower traits in *Gerbera hybrida* Compositae. I .Genetic and environmental correlations. *Theor Appl Genet*. 80, 552-558

Harding, J., T. G. Byren, and R. L. Nelson, 1981: Estimation of heritability and response to selection for cut-flower yield in gerbera. *Euphytica*. 30, 313-322.

Harding, J., T. G. Byren, and R. L. Nelson, 1981: Heritability of cut-flower vase longevity in gerbera. *Euphytica*. 30, 653-657.

Hof, L., I. G. Nieboer, and O. Dolstra, 1999: Heritability of oil content and onset of flowering in a population of *Dimorphotheca pluvialis*. *Plant Breeding*. 118, 63–69.

Hof, L., L.C.P. Keizer, I.A.M. Elberse, and O. Dolstra, 1999: A model describing the lowering of single plants, and the heritability of flowering traits of *Dimorphotheca pluvialis*. *Euphytica*. 110, 35–44.

Huang, H., and J. Harding, 1998: Quantitative analysis of correlations among flower traits in *Gerbera hybrida* Compositae. III. Genetic variability and structure of principal component traits. *Theor Appl Genet*. 97, 316–322

Huang, H., J. Harding, and T. Byren, 1983: Heritabilty of consumer preference in *Gerbera*. *Euphytica*. 32, 465-474.

Lee, C., 1998: Effect of day length on reproductive growth in *viola* species. *Journal of the Korean society for horticultural science*. 39, 66-73

Liu, L. F., S. X. Mao, and Y. Z. Huang, 1984: Crop quantitative genetics. *Agriculture Press,* Beijing, China.

Ma, Y. H., 1980: The basis of quantitative genetics for plant breeding. *Science & Technology* Press of Jiangsu Province. Nanjing, China.

Milligan, S. B., M. Balzarini , and W. H. White, 2003: Broad-sense heritabilities, genetic correlations, and selection indices for sugarcane borer resistance and their relation to yield loss. *Crop Science.* 43, 1729

Novotna, I., 1977: The number of seeds in pansy pods in relation to polliation method. *Archiv fur Zuchtungs for schung.* 4, 237-246

Novotna, I., 1978/1979: Analysis of flower size in self-pollinated and open-pollinated individuals which differ as a result of mixed pollination of garden pansy (*Viola wittrokiana*) and wild species (*V. arvensis* or *V. tricolor*). *Zbornik Uvtiz, Zahradnictvi.* 5/6, 283-290

Novotna, I., 1979: Expression of some characteristics in progeny of individual clones derived from open- and self – pollination in *viola wittrokiana. Archiv fur Zuchtungs for schung.* 9, 363-371

Novotna, I., 1981: Studies on the relationship between flower size and height in pansy in relation to pollination method or sampling date. *Preslia.* 53, 269-276

Novotna, I., 1984: Inheritance of productivity in pansies (*Viola wittrokiana*) and its use in breeding. Zbornik Uvtiz, Zahradnictvi. 11, 315-324

Novotna, I., 1985: Inheritance of flower size in garden pansies (*Viola wittrokiana*) and its use in breeding. *Zbornik Uvtiz, Zahradnictvi.* 12, 141-148

SAS Institute Inc., 2000: *SAS/STAT User Guider Version 8.1*. Cary, NC, USA.

In: Horticulture in the 21st Century
Editors: Abigail N. Sampson, pp. 223-256

ISBN: 978-1-61668-582-9

Chapter 11

DEVELOPMENTS IN APPLYING BIOTECHNOLOGY TOOLS IN ENVIRONMENTAL ENGINEERING

Bing-Gang Ma*[*], Xiao-Yan Lu, Zhi-Ping Wu and Hu-Ping Zhang

Key Oasis Eco-agriculture Laboratory of Xinjiang Bingtuan, Department of Horticulture, Agricultural College, Shihezi University, Shihezi 832000, People's Republic of China

ABSTRACT

The explosion in biological knowledge is touching almost all areas of science and engineering. Improved understanding can lead to improved design, operation, and maintenance of environmental engineered systems. The integration of molecular biology tools in environmental engineering is a challenge. We discuss our views on the following two critical issues: The opening section provides an informative introduction to Environmental Biotechnology. Like white biotechnology, environmental biotechnology, often referred to as "grey biotechnology", also focuses on sustainability. For instance, environmental biotechnology deals with the treatment of sewage water, the purification of exhaust gas or the decontamination of soils or ground water using specific microorganisms. Environmental technologies are likely to play a more important role as they involve the use of environment-friendly mechanisms. This means starting environmental protection from the very beginning of production. In 1992, at the 21st Agenda of the UN Conference on Environment and Development in Rio de Janeiro, environmental biotechnology was, for the first time, seen to have an important role to play in tackling environmental problems. Traditional biotechnological processes are an integral part of classical "end of pipe" technologies. One could also refer to this as "first generation environmental biotechnology". Traditional environmental biotechnology therefore relies on the metabolic by-products of naturally occurring organisms. The rapid progress in biotechnology, molecular biology and gene technology broadens the application spectrum of environmental biotechnology: Desired metabolic products can be provided in a targeted and inexpensive way. Apart from genetically optimized organisms, isolated enzymes or biological products are available in nearly all amounts desired. Second generation environmental biotechnology is more and more based on high-tech

[*] E-mail: mbg_agr@shzu.edu.cn

methods and processes. This section provides a critical review of recent biotechnological approaches to environment protection. The second section focuses on the application of biotechnology in the bioremediation of environment pollution. Bioremediation is an alternative to traditional remediation technologies such as land filling or incineration. Bioremediation works by either transforming or degrading contaminants to non-hazardous or less hazardous chemicals. A number of bioremediation strategies have been developed to treat contaminated wastes and sites. Several strategies are currently being developed on biological wastewater treatment including aquatic plants, membrane and microorganisms. Bioremediation is also applied in removing pollutants from contaminated soil and petroleum industry. Bioremediation has become one of the most rapidly developing fields of environmental restoration.

INTRODUCTION

This is a century of biotechnology. Many of the scientific breakthroughs we predicted more than twenty years ago are now coming out of the laboratory and into widespread commercial use, bringing us face to face for the first time with the promises and perils of this new era. Although GM technology was first developed for plants in 1983, the First GM crop (termed more appropriately as 'biotech crop') reached the markets during the mid-1990s and during the past decade the development and adoption of transgenic technology has progressed rapidly. In 2007, the global area of biotech crops increased for the twelfth consecutive year at an annual growth rate of 12%. While the technology was initially applied in developed countries, in 2007 biotech crops were grown by 12 million farmers in 23 countries covering 114.3 million hectares and over 90% of the beneficially farmers were resource-poor in developing countries, with increased incomes from biotech crops contributing to the alleviation of poverty (James, 2007). Significant progress has also been made towards the elucidation of plant genes that, as single variables, are able to improve yield (Camp, 2005). Much has been written about the various, different perceptions of transgenic crops. To some they represent an exciting opportunity to use modern biological techniques to improve the quality of agricultural plants and their products, whilst to others they are seen as an insidious attempt to extend the power of multinational companies and thereby threaten the well being of subsistence farmers, as well as the ecology of fragile ecosystems throughout the world. In addition, the rise of opposition is linked to a lack of public understanding of current agriculture. There are two perceptions: (1) 'Traditional' agriculture is well perceived (often associated with the preservation of nature, quality products, and small farming systems). (2) 'Modern' agriculture is considered to have had a negative impact on all of the above factors. Therefore, the use of GMOs in agriculture is associated with modern agriculture and, thus, motivates a negative perception.

ENVIRONMENTAL BIOTECHNOLOGY

Biotechnology could be widely applied in the technological development of environmental protection. Environmental biotechnology is newly developing technology in the 21st century. It is the sector of biotechnology that concerns microbial processes for efficient, effective and economic waste treatment, particularly the elimination of noxious pollutants; microbial processes for the remediation of previously polluted natural environments and for cleaning up dirty naturally occurring feedstock; the systematic evaluation and control of microbial mediated degradation processes in engineered environments, and the impact of both intentional and accidental contamination of natural environments with non-indigenous microbes. It is generally acknowledged that environmental pollution, resource problem, ecology destroy, and health problem and so on can be solved by environmental biotechnology. The term environmental biotechnology has a certain air of modernity when in fact it has a long history of use, if one considers the underlying principles and not the appellation. However, as part of its complex meaning, there is a dynamic new definition and purpose in this discipline with regard to bioremediation. The ability to probe the environment at the molecular level with exquisite methods, to create a new awareness of fundamental biological processes therein, has created an important new paradigm in remediation engineering design and management. Further, biological lines of evidence made extremely robust through the merger of biotechnology and environmental science are poised to be incorporated into the very fabric of site evaluation and disposition at the regulatory level. At the operational level, the field of environmental biotechnology is driven by the "omics" the common suffix for disciplines like genomics, proteomics, and metabolomics. Environmental biotechnology has made great progress in the following fields: Monitoring technology and treatment of dioxin-like chemicals and cyanobacterial toxins, biofiltration for air pollution control, solid waste treatment and reutilization, bioremediation of soil pollution, risk assessment and control of endocrine disturbing substances in environment, wastewater.

2. Gene Engineering and Environmental Safety

Plant biotechnology is founded on the principles of cellular totipotency and genetic transformation. Many experimental strategies that led to the production and commercialization of biotech (transformed or transgenic) plants expressing many useful genes, and emphasizes the beneficial effects of plant biotechnology on food security, human health, the environment, and conservation of biodiversity. In the last decade, the technique to genetically modify crop plants has gained more and more interest in terms of bioproduction of heterologous proteins. Plants have been discovered as a possible source for large amounts of cost effective recombinant protein. Main application fields are therapeutics for use in animal and human health, diagnostics, and technical enzymes.

Potential impacts of genetically modified organisms (GMOs) on the environment have been addressed in several authoritative reviews (Rissler and Mellon, 1996; Traynor and Westwood, 1999; Wolfenbarger and Phifer, 2000; Letourneau and Burrows, 2002). Evidence from recent studies is then presented with the aim of suggesting possible improvements to contemporary risk assessment.

(a) 2.1. Transgenic Biotechnology

Development of transgenic plants in an effective manner is important pre-requisites for sustainable and economic use of biotechnology for crop improvement (Sharma et al., 2002).

The process of genetic transformation involves several distinct stages, namely insertion, integration, expression and inheritance of new DNA and, in most current transformation systems, a tissue culture stage that is prerequisite for the ultimate recovery of transgenic plants. Although almost every significant crop species has been successfully transformed, in many species the development of rapid, highly efficient, and routine transformation systems is still in progress and thus represents a bottleneck in the development of stable high yielding transgenic plants (Sharma et al., 2002). During the last two decades the development and adoption of transgenic technology has progressed rapidly. Although almost every significant crop species has been successfully transformed, the absence of a rapid, highly efficient, and routine transformation system in many crop species represents a problem in the development of stable high yielding transgenic plants. Almost every type of plant tissue has been subjected to gene transformation experiments and several methods for transformation have been developed. However, only *Agrobacterium* mediated transformation and particle bombardment are being used routinely for successful development of transgenic plants and most of the efforts are focused on herbicide and insect tolerance. The main drawback of the *Agrobacterium*-mediated system is its dependence on tissue culture regeneration ability of a plant species or a variety of crop plant and the transformation ability of that species or variety by *Agrobacterium*. Although particle bombardment circumvents the problems associated with agro-infection, regeneration of a complete plant still requires tissue culture. For this reason, there is considerable interest in developing plant transformation methods that circumvent tissue culture steps. Gene targeting is still an elusive goal for the plant biotechnologist. It will continue to be an interesting area for research in the years to come because of its outstanding potential. Until an efficient method for creating precise changes at the DNA level is developed, RNAi will continue to be the major tools for gene silencing and the study of gene functions in plants. (Akhond and Machray, 2008)

Crucial steps in gene targeting in plants include: the design of the gene targeting construct, the choice of the tissue to be transformed, the selection of transformants and their regeneration, and the screen for clean, targeted transformants. In the reported rice GT strategy, a combination of all these factors probably contributed to the reported success (Hohn and Puchta, 2003). A number of different approaches have been taken to combine or 'pyramid' transgenes in one plant and have met with varying degrees of success. These approaches include sexual crossing, re-transformation, co-transformation and the use of linked transgenes. These methods have been utilized to generate the 'stacked trait' varieties now increasingly being grown in the USA. In the future it is likely that novel, alternative technologies will be developed that use single transgenes to coordinate the expression of multiple genes. Amongst the constructs being tested at present are some modeled on viral polyproteins can be used to simultaneously introduce multiple protein-coding genes into plants.

(b) 2.2. Development in Transgenic Plants

1) 2.2.1. The Development of Transgenic Plants

The various types of research programme will be categorized according to the various objectives which range from industrial applications, to alterations in seed quality traits and food constituents, to more long term strategies (Dunwell, 1999) to use crops as a source of valuable pharmaceutical products.

During the past decade the development and adoption of transgenic technology has progressed rapidly. In 2007, biotech crops were grown by 12 million farmers in 23 countries covering 114.3 million hectares. This progress can be attributed to development in molecular genetics, plant transformation and regeneration techniques and a better understanding of the underlying processes involved in DNA recombination. While almost every significant crop species has been successfully transformed, in many species the development of rapid, highly efficient, and routine transformation systems is still in progress.

The 'first generation products principally comprise soybean, maize (corn) and cotton resistant to herbicides and/or insects. The wide range of 'second generation products, under development and testing in many commercial and academic laboratories, include examples for a variety of food, medical, veterinary and industrial purposes. Most of the GM crop varieties has around the world contain introduced microbial genes providing tolerance to broad spectrum herbicides (e.g., glufosinate or glyphosate) and/or resistance to insect pests such as the European corn borer. These varieties now predominate in some areas of the USA, although they are still only undergoing relatively small scale testing in Europe, where the need for GM products and their safety are the subject of concerted opposition from some individuals and groups. Most participants in this debate accept that the lack of perceptible consumer benefit has been a significant deterrent to acceptance of this technology, particularly in the context of other food scares. Whether self interest will ever overcome this resistance may depend partly on the development of new products where the real and obvious benefit can be seen to be outweighing the assumed and non-obvious risk.

(ii) 2.2.2. Applications of Transgenic Plants

More food and feed have to be grown on less land. Fortunately, in the 1970s, new tools for more efficient plant breeding arose, one of which is biotechnology. As agrochemistry and mechanization have reached a level where further progress is neutral in relation to plant production—scientific progress helps environmental needs—only breeding may close the yield gap by a combined application of classical and molecular techniques (Gerhard Wenzel, 2006). Application of molecular plant breeding is now contributing to discoveries of genes and their functions opening new avenues for basic plant biology research (Moose and Mumm, 2008). Transgenes could also be transferred using much cheaper conventional plant breeding techniques to locally adapted crop varieties in countries with less sophisticated research facilities. The combination of traditional crop improvement techniques and modern biotechnological techniques can contribute substantially to human well-being by providing renewable sources of food, feed, chemicals, pharmaceuticals, energy and for the creation of a sustainable environment for the developed as well as for the developing countries of the world.

One of the potential environmental benefits of transgenic plants is their use in bioremediation, the process of decontamination of polluted soils or water. For example, mercury pollution is a major environmental problem accompanying industrial activities, with most of the mercury released being retained in the soil as complexes of the toxic ionic mercury (Hg^{2+}*)*, which then can be converted by microbes into the even more toxic methylmercury. Mercurydetoxification of the soil can also occur by microbes converting the ionic mercury into the least toxic metallic mercury (Hg0*)* form, which then evaporates. The remediation potential of transgenic plants carrying the *MerA* gene from *E. coli* encoding mercuricion reductase has been evaluated in several cases. For example, a codon-optimized version of the gene was introduced into tobacco and it was found that transgenic seeds were resistant to $HgCl_2$ at 50 μM, and some of them could germinate on media containing as much as 350 μM $HgCl_2$, while the control plants were fully inhibited or died on 50 μM $HgCl_2$. The rate of elemental mercury evolution from Hg^{2+} (added as $HgCl_2$) was 5-8 times higher for transgenic plants than the control with volatilization being up to 5 times higher in the roots than in shoots or the leaves. Other compounds shown to be detoxified by transgenic plants include explosives, chlorinated solvents (Doty et al., 2000) and hydrocarbons (oil and petroleum residues). Recovery of agricultural resources (vast territories polluted with pesticides, oil, or heavy metals) was yet another important issue dealt with at the section. Bioremediation techniques, as being the least expensive and risky are to be implemented to address it. The use of rhizospheric bacteria of the genus *Pseudomonas*, which act as growth stimulators in certain plants, appears to offer the greatest promise in this respect.

One of the recurring themes concern the potential of using plants as a production system for high value compounds of medical importance. Examples include the expression of antibodies that may reduce the growth of bacteria associated with dental caries or that can be used in the treatment of human tumours. The advances in the use of plants to produce vaccines (Arntzen et al., 2001) for veterinary or medical purposes have extended greatly. Much attention to GM crops is discussion of their potential as a production system for biodegradable plastics, particularly poly beta-hydroxy butyrate and poly beta-hydroxy valerate.

(c) 2.3. Achievements of Transgenic Biotechnology

(i) 2.3.1. Improvement of Nutritional Qualities

How do genetically engineered foods compare with conventional foods in nutritional quality? This is an important issue, and one for which there will probably be much research in the future, as crops that are engineered specifically for improved nutritional quality are marketed. There are many, varied applications of transgenic technology in modifying food quality. Although some of these refer to general attributes such as resistance to freezing the majority are concerned with specific changes in the amounts of beneficial or toxic compounds.

The improvement of either the quantity or quality of seed protein is at the centre of many plant breeding programmes (Vasil et al., 2001) and there is a wide diversity of approaches to the modification of specific traits. For example, it has been claimed that expression of an inositol polyphosphate phosphatase homologue improves the total amount of protein. Amongst the novel means of modifying a quantitative trait, namely grain texture, is one comprising the introduction into wheat of genes encoding puroindoline A and B. Some projects aim to alter only a specific amino acid (e.g., methionine, lysine, tryptophan) in order to improve nutritional quality whilst other studies have more general objectives, These latter studies depend on the fact that the composition of high molecular weight subunits of glutenin determines the gluten strength and influences the baking quality of bread wheat.

As well as the projects designed to modify protein quality, there are many concerned specifically with changing the profile of starch produced either in the seed or other storage organ such as the tuber of potatoes. Examples include the use of a beta-amylase gene, a rice starch granule-bound protein, a debranching enzyme, or starch synthase to produce modified starches in transgenic plants. It is also claimed that an overall increase in the level of stored carbohydrate can be achieved by over expressing dihydroorotase, a gene of pyrimidine metabolism.

(ii) 2.3.2. Insect Resistance

Losses due to insect herbivores, estimated at 10-20% for major crops, are a significant factor in limiting food production. Engineering crop plants for endogenous resistance to insect pests has been one of the real successes of GM technology, following the commercial introduction of genetically modified maize, potato and cotton plants expressing genes encoding the entomocidal-endotoxin from *Bacillus thuringiensis* (*Bt*) in the mid-1990s (Ferry et al., 2004). In 2007, 20.3 million hectares (18%) were planted with *Bt* crops making it the third most important transgenic trait.

Although the acreage of Bt-transgenic crops has increased dramatically in recent years, there are concerns that sustained efficacy of Bt crops will be endangered by the development of resistance in populations of insect pests that they are intended to control. To date, several major pest species, including *Heliothis virescens*, *Leptinotarsa decemlineata*, and *Plodia interpunctella*, have demonstrated the ability to adapt to Bt in the laboratory. Moreover, *Plutella xylostella* (diamondback moth, DBM) has evolved sufficiently high resistance to Cry1A proteins in field settings to result in field control failures. Such evidence of insect resistance to Bt proteins emphasizes the need to develop effective resistance management strategies.

(iii) 2.3.3. Herbicide Resistance

Detoxification is an important mechanism by which many plants as well as bacteria develop natural tolerance to a variety of herbicides. During the past two decades, novel and successful strategies have been developed for engineering herbicide tolerance in plants by isolating, cloning and characterizing bacterial or plant genes that detoxify specific herbicides and then transferring them to crops of choice by genetic transformation. This has made it possible to use a new generation of non-selective herbicides that are relatively inexpensive, highly potent, environmentally friendly and biodegradable, such as Roundup and Basta, by engineering the relevant resistance genes into a variety of crop species, including wheat (De Block et al., 1987; Duke, 1996; Gressel, 1996; Padgette et al., 1996; Vasil, 1996, 1999; Heck et al., 2003). The fact that nearly 75% of the biotech crops planted worldwide are engineered for herbicide resistance (mostly Roundup tolerance) is a testament to their usefulness in agriculture and their attraction to farmers.

(iv) 2.3.4. Disease Resistance

Protection against diseases caused by pathogenic organisms like fungi, bacteria and viruses are a major challenge for crop improvement. The most important class of genes that has been used by breeders for disease control are the plant resistance (*R*) genes. The introgression of *R*-genes into elite cultivars *via* traditional breeding can take up to 15-20 years. This process can be considerably shortened by introducing these genes by genetic Modification of crop plants (Rommens and Kishore, 2000). Recently cloning of the major resistance gene *RB* in *Solanum bulbocastanum* by using a map-based approach in combination with a long-range (LR)-PCR strategy has been reported. A cluster of four resistance genes was found within the genetically mapped *RB* region. Transgenic plants containing a LR-PCR product of one of these four genes conferred broad spectrum late blight resistance to potato (Song et al., 2003).

Significant advances have also been made in the development of virus-resistant crop plants since the first report of pathogen-derived resistance (PDR) to limit virus infection in the form of coat protein mediated resistance (Beachy, 1997). Although coat protein-mediated resistance (CPMR) has been the most widely applied PDR strategy for the last 20 years, it could be replaced in future by the recently developed RNA-mediated mechanisms based on short interfering RNA (siRNAnd microRNA (miRNA)-mediated gene silencing- derived strategy (Ritzenthaler, 2005). Besides, Niu et al. (2006) reported that artificial miRNAs can confer virus resistance in plants.

(v) 2.3.5 Pharmaceuticals, Proteins, Plastics Etc

For more than two decades, bio-engineered plants have produced protein therapeutics for human and animal use. Almost all proteins produced by other existing systems, including antibodies, vaccines and plasma proteins, have now been manufactured in plants (Joshi and Lopez, 2005). Several seed crops, including cereals, grain legumes and oilseeds, have been explored as production platforms, and the first commercial products—all technical proteins and enzymes—have already reached the market (Stoger et al., 2005). Fibrous proteins from spiders, proteins with synthetic multiple repeats, mammalian structural proteins such as collagen and biodegradable plastic have been produced in transgenic plants (Scheller and Conrad, 2005). Environmental pollution with metals and xenobiotics is a global problem, and the development of phytoremediation technologies for the plant-based clean up of contaminated soils is therefore of significant interest (Krämer, 2005).

(d) 2.4. Safety of Transgenic Plants

Because of the widespread public concern regarding the safety of biotech crops, rigorous testing and safety assessment is carried out before any such crops are approved for commercial cultivation. Besides the concern for marker genes, other worries with genetically engineered crops persist partially because of the perception that gaining access to a wider range of genetic diversity, crossing species barriers, and introducing foods with additional proteins present safety concerns via our diets (Cohen et al., 2003). The likelihood that a transgene will spread in the environment depends on its potential fitness impact. Mathematical models and empirical experimental evidence suggest that genetic approaches have the potential to effectively prevent transgenes from escaping into wild relatives and becoming established in wild populations that are not reproductively isolated from genetically engineered crops (Lee and Natesan, 2006). In regards to food safety issues, most biotech products are considered substantially equivalent to traditional counterparts, with exceptions for certain well-defined differences. Safety evaluations usually focus on these defined differences (Cohen et al., 2003).

In the US, according to the National Academy of Sciences, the trait of a GM plant should be the focus of the risk assessment, and not the process by which it has been produced (NAS, 2000). As a result the Environmental Protection Agency (EPA), requires that risk assessment tests are performed on non-target organisms to assess toxicity of so called plant-incorporated protectants. This is in contrast to EU regulatory procedures which consider GM plants as something new, for which both direct (toxic) and indirect effects should be assessed prior to their commercial release. Since in the US the major concern regarding the potential effects of GM plants expressing insecticidal proteins on non-target organisms is that related to the toxicity of the protein contained in the plant, test requirements on non-target organisms remained similar to those required for plant protection products. This means testing the effects on a set of representative insects such as bees, ladybirds and lacewings (NAS, 2002; EPA, 2002) using a tiered testing system in which the first tier reflects a maximum hazard approach. Organisms are exposed to the pure insecticidal protein at a range of doses and concentrations 10 to 100 times higher than those expected in the environment. Since these tests do not provide any information on the possible non-intended indirect effects that could occur once the plants are released in the field, the EPA often requires information from field studies (EPA, 2003). Field studies, however, have often the drawback of being expensive and lack appropriate statistical power.

(i) 2.4.1. Effects of Transgenic Plants on Natural Ecosystems

Genetic pollution is undesirable gene flow into wild populations. The term is usually associated with the gene flow from a genetically engineered (GE) organism (or genetically modifed organism (GMO) to a non GE organism; however, conservation biologists and conservationists are using it to describe gene flow from a domestic, feral, non-native or invasive species to a wild indigenous population. Hybridization of crops with nearby weeds may enable weeds to acquire traits we wish they didn't have, such as resistance to herbicides. Research results indicate that crop traits may escape from cultivation and persist for many years in wild populations. Genes that provide a competitive edge, such as resistance to viral disease, could benefit weed populations around a crop field. Many cultivated crops have sexually compatible wild relatives with which they hybridize under favorable circumstances. The likelihood that transgenes will spread can be different for each crop in each area of the world. Thus each crop must be evaluated individually for the risk of gene flow in the area where it will be grown.

Many factors influence the potential for gene flow from crop to crop. Some crops are highly out crossing, with pollen carried to other fields by wind and by insects. Other species are highly self-pollinating, with little potential for pollen transfer to neighboring plants. Because of the differences among crops species, every case must be evaluated individually for potential to contribute to gene flow from transgenic to conventional crops.

(ii) 2.4.2. Effects of Transgenic Plants on Human Health

Genetically modified (GM) plants of the ‘‘first generation were conferred with traits beneficial to producers, whereas GM plants in subsequent ‘‘generations’’ are intended to provide beneficial traits for consumers. Golden Rice is a promising example of a GM plant in the second generation, and has overcome a number of obstacles for practical use. Furthermore, consumer-acceptable plants with health-promoting properties that are genetically modified using native genes are being developed. The emerging technology of metabolomics will also support the commercial realization of GM plants by providing comprehensive analyzes of plant biochemical components.

Agronomic traits such as herbicide tolerance, insecticides, virus resistance, disease-tolerance, and delayed ripening, were the initial focus of genetic engineering in plants, which are primarily beneficial to growers and the seed companies that pioneered the work. These are the so-called ''first generation genetically modified (GM) plants''. In 2005, almost all GM acreage was occupied by first generation plants, dominated by herbicide-tolerance (71%), followed by insect resistance (18%) and a combination of these traits (11%) (ISAAA). The last decade has largely been dedicated to the development of ''second generation'' GM plants with traits beneficial to consumers such as high nutritional value, premium quality and low allergenicity. The second generation includes soybeans with high oleic acid- or laurate-content and rice with high vitamin A or iron content, or low allergenicity. Development of a ''third generation'' as protein factories of vaccines and therapeutic medicines is in progress. First generation GM plants mainly benefited producers by improving crop productivity, reducing agrichemical inputs and subsequently reducing labor costs. However, GM plants provided no obvious benefits for consumers over traditional plants. In contrast, the next generation of GM plants has been developed based on consumer benefits. Researchers and developers have been carefully choosing the traits to induce, including their origin, expression systems and selectable markers. The second and third generation GM crops which can produce substances for promoting human health, including medicines and vaccines have great potential for commercial application (Keiko Yonekura-Sakakibara and Kazuki Saito, 2006).

Fears and exaggerated claims of possible adverse effects of biotech foods and crops on humans and the environment are unfounded and have no basis in fact. Indeed, it is remarkable that no such ill effects have been documented after 12 years of extensive cultivation in diverse environments, and the consumption of biotech foods by more than a billion humans and even a larger number of animals. Many practical applications of plant biotechnology, including the introduction into the market place of other biotech crops with useful and highly desirable characteristics, have been prevented by political and ideological opposition from vested interests—including the European Union, a few developing countries, and some self-serving environmental groups—and by the many redundant, overly stringent, time-consuming and exorbitantly expensive regulatory requirements (Vasil, 2003).

The potential human health hazards from transgenic crops are also generally recognized by the scientific and regulatory communities (NRC, 2001). The potential risks generally relate to 'the possibility of introducing new allergens or toxins into food-plant varieties, the possibility of introducing new allergens into pollen, or the possibility that previously unknown protein combinations now being produced in food plants will have unforeseen secondary or pleiotropic effects.' Scientific studies have found transgenic crops that contain allergens (Nordlee et al., 1996) as well as crops with unintended effects. As with the potential environmental risks from transgenic crops, the likelihood that a particular transgenic crop may have a harmful health effect may be small but if it can occur, the government should step in to make sure that any transgenic crop with those harmful characteristics is not commercialized.

According to the information reported by the WHO, the genetically modified (GM) products that are currently on the international market have all passed risk assessments conducted by national authorities. These assessments have not indicated any risk to human health. For transgenic crops, protecting humans and the environment is the primary, but not exclusive, reason for government oversight and regulation. To date, there is little evidence of specific harmful environmental effects from the millions of acres of transgenic crops that have been planted worldwide nor any evidence of harm from the many foods that humans have eaten containing transgenic crop ingredients. There is general scientific consensus, however, that individual transgenic crops could potentially present risks to humans or the environment. For example, the United States National Research Council identified four categories of potential environmental hazards from the release of a transgenic crop: '(1) hazards associated with the movement of the transgene itself with subsequent expression in a different organism or species, (2) hazards associated directly or indirectly with the transgenic plant as a whole, (3) non-target hazards associated with the transgene product outside the plant, and (4) resistance evolution in the targeted pest population. That one or more of those risks could result from applying transgenic technology to a crop is generally recognized throughout the scientific community (Snow and Moran-Palma, 1997; NRC, 2001)

(iii) 2.4.3. Effects of Biotechnology in Industries Product

The development of novel bioindustries is dependent upon the successful tapping of the invisible reservoir of microbial diversity. Naturally occurring or genetically engineered microorganisms are constituent components of bioactive processes that are conducive to the environment and the generation of market-competitive technologies. The role of microbial intervention in every sphere of human existence and endeavour is so integral that it is always glossed over. However, the role of microorganisms, and those that have been genetically engineered for specific purposes, in globalized biotechnologies is significant especially since they are economic agents of change, for example in fields such as: bioremediation bacterial using of aquatic and terrestrial environments; bioconversion technologies conversion of biowastes into utility products; biofuels and bioenergy generation biodiesel, alcohols, and ethers; biocleaning technologies desulphurization of coal; biopharmaceutical production gene therapy; bioproducts for agriculture biofertilizers, biopesticides; bioinformatics DNA-based computers, gene sequencing, novel plants and oral ornamentals.

(iv) 2.4.4. Effects of Transgenic Biotechnology on Biodiversity

Fear for the loss of biodiversity is the focal point of opposition of several influential environmentalists groups against GM crops. It was hypothesised that GM crops could be a threat to the crop diversity or outgrow a local flora to the detriment of native species in Mexican maize landraces (Rissler and Mellon, 1993). However, a large-scale systematic survey showed no evidence for transgene introgression into maize landraces of Oaxaca, Mexico and is an important reference for local farmers, government agencies. The study also addresses socioeconomic and ethical implications of use of GM crops thereon (Garcia et al., 2005).We recognize that non-target herbivorous arthropods feeding on the transgenic crop may also be affected (lethally or sub-lethally). These effects may lead to a reduction in populations which may affect organisms in higher trophic levels. For registration of pesticide products, tests on non-target herbivorous arthropods are generally not required, despite the fact that most insecticides affect to some extent non-target herbivores that occur in the crop. By adopting the same principle, one can argue that risk assessment for non-target herbivores should not be required for registration of insect resistant transgenic plants. However, given the season-long expression of insecticidal proteins in transgenic plants it would be useful to have information on the effects of transgenic plants on some non-target herbivores.

One of the primary concerns related to the adoption of insect resistant transgenic plants in the environment is the detrimental effect that these may pose on non-target organisms, including entomophagous arthropods (parasitoids and predators) which have an important function in regulating pests. Despite the fact that regulatory bodies require information regarding the potential risk of releasing transgenic plants in the environment, to date, no specific protocols have been designed for assessing the risks of insect resistant transgenic crops on entomophagous arthropods. The only insect resistant transgenic plants that are commercially available are those expressing genes which code for *Bacillus thuringiensis* 612 ANNA DUTTON ETAL (Bt) toxins. The amount of area being cultivated with these crops is rapidly increasing (James, 2002), and other genes coding for new Bt-toxins, lectins, proteinase or α-amylase inhibitors, and other insecticidal products have been successfully engineered in plants. Some of these plants are being tested at the field scale, such as peas (*Pisumsativum*) expressing the gene coding for common bean α-amylase inhibitors (αAIs).

Future agricultural and, ultimately, also natural ecosystems will be challenged by the large-scale introduction of entirely novel genes and gene products in new combinations at high frequencies all of which will have unknown impacts on their associated complex of non-target organisms, i.e. all organisms that are not targeted by the insecticidal protein. In times of severe global decline of biodiversity, pro-active precaution is necessary and careful consideration of the likely expected effects of transgenic plants on biodiversity of plants and insects is mandatory. The future challenge in pest management will be to explore how transgenic plants can be incorporated as safe and effective components of IPM systems and what gene technology can contribute to the needs of a modern sustainable agriculture that avoids or reduces adverse impacts on biodiversity? For mainly economically motivated resistance management purposes, constitutive high expression of Bt-toxins in transgenic plants is promoted seeking to kill almost 100% of all susceptible target pest insects. However, for pest management this is usually not necessary. Control at or below an established economic injury level is sufficient for most pests and cropping systems. It is proposed that partially or moderately resistant plants expressing quantitative rather than single gene traits and affecting the target pest sub-lethally may provide a more meaningful contribution of agricultural biotechnology to modern sustainable agriculture. Some examples of such plants produced through conventional breeding are presented. Non-target effects may be less severe allowing for better incorporation of these plants into IPM or biological control programs using multiple control strategies, thereby, also reducing selection pressure for pest resistance development.

(v) 2.4.5. Un-Predictability of Transgenic Biotechnology

Transgenic biotechnology presents an exciting range of possibilities, from feeding the hungry to preventing and treating diseases; however, these promises are not without potential peril. Genetic engineering is changing the terrain of development studies. Societies have reacted with indifference or appreciation to genetically engineered pharmaceuticals, beginning with insulin; yet for food and agriculture, a globally contentious politics and unprecedented policy dilemmas have arisen. Transgenic organisms raise questions of property, ethics and safety unimaginable a generation ago: what can be owned and with what responsibility? Are we blurring the lines between species by creating transgenic combinations? What ethical, social, and legal controls or reviews should be placed on such research? Much turns on science: how one conceptualizes evidence, knowledge, uncertainty and risk. Both opponents and proponents of frontier applications in biotechnology have a poverty story to tell, but with divergent implications. The balance in this global debate has perceptibly shifted; a new developmentalist consensus concludes that the world's poor may benefit from genetic engineering: the question is 'under what conditions'? The new consensus assumes much about biosafety, bioproperty and biopolitics that is contrary to ground realities

- the actual capacity of firms and states to monitor and control biotechnology - but raises new questions at the frontiers of development studies. Transgenic and genetic engineering present intriguing and difficult challenges for 21st century scientists and ethicists. Until we as a society or, perhaps, as a global entity can agree on what beings, human or otherwise, are worthy of moral and legal status and respect, we can expect intense cross-disciplinary debate and discussion as new intelligent life is created through science and medicine.

(vi) 2.4.6. Safety Assessment and Management of Transgenic Plants

Transgenic crops have the potential to benefit both developed and developing countries. To ensure safe crops to humans and the environment, a strong, but not stifling, regulatory system needs to be established and properly implemented. Compositional analysis represents one component of the risk assessment process for GMOs (EFSA 2004). In Europe, the risk assessment strategy for GMOs focuses on comparison between the GMO and derived products and their non-GM counterparts. This comparative assessment approach is based on the premise that traditionally cultivated crops have gained a history of safe use as food/feed for human or animals and with regard to any adverse environmental impact. These crops serve as a baseline for the environmental and food/feed safety assessment of GMOs. The concepts of familiarity and substantial equivalence were developed by the OECD (EFSA 2004) and further elaborated by WHO/FAO. This comparison is the starting point of the safety assessment which then focuses on the environmental or food/feed safety and nutritional impact of any intended or unintended differences identified.

"Novelty" and "societal benefit" are two central criteria for the possible significance of technological developments. The present report on transgenic plants of the second and third generation (with the emphasis on molecular farming) is devoted to a balanced consideration of these two perspectives: In terms of novelty, the aim is to focus on new questions of evaluation, while the aspect of societal benefits will focus on the societal context as a whole, rather than one-sided consideration of the risk dimension or specific economic interests. At present, transgenic plants of the safety evaluation focused on two aspects, one is the safety of the environment, and the other is food security. There is little reliable information on the individual aspects of possible ecological and health risks of the various groups (or even individual transgenic properties). The risk discussion for FF GMP is focusing on the basic question of safety evaluation of innovative and primarily functional foods, while for PMP GMP the emphasis is on possible release into the environment and food and preventing this (depending on the individual instance, PMI can be closer to FF or PMP GMP). Particularly with regard to PMP, the risk debate on molecular farming generally has so far concentrated almost entirely on the question of reliable sequestration and containment/confinement of GMP. The discussion of benefits is taking on new priority compared with the first generation of GMP, including risk evaluation and regulation. So far, it has been possible to ignore doubts about the benefits of the genetically introduced properties e.g. from the regulatory point of view (because no concrete risks to health and the environment were established as a prerequisite for approval), and to leave evaluation to market forces. In future, however, the desired benefit (e.g. of drug production) is likely to play a greater role - at least in individual cases - in risk evaluation, including in the approval decision□

To properly utilize agricultural biotechnology, a strong regulatory system is needed to ensure products are safe to humans and the environment. There is currently no country with an ideal regulatory system, although portions of many different country's systems are worth using as models. It is important that every country be given the opportunity to establish a biosafety system so that it can make its own determination whether a particular product meets its safety standards. Once a system is in place, two countries may rationally reach different conclusions on approving the same product, due to a different balancing of the risks and benefits of that product to that country's population and environment. Each country needs a strong regulatory structure, however, to be able to make those decisions. It is essential that regulatory systems be transparent and allow for public participation. Unless the public is provided access to the system as well as detailed information about the products that go through it, there will be little public trust in the regulatory process and no consumer confidence in the marketed products. The key to reaping the long-term benefits of agricultural biotechnology is a strong, but not stifling, regulatory system that independently reviews and approves each product for human and environmental safety before it is commercialized.

3. Bioremediation in Environment Pollution

Bioremediation technique can be considered a promising alternative to reduce the toxicity of water, soil and petroleum pollutants. The term "bioremediation has been used to describe the process of using microorganisms to degrade or remove hazardous components of the wastes from the environment (Glazer and Nikaido, 1995). Here, recent advances of bioremediation world water resources, soils and petroleum industry are discussed.

(e) 3.1. Bioremediation Technology and Bioreactors

A narrow definition of bioremediation might focus on the conversion of contaminating organic molecules to CO_2, water and inorganic ions, and the oxidation or reduction of contaminating inorganic ions. A broader definition would include biological processes for ameliorating extremes of pH, concentrating contaminants so that they can be more easily removed by physical techniques, converting toxic species to less toxic forms, thereby restoring functional ecosystems in contaminated or disturbed sites when the contaminants or disturbance cannot be removed (Adhya, 2008).

Bioremediation technology uses microorganisms to reduce, eliminate, contain and transform to benign products, contaminants present in soils, sediments, water or air. Micobial processes are beginning to be used in the cleanup of radioactive and metallic contaminants, though these contaminants present special problems since they cannot be destroyed, only transformed or contained. There is an increase in the types of contaminants to which bioremediation is being applied, including solvents, explosives, polycyclic aromatic hydrocarbons (PAHs), and polychlorinated biphenyls (PCBs). Bioremediation is an alternative to traditional remediation technologies such as landfilling or incineration. Bioremediation works by either transforming or degrading contaminants to non-hazardous or less hazardous chemicals. Bioremediation and biodegradation are of biofilm formation, bacterial adhesion and xenobiotic compounds. Environmental Biotechnology worked at the design phase and deals with mass transfer mechanisms in bacterial biofilms, surface and groundwater flow and contaminant transport, optimization of novel catalytic materials and absorbents, fluid mechanics and heat transfer of reacting single and multi-phase flows, forest micrometeorology and microclimatology, estuarine restoration and many other areas of application.

The fungus *Trichoderma harzianum produces homothallin* II, a nitrile with a broad activity against fungi and Gram-positive and Gram-negative bacteria (Faull et al.,1994), *Chromobacterium violaceum* ATCC 53434 produces the isonitrile antibiotic aerocyanidin that is primarily active against Gram-positive bacteria (Parker et al.,1988).A number of fungi and bacteria utilize the toxicity of cyanide compounds in a more overt way by producing and secreting antibiotic cyanide compounds to inhibit competitive organisms.

Polycyclic aromatic hydrocarbons (PAHs) are compounds of intense public concern due to their persistence in the environment and potentially deleterious effects on human, environmental and ecological health. Bioremediation, which exploits the metabolic potential of microbes for the clean-up of recalcitrant xenobiotic compounds, has come up as a promising alternative. Microorganisms have been found to degrade PAHs either via metabolism or co-metabolism. In spite of their xenobiotic properties, a variety of genera of gram-positive and -negative bacteria, fungi and algae have been isolated and characterized for their ability to utilize PAHs. Different regulatory mechanisms have been identified in different PAH degrading bacteria. Analysis of the PAH catabolic genes is of interest in both fundamental and applied contexts. Approaches such as improvement in PAH solubilization and entry into the cell, pathway and enzyme engineering and control of enzyme expression etc. are in development. Several strategies could be employed to improve the induction of PAH catabolic operons under field conditions. One strategy could be the expression of the PAH degrading genes under the control of a constitutive promoter. For this the novel approaches of promoter implantation by homologous recombination developed by Ohtsubo et al. (2003). Another strategy could be the identification of a cheap and non-toxic inducer, as demonstrated by Gilbert and Crowley (1997) for PCB degrading *Arthrobacter sp.* B1B. Genetic engineering of well characterized enzymes is a strategy for the creation of the enzymes with the desired properties. Biphenyls and PCBs could be exploited (a new naphthalene dioxygense from *Sphingomonas* CHY-1 which has the ability to attack PAHs composed of up to five rings. The use of such strains having broad substrate specificities is one strategy to improve the degradation of PAHs establishment of a high level of competence of the introduced bacteria. The inoculation of pollutant-degrading bacteria on plant seed can be an important additive to improve the efficiency of phytoremediation or bioaugmentation in the near future (Archana Chauhan et al., 2008).

Sarkar et al. (2006) has reported that the use of molecular biomarkers as diagnostic and prognostic tools for marine pollution monitoring. In order to assess the impact of highly persistent pollutants such as polychlorinated biphenyls (PCB), polychlorinated dibenzo-dioxins (PCDD), polychlorinated dibenzo-furans (PCDF), polynuclear aromatic hydrocarbons (PAH), tributyltin (TBT) and other toxic metals on the marine ecosystem a suite of biomarkers are being extensively used worldwide. Among the various types of biomarkers, the following have received special attention: cytochrome P4501A induction, DNA integrity, acetylcholinesterase activity and metallothionein induction. These biomarkers are being used to evaluate exposure of various species of sentinel marine organisms (e.g. mussels, clams, oysters, snails, fishes, etc.) to and the effect of various contaminants (organic xenobiotics and metals using different molecular approaches biochemical assays, enzyme linked immuno-sorbent assays (ELISA, spectrophotometric, fluorometric measurement, differential pulsed polarography, liquid chromatography, atomic absorption spectrometry. The induction of the biotransformation enzyme, cytochrome P4501A in fishes (*Callionymus lyra, Limanda limanda, Serranus sp., Mullus barbatus*) and mussels (*Dreissena polymorpha*) by various xenobiotic contaminants such as PCBs, PAHs, PCDs is used as a biomarker of exposure to such organic pollutants. The induction of cytochrome P4501A is involved in chemical carcinogenesis through catalysis of the covalent bonding of organic contaminants to a DNA strand leading to formation of DNA adduct. Measurement of the induction of cytochrome P4501A in terms of EROD (7-ethoxy resorufin O-deethylase) activity is successfully used as a potential biomarker of exposure to xenobiotic contaminants in marine pollution monitoring. In order to assess the impact of neurotoxic compounds on marine environment the evaluation of acetylcholinesterase activity in marine organisms is used as a biomarker of exposure to neurotoxic agents such as organophosphorus, carbamate pesticides etc. Metallothioneins (MTs) are induced by toxic metals such as Cd, Hg, and Cu by chelation through cysteine residues and are used in both vertebrates and invertebrates as a biomarker of metal exposure. The measurement of the levels of DNA integrity in marine organisms such as Sea stars (Asterias rubens) from the North Sea and the marine snails (Planaxis sulcatus) from the Arabian Sea along the Goa coast exposed to environmental xenobiotic contaminants clearly indicated the extent and the nature of pollution at the sampling sites along coastal environment.

Bioreactor systems are becoming more important for the production of large quantities of proteins, particularly in cases where traditional sources are limited due to cost and/or availability. Bioreactors based on mammalian cell lines are currently the system of choice for the production of proteins for therapeutic use and 21 different diagnostic and therapeutic proteins have been approved in the USA (Kelley, 2001). Plant-based bioreactors to be commercially viable, plants are not hosts to major human pathogens that can contaminate bacterial and mammalian systems (Herbers and Sonnewald 1999). PMP which was for the vaccine against ''Newcastle Disease Virus'' was a landmark success in the manufacturing of plant cell culture based proteins. To date, about 20-plant made proteins (PMPs) have been produced in proof-of principle studies. For plant cell and tissue cultures, a variety of bioreactor types providing growth and expression of bioactive substances is available today. In contrast to traditionally grown ''whole wild plants'' or ''whole transgenic plants'', their production in bioreactors guarantees defined controlled process conditions and therefore minimizes or even prevents variations in product yield and quality, which simplifies process validation and product registration. Most of the plant cell suspension cultures and hairy roots used are grown in submerged stirred bioreactors, bubble columns, airlift reactors and their modifications, gas-phase bioreactors and their combinations, or socalled hybrid systems up to m^3-range. The development of a number of novel bioreactor systems based on a diverse variety of living cells and organisms have been seen recently. Choice of bioreactor system for the production of a particular protein will help not only to enhance and sustain high productivity, but also to reduce process costs (Regine Eibl and Dieter Eibl, 2008).

(f) 3.2. In Situ Bioremediation

Bioremediation is a biodegradation process in which sites contaminated with xenobiotics are cleaned up by means of microbial bio-geochemical processes.

In general, remediation techniques can be divided into two classes depending on whether the pollutant is directly removed/degraded in-place or not, i.e., in situ or ex situ. One of the main problems associated with ex situ remediation is its high operating cost for activities like soil excavation and groundwater pumping. Ex situ technologies for the bioremediation of soil contaminated with toxic organic compounds include land farming, composting and bioreactors. Operation of these bioremediation strategies involves the excavation and /or agitation of soil, which often leads to enhanced abiotic loss of contaminants. In recent years, in situ techniques have become popular. In situ treatments, including natural attenuation - defined as *in situ* attenuation processes that reduce the hazard associated with the release of contaminant (Soo Cho et al.,1997), biostimulation and bioaugmentation, usually avoid excavation and emission control costs, and may therefore be considered as cheap and clean technologies. In situ biodegradation has been shown to be effective at degrading or transforming a large number of organic compounds to environmentally acceptable or less mobile compounds. Soluble organic contaminants are particularly amenable to biodegradation. Relatively insoluble contaminants may also be degraded if they are accessible to microbial degraders (Chena et al., 2003) In situ and on site bioremediation of contaminated soils and groundwater has recently received widespread attention. A number of structure-biodegradability relationships have been developed for aerobic microbial metabolism which aid the assessing the potential of specific contaminants for bioremediation

Potential evidence for in-situ biodegradation is often gathered by comparing geochemical and microbial characteristics in the contaminated zone to those of uncontaminated background areas. Some types of evidence for in- situ biodegradation includes the following observations: (i) decreasing contaminant concentration over time, (ii) consumption of electron acceptors and production of their reduced byproducts, (iii) production of signature degradation intermediates and products, (iv) stable isotopic ratios and radioisotopic content of contaminants and degradation products, and (v) distinct patterns of elevated biomass concentrations and activities. While these observations can provide valuable information supportive of in-situ biological activities, each has inherent interpretational limitations, and, when used alone, can yield ambigious results. However, when used in combination, these observations can provide convincing evidence for the occurrence of in-situ bioremediation .in-situ bioremediation can be an effective method for treating fuel and other hydrocarbon contamination in soil and groundwater (Vishvesh K, 2002).

Sanjeet (2001) validates the large-scale use of a carrier-based bacterial consortium and nutrients for the treatment of land contaminated with oily sludge, a hazardous hydrocarbon waste generated by petroleum industry. Improper disposal and handling of oily sludge contaminates soil and may pose a serious threat to groundwater, and in situ bioremediation offers a promising means to reclaim such contaminated soil .The process treats the contaminants on site (in situ) instead of merely transporting them elsewhere.

(g) 3.3. Bioremediation of World Water Resources

Water plays without a doubt critical role in human health, agricultural production and modulating climate. Water also provides planetary biogeochemical processes with a medium for the distribution and mineralization of pollutants. Risks of water contamination from current and past pollutant discharges abound, for example, it is estimated that the world industrial output of organic chemicals to freshwaters alone totals over 19,000 metric tons per day. Population growth and economic development without industrial efficiency improvements and sound sewage treatment and management lead to increased pollutant loads to surface and groundwater. Water is a sustainable resource if properly managed. Several strategies are currently being developed on biological wastewater treatment.

There is a growing interest in the use of aquatic or semi-aquatic plants for the removal of heavy metals from contaminated aqueous streams. Such plants could be utilized for the improvement of water quality and for reducing the pollution load in water bodies. There has been reported that water hyacinths (*Eichornia crassipes*), pennywort *(Hydrocotyle umbellate* L.) (Dierberg et al., 1987), duckweed (*Lemna minor*) and water velvet (*Azolla pinnata*) (Jain et al., 1989) take up Pb, Cu, Cd, Fe and Hg from contaminated solutions. Kara (2004) analyzed bioaccumulation of copper from contaminated wastewater by using *Lemna minor* under laboratory conditions. A substantial decrease of the exposure in the somes River could be observed due to the upgrade of the municipal wastewater treatment plant (WWTP) in Cluj-Napoca. After the upgrade of the WWTP Cluj-Napoca, the concentrations in the somes of caffeine, pentoxifylline, cyclophosphamide, galaxolide, and tonalide were significantly reduced (over 75%) (Zaharie et al., 2008).Large-scale plants usually apply coagulation with aluminium or iron salts followed by filtration but a number of anions (e.g. nitrate) have very little tendency to coordinate with metal ions and low potential for co-precipitation (Duan and Gregory, 2003).

The use of membranes is particularly attractive for separating ions between two liquid phases (purified and concentrated water streams). Because many of the difficulties associated with precipitation, coagulation or adsorption and phase separation can be avoided. The nitrate removal efficiency was close to 98% (Svetlozar et al., 2004). RO and NF can be used for removal of inorganic anions as single processes. It was demonstrated that Reverse osmosis (RO) could be effectively applied for removal of nitrate along with water desalination in a rural area (Schoeman and Steyn, 2003).

Anaerobic processes have been extensively applied for the digestion of primary and secondary sludge in wastewater treatment plants. Aanaerobic processes have played an important role in the organic load abatement and sludge manageability, mainly in large conventional aerobic treatment plants. From 1950 to 1980, several anaerobic reactor configurations were developed for industrial wastewater treatment (McCarty, 1982). Domestic anaerobic technology acts as a tool to promote a sustainable sewage treatment (DST). The ideal situation for DST systems would be the complete removal of pathogens (health protection) and the highest removal of COD (environmental protection) with recovery of energy (methane or hydrogen) and compounds of interest: nitrogen (as NH_4^+, NO_2), and NO_3), phosphorus (as phosphate) and sulfur (as S^0). The increasing use of anaerobic reactors as the first unit of DST systems is mainly due to the successful use of UASB reactors for industrial wastewater treatment. The earlier reports on the application of UASB reactors for DST are from the beginning of the eighties (Lettinga et al., 1982). At the moment, the UASB reactor is undoubtedly the most successful reactor for the treatment of raw domestic sewage, due to both the absence of a fixed bed (avoiding clogging) and the presence of a SLG separator on the reactor's top that prevents excessive solid losses. The recovery of nitrogen is quite complex due to the high solubility of all nitrogen forms of interest (NH_4^+, NO_2), NO_3). Scientists and practitioners pursue ways to remove nitrogen in its dissolved form from anaerobic effluents. Adsorption seems to be the most appropriated method for nitrogen recovery adsorption of nitrogen in its more reduced form (NH_4^+) is advantageous because this form predominates in anaerobic reactor effluents. For nitrogen removal from anaerobic effluents of DST in the traditional nitrification/denitrification design, there is a need for an external electron donor. This would open the opportunity to explore the use of electron donors produced in anaerobic reactors. Among such electron donors, VFA, methane, ammonia and sulfide are the natural candidates. Enhanced biological phosphorus removal (EBPR) has been achieved under alternating anaerobic aerobic conditions using polyphosphate-accumulating organisms (PAO) in the presence of an easily assimilating organic carbon source (Comeau et al., 1986; Eugenio et al., 2006).Based on conventional aerobic systems. The characterization and quantification of microorganisms in bioreactors is essential for describing and optimizing the bioconversion process. Most of the effort is done on the removal of more specific contaminants such as nutrients (N and P) and sulfurous compounds since they are becoming of great concern due to its impact on the quality of water bodies. There have been two recent discoveries of microbial conversions of nitrogenous compounds. One depends on the capability of ammonia oxidizers of denitrify under certain conditions resulting in a new one-step method for the removal of N-compounds. The second has been named the ANAMMOX process, wherein ammonium is oxidized to dinitrogen gas with nitrite as the electron acceptor; Biological phosphorous removal are based on the capacity of some denitrifying microorganisms to store ortho-phosphate intracellular as poly-phosphate in the presence of nitrate; The microbial conversions of sulfurous compounds involve the metabolism of several different specific groups of bacteria such as sulfate reducing bacteria, sulfur and sulfide oxidizing bacteria, and phototrophic sulfur bacteria (Santiago et al., 2004). *Nitrosomonas eutropha* was able to grow under anoxic conditions using hydrogen or organic compounds (acetate) as electron donor and nitrite as electron acceptor. The anaerobic biological treatment of industrial wastewater highly loaded with organic matter, nitrogenous and sulfurous compounds may induce manifold microbial scenarios. In consequence, very complex microbial consortia may grow within anaerobic reactors treating this type of complex matrixes of substrates (Lettinga, 1995). The use of anaerobic processes as the core technology can improves the sustainability of DST (Eugenio et al., 2006).

Cyanides are a group of compounds that present in the environment in several forms including; hydrogen cyanide; simple inorganic salts including NaCN and KCN; complex metal cyanides, thiocyanates, and as nitriles where the CN binds to an organic radical (Kjeldsen 1999). According to the degree of cyanide liberation by these compounds, the toxicity of metal-cyanide complexes or nitriles vary .Free cyanide (HCN or CN)) is an extremely potent metabolic poison in toxicity, (Figueira et al. 1995). As a consequence of its widespread use in industry, and because many cash and food crops naturally synthesize cyanide compounds, there are large volumes of waste generated during industrial and agricultural production and processing that contain significant concentrations of cyanide. Associated with the discharge of municipal wastewater by sewage treatment plants, particularly from facilities that serve industrialized areas, where cyanide can reach 0.1 mg l^{-1} (Lordi et al., 1980).These must be treated before being discharged to prevent environment degradation. There are several widely used physical and chemical treatments of cyanide wastes that have achieved acceptance in commercial applications (Ebbs, 2004). A number of fungi and bacteria utilize the toxicity of cyanide compounds in a more overt way by producing and secreting antibiotic cyanide compounds to inhibit competitive organisms. The biological treatment of metal-cyanide contaminated wastewaters has been successfully demonstrated in laboratory scale experiments under aerobic and anaerobic conditions. Relatively simple batch reactor set ups have exploited fungal biomass to degrade model metal cyanide containing wastewaters (Ezzi and Lynch, 2005).The biocatalyst 'Cyanidase', developed by Novo Nordisk, utilizes an immobilized preparation of Alcaligenes denitrificans and has been developed for optimum cyanide degradation and stability with respect to the pH, ionic strength and presence of metal ions in cyanide contaminated wastewaters (Basheer et al., 1992). Microorganisms have been shown to possess a variety of catabolic activities that can be harnessed to remediate inorganic cyanide and nitriles. Further work is required to optimize the use of such biological systems both in wastewater .In particular, the development of microbial processes that are effective and robust in the face of extreme environmental conditions, such as low pH and co-contaminant toxicity, is required to ensure a technology that is competitive with the current chemical and physical remediation strategies currently practiced to combat cyanide pollution (Joanne Baxter and Stephen Cummings, 2006).

A considerable amount of agricultural waste such as coconut husk is a highly lignocellulosic material. Parimala et al. (2004) have reported the removal of chromium from coastal water using five different products prepared from coconut husk. Nitrogenous contaminants such as ammonia and nitrite have been found to appear frequently in the source water used for coastal aquaculture (Krishnani et al., 2003). Parimala et al. (2004) has been carried out to investigate the use of coconut husk in the remediation of ammonia and nitrite from coastal water, and to examine their effect on other water quality parameters. Their result shows that coconut husk can be used for the effective removal of ammonia and nitrite from coastal water. The removal of ammonia was found to depend on materials, dose and time. Oven-dried coconut husk powder (COP) and charred coconut husk powder (CCP) were also found to be effective. Other water quality parameters, such as salinity, did not show much change in the presence of CRF, CDF, COP, CCP and CSP throughout the course of the experiment.

Bioaccumulation of selenium (Se) in wetlands and evaporation ponds has created serious hazards to fish and waterfowl. Agricultural drainage water is the main source for Se contamination in this region. Treatment technologies currently being evaluated to reduce the Se load in agricultural drainage water include physical, chemical and biological methods. The concept of the ABSR process is to grow microalgae in drainage water at the expense of nitrate and then to utilize the naturally settled algal biomass as a carbon source for native bacteria which, in the absence of oxygen, reduce the remaining nitrate to nitrogen gas and reduce selenate to Se. The insoluble Se is then removed from the water by sedimentation in deep ponds and, as needed, by dissolved air flotation and sand filtration. Supplemental carbon sources such as molasses can be employed as a reductant in addition to algal biomass. Several species of microalgae isolated from agricultural evaporation ponds transform Se oxyanions into volatile alkylselenides, selenonium ions, and proteinaceous selenomethionine (Fan et al.,

1997). In particular, two cyanophyte strains isolated from the evaporation basins of the Tulare Lake Drainage District (TLDD), California were very active in Se volatilization. Several pilot-scale projects that use anaerobic reduction reactions to precipitate Se from drainage water have been tested. These treatment processes vary based on reactor design and the type of carbon source and nutrients provided to the microorganisms. Reactor designs include ground tanks containing sludge beds, fluidized beds, or fixed films. Constructed wetlands constitute a complex ecosystem, the biological and physical components of which interact to provide a mechanical and biogeochemical filter capable of removing many different types of contaminants from water. Constructed wetlands remove Se by reduction to insoluble forms which are deposited in the sediments, by accumulation into plant tissues, and/or volatized to the atmosphere, i.e., through biological volatilization of plants, plant/microbe associations and microbes alone. Allen (1991) found that the rate of Se removal by the wetland exceeded 90% of the total Se introduced via biological volatilization as an important component of this loss. Actual field measurements of Se volatilization performed by Terry and his co-workers have shown that biological volatilization may have accounted for as much as 10-30% of the total Se removed from the San Francisco Bay constructed wetland (Hansen et al., 1996). Constructed subsurface flow (SSF) treatment wetlands can also be used to remove Se from irrigation drainage water. The wetland would consist of gravel beds that serve as the rooting medium for salt-tolerant emergent wetland plants such as bulrush, with the agricultural drainage water flowing through the gravel. As the water flows through the gravel bed, oxidation-reduction processes in the plant root zone (especially due to the interactions of the microbial community and the plant roots) would remove Se through a combination of uptake and volatilization. The primary advantage of SSF treatment wetlands is that they reduce the potential for an aquatic food chain pathway for Se exposure to fish or birds.

There are a number of reports that one, more or algal assemblage could be used as organism indicative of water quality (Mohapatra and Mohanty, 1992). Use of such organism offers a great promise in ensuring their success as qualitative and quantitative indicator suitable for a wide variety of pollutants. Rai et al. (2005) have shown that different groups of algae growing on tannery effluent exhibited high Cr accumulation and have been found useful for bioremediation of Cr pollution Algae constitute important components in polluted and unpolluted fresh water bodies and are sensitive to metallic pollutants (Whitton et al. 1982), the adverse impact of any pollutant can be adjudged by their presence/absence and metal accumulation potential. Rai et al. (2008) demonstrates a relationship between physicochemical characteristics of these water bodies and metal bioaccumulation in major algal forms. Results showed that some of these forms growing in polluted environment and accumulating high amounts of toxic metals may be used as bioindicator species, their performance in metal contaminated water under different ecological niche is to be ascertained.

(h) 3.4. Bioremediation of Soils

Bioremediation is considered to be a safe, efficient, eco-friendly and economic means of removing pollutants from contaminated soil.

Endosulfan is used extensively throughout the world to control the insect pests of a wide range of crops. It can bind to soil particles and persist for a relatively long period with half-life of 60-800 days (Siddique et al., 2003). Endosulfan is extremely toxic to fish and aquatic invertebrates and has been implicated increasingly in mammalian gonadal toxicity, genotoxicity, and neurotoxicity (Sutherland, 2000). Several microorganisms both bacteria and fungi, which degrade endosulfan have been isolated and characterized report on the diversity of endosulfan degrading bacterial systems, evaluate the capability of the enriched bacterial isolates to *in situ* degrade endosulfan and endosulfan sulfate in the complex environment of soil and to assess the survival of the introduced population in the soil based on ERIC - PCR. Molecular characterization based on 16s rDNA gene sequence analysis of bacterial colonies isolated from endosulfan contaminated soil showed the presence of *Ochrobacterum* sp, *Burkholderia* sp, *Pseudomonas alcaligenes, Pseudomonas* sp and *Arthrobacter* sp which degraded 57-90% of α-endosulfan and 74-94% of β-endosulfan after 7days (Mohit et al., 2008).

Earthworms maintain aerobic conditions through the continuous mixing of the soil (Schack-Kirchner and Hildebrand, 1998). They also ingest soil and expel a partially stabilized product (casting), in this way they ensure the availability of organic substrates for proliferation of the autochthonous microorganisms in the soil, thus increasing the microbiological and biochemical soil activity (Wolters, 2000).The use of earthworms, as an efficient method to support the bioremediation of a soil has been experimented (Singer et al., 2001). The bioremediation treatments were effective in the degradation of hydrocarbon polluted compounds. In particular, the use of just mineral nutrients to stimulate soil microbial activity caused a high decrease in total organic carbon, thus suggesting an attack of the native soil organic matter (priming effect). On the other hand, the addition of compost (especially with earthworms) had the double effect of adding nutrients and labile organic matter for the soil authoctonous microrganisms, but also of introducing new microrganisms in the soil system. Finally, the compost treatments, stimulating soil metabolic processes, achieved the best result in the hydrocarbon degradation preserving the biochemical quality of the soil (Brunello, 2006).

Katia et al. (2005) reported the ability of tropical Brazilian basidiomycetes to degrade pentachlorophenol (PCP) in soils from areas contaminated with organochlorine industrial residues. The study shows the potential of Brazilian fungi for the biodegradation of toxic and persistent pollutants and it is the first to report fungal growth and PCP depletion in soils with high pentachlorophenol concentrations.

Polycyclic aromatic hydrocarbons (PAHs) represent a large and diverse group of organic molecules with low bioavailability, hydrophobic nature, strong sorption phenomena, and high persistence in soil. PAH-contaminated soil employing various remediation technologies such as solid phase treatment, land treatment/farming, composting, bioreactors, phytoremediation, enzyme catalyzed bioremediation and combined methods (chemical pre-treatment followed by bioremediation) has been documented, in addition to the application of strategies such as biostimulation, microbial adaptation, bioaugmentation, bacterial chemotaxis, etc. (Venkata et al., 2006).

Phytoremediation, or rhizoremediation, is the use of plants and/or the associated rhizosphere to decontaminate polluted sites, and is considered today to be a realistic, low-cost alternative for treating extensive areas of pollution by organic chemicals (Susarla et al., 2002). The advantages of phytoremediation are that it preserves the natural structure and texture of the soil (Huang et al., 2004). Two reactor configurations, namely soil column bioreactors and a soil slurry phase bioreactor were studied in the application of bioreactor technology for the remediation of PAH-contaminated soil (Kumada et al., 2005).

Metal contamination of soil is especially problematic because of the strong adsorption of many metals to the surfaces of soil particles. Metals and radionuclides cannot be biodegraded. Bioremediation of metals and radionuclides relies on a complex interplay of biological, chemical and physical processes. There are at least three types of microbial processes that can influence toxicity and transport of metals and radionuclides: biotransformation, biosorption and bioaccumulation (Henry et al., 2005). Bioremediation of metals and radionuclides has had many field tests, Field research in this area has occurred for many different metals and radionuclides using different strategies characterized in six major categories: biotransformation, bioaccumulation/bisorption, biodegradation of chelators, volatilization, treatment trains, and natural attenuation (Terry et al., 2005). The biotransformation of metals is an exciting, developing strategy to treat metal contamination, especially in environments that are not accessible to other remediation technologies. Remediation of contaminated sediments and soil requires detailed in situ characterization of the speciation of the toxic substances and their transformations with respect to time and spatial distribution (Eric et al., 2005).

Siderophores are small molecular weight extracellular organic compounds secreted by microorganisms under iron-starved conditions, used by them to chelate and solubilize iron. Though they are specific ferric iron chelator, but is reported that they bind other metals also, such as divalent heavy metals and actinides because of potentially high metal-siderophore stability constants. Thus metal contaminant fate and transport in subsurface environment can be heavily influenced by siderophores. This approach can be successfully used in removing many toxic metals off the soil which poses a serious health threat. Anupa et al.(2007) reported that the siderophores produced by P. azotoformans was detected by CAS assay and determination of its functional groups has revealed that it is of mixed type i.e., catecholate as well as hydroxamate type; Siderophore showed high affinity for complexation with several metals in the following order —Cd^{2+}=Ni^{2+}=Pb^{2+}=Sn^{2+}=$AsO_4{}^{-2}$=$AsO_2{}^{-1}$>Mn^{2+}=Co^{2+}=-Cu^{2+}=Fe^3 and very weakly binds Al and Cr. The batch studies reported herein provides encouraging and promising results and presents an ecofriendly approach for removing the arsenic by siderophore from contaminated soils. The use of siderophore for decontamination of arsenic off the soil showed no toxic effect on soil microbial population. These results confirm the potential applicability of siderophore for metal bioremediation. The contaminated soil has arsenic above the legislative limit and the treatment with siderophore could bring it down to a lower level, but still complete removal needs extended washings with the siderophore. Bonding structure studies by FTIR spectroscopy revealed hydrogen bonding between soil, arsenic and siderophore.

George et al. (2006) propose the use of solid polymer beads for the removal of a semi-volatile model contaminant (phenol) from soil and its subsequent destruction in a bioreactor via a microbial consortium isolated for its ability to degrade phenol. Bioremediation of pesticide in soil was documented by Dileep (2008). Microbes were used to degrade the pesticides in situ. A successful bioremediation technique requires an efficient bacterial strain that can degrade largest pollutant to minimum level. The rate of biodegradation in soil depends on four variables i.e. availability of pesticide or metabolite to the microorganisms; physiological status of the microorganisms; survival and/or proliferation of pesticide degrading microorganisms at contaminated site and sustainable population of these microorganisms. Biodegradation is a natural process, where the degradation of a xenobiotic chemical or pesticide by an organism is primarily a strategy for their own survival. Most of these microbes work in natural environment but some modifications can be brought about to encourage the organisms to degrade the pesticide at a faster rate in a limited time frame. This capability of microbe is some times utilized as technology for removal of contaminant from actual site. Knowledge of physiology, biochemistry and genetics of the desired microbe may further enhance the microbial process to achieve bioremediation with precision and with limited or no scope for uncertainty and variability in microbe functioning. Several bacteria competent of degrading either carbofuran, carbarly, baygon or aldicarb have been isolated and characterized. The degradation of HCH and distribution of *lin* gene in Sphingomonads. *S. indicum* B90A was found to contain two non-identical *lin*A genes (Lal et al., 2006).The gene Ese encoding enzyme from monooxygenase family capable of degrading both endosulfan α and β from *Arthrobacter* sp (Weir et al., 2006). Esd gene that had same sequence homology to monooygenase family that use reduced fl avin, provided by a separate flavin reductase enzyme, as cosubstrates in *Mycobacterium smegmatis* (Sutherland et al., 2002).Gene encoding for enzyme has been identified for several pesticides, which will provide a new inputs in understanding the microbial capability to degrade a pesticide and develop a super strain to achieve the desired result of bioremediation in a short time. Endosulfan is extremely toxic to fish and aquatic invertebrates and has been implicated increasingly in mammalian gonadal toxicity, genotoxicity, and neurotoxicity (Sutherland et al., 2000). Molecular characterization based on 16s rDNA gene sequence analysis of bacterial colonies isolated from endosulfan contaminated soil showed the presence of *Ochrobacterum* sp, *Burkholderia* sp, *Pseudomonas alcaligenes, Pseudomonas* sp and *Arthrobacter* sp which degraded 57-90% of α-endosulfan and 74-94% of β-endosulfan after 7days. Whole cells of *Pseudomonas sp* and *Pseudomonas alcaligenes* showed 94 and 89% uptake of α-isomer and 86 and 89% of β-endosulfan respectively in 120 min. Endosulfan sulfate formed during endosulfan degradation was subsequently degraded to unknown metabolites. ERIC-PCR analysis indicated 80% survival of introduced population of *Pseudomonas alcaligenes* and *Pseudomonas* sp in treated plots (Mohit Kumar et al., 2008).

The capacity of basidiomycetes to degrade Pentachlorophenol (PCP), as well the possibility of using these organisms in processes of reclamation of PCP-contaminated soils, have been the subject of much previous research. Thirty-six basidiomycetes isolated from different tropical ecosystems were tested for tolerance to high PCP concentrations in soil. The present study shows the potential of Brazilian fungi for the biodegradation of toxic and persistent pollutants and it is the first to report fungal growth and PCP depletion in soils with high pentachlorophenol concentrations(Ka′ tia Maria et al.,2005).

On the efficacy of in situ and on-site bioremediation of contaminated soils and determine the attainable treatment end-points. Two kinds of compounds have been studied: phenol and alkyl phenols, which represent hydrophilic compounds, exhibiting high water solubility and moderate to low soil partitioning; and polycyclic aromatic hydrocarbons which are hydrophobic compounds with low water solubility and exhibit significant partitioning in soil organic carbon. A three-step experimental protocol for determining important kinetics parameters for the *in situ* biodegradation of toxic chemicals in soils was developed using phenol and naphthalene as the test compounds. The protocol was developed so that the experimental schemes used grew in complexity toward the actual *in situ* case, but remained simple enough to allow them to be adequately modeled. The data gained for each of the schemes agreed with expectations. Model predictions agreed very well with experimental data. Further, the model parameters were useful in simulating treatment endpoints for these types of soil reactors. The application of this protocol to other chemicals is feasible with only minor alterations in methodology (Tabak et al., 1997).

(i) 3.5. Bioremediation of Petroleum Products

Petroleum industry is one of the major economic sectors in the world. Bioremediation of pollution caused by petroleum industry has been considered as an important environmental topical study worldwide. The most widely distributed environmental pollution can be attributed to contamination with petroleum hydrocarbons, caused by tanker accidents, storage tank ruptures, pipeline leaks, and transport accidents. Bioremediation of soils contaminated with petroleum hydrocarbons has been established as an efficient, economic, versatile, and environmentally sound treatment (Singh et al., 2004; Van et al., 2003)

Two hydrocarbon-degrading bacterial isolates, an *Arthrobacter* sp. and a Gram-negative bacillus isolated from Kuwait oil lakes, exhibited considerable cell-surface hydrophobicity without production of exopolysaccharides in complex media. However, the bacteria produced copious amounts of exopolysaccharides in a low nutrient medium. When incubated with sawdust, Styrofoam or wheat bran, as carriers, under low nutrient conditions, stable exopolysaccharide-mediated immobilized cultures were formed. (Obuekwe and Esraa,2001).

The effects of biosurfactants on the biodegradation of petroleum compounds were investigated. Candida antarctica T-34 could produce extracellular biosurfactant mannosylerythritol lipids (MELs) when it was cultured in vegetable oil. In addition, it was found that this strain could also produce a new type of biosurfactant while it grew on n-undecane ($C11H24$), and the biosurfactant was named as BS-UC. In flask culture of Candida antarctica, the addition of BS-UC could improve the biodegradation rate of some n-alkanes, a mixture of n-alkanes and kerosene. By comparing the effects of the biosurfactants BS-UC and MEL and chemical surfactants on the biodegradation of crude oil, it was found that biosurfactants could be used to enhance the degradation of petroleum compounds instead of chemical surfactants. In a laboratory scale immobilized bioreactor, the addition of biosurfactant improved not only the emulsification of kerosene in simulated wastewater but also its biodegradation rate. The results showed that the biosurfactant BS-UC was highly promising for work on biodegradation of hydrophobic contaminants (Zhao et al., 2004).

The Indigenous bacteria, *Stenotrophomonas multophilia*, were isolated from the contaminated sites near to Jordan Petroleum Refinery and used further in the bioremediation experiments. (Bassim et al., 2008).

Microbiological and chemical analyses and a suitable bioreactor design were very useful for suggesting the best ways to improve biodegradation extents in a diesel-enriched soil. Samples taken in polluted soil and water revealed that bacteria from the genus *Acinetobacter* were predominant. In plate studies, *Acinetobacter* colonies produced a whitish substance with the characteristics of a biosurfactant. The study of the similarities between laboratory results and the diesel spill site strongly suggested that natural conditions at the field site allowed the implementation of *in situ* bioremediation after physical removal of LNAPL (light nonaqueous-phase liquids) (Jose et al., 2001).

Hwang et al. (2001) investigated the bioremediation of diesel-contaminated using composting techniques. Simoncyril (2001) suggested that inoculation of *Pseudomonas putida* (PP) with addition of appropriate inorganic nutrients may be a suitable method for a rapid rehabilitation of agricultural land upon pollution with crude petroleum. Odd et al. (2006) investigated biodegradation of hydrocarbons in thin oil films in seawater at low temperatures, 0 and 5□. The results showed that marine microbial communities from cold seawater have potentials for oil film HC degradation at temperatures ≤5□, and *psychrotrophic* or *psychrophilic* bacteria may play an important role during oil HC biodegradation in seawater close to freezing point. Obbard et al. (2004) reported on the ability of crude palm oil (CPO) and several component fatty acids to stimulate the indigenous microbial biomass and degrade selected petroleum hydrocarbons in oil-spiked beach sediments under laboratory conditions as a prelude to further field trials in Singapore. A combination of geochemical, microbiological and isotopic methods were used to evaluate in-situ bioremediation of petroleum hydrocarbons at one site contaminated with refinery waste and a second site contaminated with aviation gasoline at Alameda Point, California. At each site, geochemical and microbiological characteristics from four locations in the contaminated zone were compared to those from two uncontaminated background locations. At both sites, the geochemical indicators of in-situ biodegradation included depleted soil gas and groundwater oxygen, elevated groundwater alkalinity, and elevated soil gas carbon dioxide and methane in the contaminated zone relative to the background. Radiocarbon content of methane and carbon dioxide measured in soil gas at both sites indicated that they were derived from hydrocarbon contaminant degradation. Direct microscopy of soil core samples using cell wall stains and activity stains, revealed elevated microbial numbers and enhanced microbial activities in contaminated areas relative to background areas, corroborating geochemical findings. While microbial plate counts and microcosm studies using soil core samples provided laboratory evidence for the presence of some microbial activity and contaminant degradation abilities, they did not correlate well with either contaminant location, geochemical, isotopic, or direct microscopy data (Vishvesh et al., 2002).

A field-scale study was conducted by Sanjeet et al. (2001) in a 4000 m^2 plot of land contaminated with an oily sludge by use of a carrier-based hydrocarbon-degrading bacterial consortium for bioremediation. The land belonged to an oil refinery. The site selected for bioremediation contained approximately 300 tons of oily sludge. This study validates the large-scale use of a carrier-based bacterial consortium and nutrients for the treatment of land contaminated with oily sludge, a hazardous hydrocarbon waste generated by petroleum industry.

In-situ bioremediation of petroleum waste sludge in landfarming sites of Motor Oil Hellas (petroleum refinery) was studied by monitoring the changes of the petroleum composition of the waste sludge, as well as the changes in the structure of the microbial community, for a time period of 14 months. The analyses indicated an enhanced degradation of the petroleum hydrocarbons in the landfarming areas. A depletion of n-alkanes of approximately 75-100% was obtained. Marked changes of the microbial communities of the landfarms occurred concomitantly with the degradation of the petroleum hydrocarbons. (Katsivela et al., 2005).

The effects of bioremediationrogram of sterile agricultural soils contaminated with crude petroleum were determined with a view to developing a suitable technique for rehabilitation of similar environments upon pollution by oil spillage. Sterile soils inoculated with the soil bacterium, *Pseudomonas putida* (PP), with inorganic nutrients monitoring and supplementation constituted the experimental set-ups (ESU). results suggest that phylogenetic analysis of *dsr* transcript sequences may provide insight into the active sulfate reducers in marine sediments and that quantifying levels of *dsrA* transcripts can indicate whether sulfate reducers are active in particular sediment. Furthermore, it may only be possible to use *dsrA* transcript levels to compare the relative rates of sulfate reduction in sediments when sulfate concentrations, and possibly other environmental conditions, are comparable (Kuk-Jeong Chin et al., 2008).

Conclusion

Biotechnology, from a global perspective, can be expected, within the next 20 years, to transform the sectors of health, pharmaceuticals, agriculture, food and the environment. As a tool and a social process, biotechnology has influenced the development of societies and the advancement of civilization. Located and rooted within national borders, its productive, scientific and commercial aspects are generally a function of the multinationals. Sensitive to the dictates of global markets biotechnology (Callan, 1997) has a transformative capacity in fostering and driving the expansion of the world's economy. It permeates national and regional frontiers and, is at the basis of new market-oriented processes and products (Oman, 1994). Globalization of markets involves an unavoidable interdependency of the world's economies, increasing competition and participation in cross border financing and transactions and a concern for the protection of intellectual property. Advances in environmental health science and technology can be exploited to prevent long-term disaster and irreparable damage due to environmental negligence. The harmful effects of chemicals can be prevented or at least minimized. There is an urgent need for governments and scientists to resolve these emerging problems and to plan for future prevention (DaSilva, 1998). A number of bioremediation strategies for world water resources, soils and petroleum products have been developed to degrade highly stable, toxic matter. With the help of advances in bioinformatics, biotechnology holds a bright future for developing bioprocesses for environmental applications. However, there are still many pollutants persist in the environment.

References

Adhya TK (2008) Biodegradation and Bioremediation. *Indian J. Microbiol.* 48:1-2.

Akhond MAY, Machray GC (2009) Biotech crops: technologies, achievements and prospects. *Euphytica.* 166:47-59.

Allen KN (1991) Seasonal variation of selenium in outdoor experimental stream-wetland systems. *Journal of Environmental Quality.* 20: 865-868.

Anupa Nair, Asha A Juwarkar, Sanjeev K Singh (2007) Production and Characterization of Siderophores and its Application in Arsenic Removal from Contaminated Soil. *Water, Air, & Soil Pollution.* 180:199-212.

Archana Chauhan, Fazlurrahman, John G. Oakeshott, Rakesh K. Jain (2008) Bacterial metabolism of polycyclic aromatic hydrocarbons:strategies for bioremediation. *Indian J. Microbiol.* 48:95-113.

Arntzen CJ, Mason HS, Haq TA (2001) Oral immunization with transgenic plants. US Patent 6,194,560.

Basheer S, Kut OM, Prenosil JE, Bourne JR (1992) Kinetics of enzymatic degradation of cyanide. *Biotechnol. Bioeng.* 39:629-634.

Bassim E Abbassi, Walid D Shquirat (2008) Kinetics of Indigenous Isolated Bacteria used for Ex-Situ Bioremediation of Petroleum Contaminated Soil. *Water, Air, & Soil Pollution.* 192:221-226.

Beachy RN (1997) Mechanisms and applications of pathogenderived resistance in transgenic plants. *Curr. Opin. Biotechnol.* 8:215-220.

Brunello Ceccanti, Grazia Masciandaro, Carlos Garcia, Cristina Macci, Serena Doni (2006) Soil bioremediation: combination of earthworms and compost for the ecological remediation of a hydrocarbon polluted soil. *Water, Air, & Soil Pollutio.* 177: 383-397.

Callan, B (1997) Why production technology is not a measure of competitiveness in the biotechnologies. *Science and Public Policy.* 24, 146-160.

Camp WV (2005) Yield enhancement genes: seeds for growth. *Curr. Opin. Biotechnol.* 16:147-153.

Chen Z, Huang GH, Chan CW, Geng LQ, Xia J (2003) Development of an expert system for the remediation of petroleum-contaminated sites. *Environmental Modeling and Assessment.* 8:323-334.

Cohen JI, Quemada H, Frederick R (2003) Food safety and GM crops: implications for developing-country research. In: Food safety in food security and food trade. Focus 10, Brief 16 of 17. International Food Policy Research Institute, Washington, DC, USA.

Comeau Y, Hall K, Hancock R, Oldham W (1986) Biochemical model for enhancing biological phosphorus removal. *Water Res.* 20(2):1511-1521.

DaSilva E (1998) Review: Biotechnology: developing countries and globalization. World *Journal of Microbiology & Biotechnology.* 14, 463-486

De Block M, Botterman J, Vandeweile M, Dockx J, Thoen C, Gossele V, Movva NR, Thompson C, van Montagu M, Leemans J (1987) Engineering herbicide resistance in plants by expression of a detoxyfying enzyme. *EMBO J.* 6:2513-2518

Dierberg FF, DeBusk TA, Goulet Ja NA (1987) Removal of copper and lead using a thin film technique, In: Reddy KB, Smith WH(eds). Aquatic plants for water treatment and resource recovery. Magnolia Publishing Inc, Florida.

Dileep K Singh (2008) Biodegradation and bioremediation of pesticide in soil: concept, method and recent developments. *Indian J. Microbiol.* 48:35-40.

Doty SL, Shang TO, Wilson AM, Tangen J, Westergreen AD, Newman LA, Strand SE, Gordan MP (2000) Enhanced metabolism of halogenated hydrocarbons in transgenic plants containing mammalian cytochrome P450 2E1. *Proc. Natl. Acad. Sci. U.S.A.* 97:6287-6291

Duan J, Gregory J (2003) Coagulation by hydrolysing metal salts. *Adv. Coll. Inter. Sci.* 100-102: 475-502.

Duke SO (ed) (1996) Herbicide-resistant crops: agricultural, environmental, economic, regulatory, and technical aspects. CRC Press, Boca Raton

Dunwell JM (1999) Transgenic crop plants: the next generation or an example of 2020 vision. *Ann. Bot.* 84: 269-277.

Ebbs S (2004) Biological degradation of cyanide compounds. *Environ. Biotechnol.* 15:231-236.

Eric D van Hullebusch, Piet NL Lens, Henry H Tabak (2005) Developments in bioremediation of soils and sediments polluted with metals and radionuclides. 3. Influence of chemical speciation and bioavailability on contaminants immobilization/mobilization bio-processes. *Reviews in Environmental Science & Biotechnology.* 4:185-212.

Eugenio Foresti, Marcelo Zaiat, Marcus Vallero (2006) Anaerobic processes as the core technology for sustainable domestic wastewater treatment: Consolidated applications, new trends, perspectives, and challenges. *Reviews in Environmental Science and Biotechnology.* 5:3-19.

Ezzi M, Lynch JM (2005) Biodegradation of cyanide by Trichoderma spp. and Fusarium spp. *Enz. Microb. Technol.* 36:849 -854.

Fan TWM, Lane AN, Higashi RM (1997) Selenium biotransformations by a euryhaline microalga isolated from a saline evaporation pond. *Environmental Science & Technology.* 31:569-576.

Faull JL, Graeme-Cook KA, Pilkington BL (1994) Production of an isonitrile antibiotic by an UV-induced 14 mutant of Trichoderma harzianum. *Phytochemistry.* 36:1273-1276.

Ferry N, Edwards MG, Gatehouse JA, Gatehouse AMR (2004) Plant-insect interactions: molecular approaches to insect resistance. *Curr. Opin. Biotechnol.* 15:155-161.

Figueira MM, Ciminelli VST, Linardi VR (1995) Bacterial degradation of metal cyanide complexes. In: Jerez C.A.,Vargas T., Toledo H., Wiertz J.V. (eds), Biohydrometallurgical Processing, University of Chile, Chile, pp. 333-339.

Garcia S, Ezcurra E, Schoel B, Acevedo F, Soberon J, Snow AA (2005) Absence of detectable transgenes in local landraces of maize in Oaxaca, Mexico. *Proc. Natl. Acad. Sci. U.S.A.* 35:12338-12343

George P Prpich, Rachel L Adams, Andrew J Daugulis (2006) Ex situ bioremediation of phenol contaminated soil using polymer beads. *Biotechnol. Lett.* 28:2027-2031.

Gerhard Wenzel(2006) Molecular plant breeding: achievements in green biotechnology and future perspectives. *Appl. Microbiol. Biotechnol.* 70: 642-648

Gilbert ES, Crowley DE (1997) Plant compounds that induce polychlorinated biphenyl biodegradation by *Arthrobacter* sp. strain B1B. *Appl. Environ. Microbiol.* 63:1933-1938.

Gressel J (1996) The potential roles for herbicide-resistant crops in world agriculture. In: Duke SO (ed) Herbicide-resistant crops. CRC Press, Boca Raton, pp 231-250

Hansen HCB, Koch CB, Nancke-Krogh H, Borggaard OK, Sorensen J (1996) Abiotic nitrate reduction to ammonium: key role of green rust. *Environmental Science & Technology.* 30:2053-2056.

Heck GR, CaJacob CA, Padgette SR (2003) Discovery, development, and commercialization of Roundup Ready crops. In: Vasil IK (ed) Plant biotechnology 2002 and beyond. Kluwer, Dordrecht, pp 139-142.

Henry H Tabak, Piet Lens, Eric D van Hullebusch, Winnie Dejonghe (2005) Developments in bioremediation of soils and sediments polluted with metals and radionuclides - 1. Microbial processes and mechanisms affecting bioremediation of metal contamination and influencing metal toxicity and transport. *Reviews in Environmental Science and Biotechnology.* 4:115-156.

Herbers K, Sonnewald U (1999) Production of new/modified proteins in transgenic plants. *Curr. Opin. Biotechnol.* 10:163-168.

Hohn B, Puchta H (2003) Some like it sticky: targeting of the rice gene *Waxy*. *Trends Plant. Sci.* 8:51-53.

Huang XD, Alawi YE, Penrose DM, Glick BR, Greenberg BM (2004) A multi-process phytoremediation system for removal of polycyclic aromatic hydrocarbons from contaminated soils. *Environ. Pollut.* 130:465-476.

Hwang EY, Wan N, Park JS (2001) Recycling of remediated soil for effective composting of dieselcontaminated soil. *Compost. Science and Utilization.* 9 (2), 143-149.

Jain SK, Vasudevan P, Jha NK (1989) Removal of some heavy metals from polluted waters by aquatic plants: studies on duckweed and water velvet. *Bio Wastes.* 28:115-126.

Joanne Baxter, Stephen P. Cummings (2006) The current and future applications of microorganism in the bioremediation of cyanide contamination. *Antonie van Leeuwenhoek.* 90:1-17.

Jos´e LR Gallego, Jorge Loredo, Juan F Llamas, Fernando V´azquez, Jes´us S´anchez (2001) Bioremediation of diesel-contaminated soils: Evaluation of potential *in situ* techniques by study of bacterial degradation. *Biodegradation.* 12: 325-335.

Joshi L, Lopez LC (2005) Bioprospecting in plants for engineered proteins. *Curr. Opin. Plant. Biol.* 8:223-226.

Ka′ tia Maria Gomes Machado, Da′ cio Roberto Matheus, Regina Teresa Rosim Monteiro, Vera Lu′ cia Ramos Bononi (2005) Biodegradation of pentachorophenol by tropical basidiomycetes in soils contaminated with industrial residues. *World Journal of Microbiology & Biotechnology.* 21:297-301.

Kara Y (2004) Bioaccumulation of Copper from Contaminated Wastewater by Using Lemna minor. *Bull. Environ. Contam. Toxicol.* 72:467-471.

Katsivela E, Moore ERB, Maroukli D, Strompl C, Pieper D, Kalogerakis N (2005) Bacterial community dynamics during in-situ bioremediation of petroleum waste sludge in landfarming sites. *Biodegradation.* 16:169-180.

Keiko Yonekura-Sakakibara, Kazuki Saito(2006) Review: genetically modified plants for the promotion of human health. *Biotechnol. Lett.* 28:1983-1988

Kelley BD (2001) Biochemical engineering: bioprocessing of therapeutic proteins. *Curr. Opin. Biotechnol.* 12:173-174.

Kjeldsen P (1999) Behaviour of cyanides in soil and groundwater: a review. *Water, Air, & Soil Pollution.* 115:279-308.

Krämer U (2005) Phytoremediation: novel approaches to cleaning up polluted soils. *Curr. Opin. Biotechnol.* 16:133-141.

Krishnani KK, Gupta BP, Joseph KO, Muralidhar M, Sarda C, Nagavel A, Parimala V (2003) Decontamination of nitrogenous toxicants from brackish water using plant and animal extracts. *Bull. Environ. Contam. Toxicol.* 71:196-203.

Kuk-Jeong Chin, Manju L Sharma, Lyndsey A Russell, Kathleen R O'Neill, Derek R Lovley (2008) Quantifying Expression of a Dissimilatory (bi) Sulfite Reductase Gene in Petroleum-Contaminated Marine Harbor Sediments. *Microb. Ecol.* 55:489-499.

Kumada Y, Kisa T, Kuwano Y, Koshikawa H, Tashiro,Yasukagawa E, Shimizu Y (2005) Microbial community analysis of PAHs degradation in soil. Proc 5th Korea-Japan joint seminar on geo-environmental engineering, 4th July, Seoul National University, Seoul.

Lal R, Dogra C, Malhotra S, Sharma P, Pal R (2006) Diversity, Distribution and Divergence of *lin* genes in hexachlorocyclohexane degrading sphingomonads. *Trends in Biotechnology.* 24:121-130.

Lee D, Natesan E (2006) Evaluating genetic containment strategies for transgenic plants. *Trends Biotechnol.* 24(3):109-114.

Letourneau, DK, Burrows BE(2002) Genetically Engineered Organisms. Assessing Environmental and Human Health Effects. CRC Press, Boca Raton.

Lettinga G (1995) Anaerobic digestion and wastewater treatment systems. *Antonie van Leeuwenhoek.* 67:3-28.

Lettinga G, Roersma R, Grin P, de Zeew W, Hulshof Pol LL,van Velsen L, Hobma S, Zeeman G. (1982) Anaerobic treatment of low strength waste waters. In: Hughes E, Stafford DA, Wheatley BI, Baader W, Lettinga G, Nyns EJ, Verstraete W, Wentworth RL (Eds) Anaerobic Digestion 1981 (pp 271-291). Elsevier Biomedical Press BV, Amsterdam.

Lordi DT, Lue-Hing C, Whitebloom SW, Kelada N, Dennison S (1980) Cyanide problems in municipal waste water treatment plants*J. Wat. Poll. Cont. Fed.,* 52:597 -609.

Lovley DR (1997) Potential for anaerobic bioremediation of BTEX in petroleum-contaminated Aquifers. *Journal of Industrial Microbiology & Biotechnology.* 18:75-81.

McCarty PL (1982) One hundred years of anaerobic treatment. In: Hughes DE, Stafford DA, Wheatley BI, Baader W, Lettinga G, Nyns EJ, Verstraete W, Wentworth RL(Eds) Anaerobic Digestion (pp 3-22). Elsevier Biomedical Press BV, Amsterdam.

Mohapatra, PK, Mohanty, RC (1992) Determination of water quality of two water bodies using algal bioassay method. *Phycos.* 31(1,2):77-84.

Mohit Kumar C, Vidya Lakshmi, Sunil Khanna (2008) Microbial biodiversity and in situ bioremediation of endosulfan contaminated soil. *Indian J. Microbiol.* 48:128-133.

Moose SP, Mumm RH (2008) Molecular plant breeding as the foundation for 21st century crop improvement. *Plant Physiol.* 147:969-977.

Niu QW, Lin SS, Reyes JL, Chen KC, Wu HW, Yeh SD (2006) Expression of artiWcial microRNAs in transgenic *Arabidopsis thaliana* confers virus resistance. Nat. Biotechnol. 24(11):1420-1428.

Nordlee JA, Taylor SL, Townsend JA, Thomas LA, Bush RK (1996) Identification of a Brazil-nut allergen in transgenic soybeans. *New Eng. J. Med.* 334(11): 688-692.

Obbard JP, NG KL, XU R. (2004) Bioremediation of petroleum contaminated beach sediments: use of crude palm oil and fatty acids to enhance indigenous biodegradation. *Water, Air, and Soil Pollution.* 157:149-161.

Obuekwe CO, Esraa M Al-Muttawa (2001) Self-immobilized bacterial cultures with potential for application as ready-to-use seeds for petroleum bioremediation. *Biotechnology Letters.* 23: 1025-1032.

Odd G Brakstad, Kristin Bonaunet (2006) Biodegradation of petroleum hydrocarbons in seawater at low temperatures (0-5□) and bacterial communities associated with degradation. *Biodegradation.* 17:71-82.

Ohtsubo Y, Shimura M, Delawary M, Kimbara K, Takagi M, Kudo T, Ohta A, Nagata Y (2003) Novel approach to the improvement of biphenyl and polychlorinated biphenyl degradation activity: promoter implantation by homologous recombination. *Appl. Environ. Microbiol.* 69:146-153.

Padgette SR, Re DB, Barry GF, Eichholtz DE, Delannay X, Fuchs RL, Kishore GM, Fraley RT (1996) New weed control opportunities: development of soybeans with a Roundup Ready gene. In: Duke SO (ed) Herbicide-resistant crops. CRC Press, Boca Raton, pp 53-84

Parimala V, Krishnani KK, Gupta BP, Jayanthi M, Abraham M (2004) Phytoremediation of chromium from seawater using five different products from coconut husk. *Bull. Environ. Contam. Toxicol.* 73:31-37.

Parker WL, Rathnum ML, Johnson JH, Wells JS, Principe PA, Sykes RB (1988) Aerocyanidin, a new antibiotic produced by Chromobacterium violaceum. *J. Antibiot,* 41:454 -460.

Rai UN, Dwivedi S, Tripathi RD, Shukla OP, Singh NK (2005) Algal biomass: An economical method for removal of chromium from tannery effluent. *Bulletin of Environmental Contamination and Toxicology.* 75, 297- 303.

Rai UN, Smita Dubey, Shukla OP, Dwivedi S, Tripathi RD (2008) Screening and identification of early warning algal species for metal contamination in fresh water bodies polluted from point and non-point sources. *Environ. Monit. Assess.* 144:469-481.

Regine Eibl , Dieter Eibl (2008) Design of bioreactors suitable for plant cell and tissue cultures. *Phytochem. Rev.* 7:593-598.

Ritzenthaler C (2005) Resistance to plant viruses: old issue, new answers? *Curr. Opin. Biotechnol.* 16:118-122.

Rommens C, Kishore GM (2000) Exploiting full potential of disease-resistance genes for agricultural use. *Curr. Opin. Biotechnol.* 11:120-125.

Sanjeet Mishra, Jeevan Jyot, Ramesh Chander Kuhad, Banwari Lal (2001) In Situ Bioremediation Potential of an Oily Sludge-Degrading Bacterial Consortium. *Current Microbiology.* 43:328-335.

Santiago Villaverde (2004) Recent developments on biological nutrient removal processes for wastewater treatment. *Reviews in Environmental Science and Bio/Technology.* 3:171-183.

Sarkar AD, Ray Amulya N, Shrivastava Subhodeep Sarker (2006) Molecular Biomarkers: Their significance and application in marine pollution monitoring. *Ecotoxicology.* 15:333-340.

Schack-Kirchner H, Hildebrand EE (1998). Changes in soil structure and aeration due to liming and acid irrigation. *Plant and Soil.* 199:167-176.

Scheller J, Conrad U (2005) Plant-based material, protein and biodegradable plastic. *Curr. Opin. Plant. Biol.* 8:188-196

Schoeman JJ, Steyn A (2003) Nitrate removal with reverse osmosis in a rural area in South Africa. *Desalination.* 155: 15-26.

Sharma HC, Crouch JH, Sharma KK, Seetharama N, Hash CT (2002) Applications of biotechnology for crop improvement: prospects and constraints. *Plant Sci.* 163:381-395.

Siddique Tariq, Benedict C Okeke, Arshad Muhammad, Frankenberger William T Jr (2003) Enrichment and Isolation of Endosulfan-Degrading Microorganisms. *J. Environ. Qual.* 32:47-54.

Simoncyril U Nwachukwu (2001) Bioremediation of Sterile Agricultural Soils Polluted with Crude Petroleum by Application of the Soil Bacterium, *Pseudomonas putida*, with Inorganic Nutrient Supplementations. *Current Microbiology.* 42:231-236.

Singer AC, Jury W, Luepromchai E, Yahng CS, Crowley DE (2001) Contribution of earthworms to PCB bioremediation. *Soil Biol. Biochem.* 33:765-776.

Singh A, Ward OP (2004) Applied Bioremediation and Phytoremediation. Soil Biology, Vol. 1. Springer Verlag, Berlin, 281.

Snow AA, Moran-Palma P (1997) Commercialization of transgenic plants: potential ecological risks. *BioScience.* 47: 86-96.

Song J, Bradeen JM, Naess SK, Raasch JA, Wielgus SM, Haberlach GT (2003) Gene RB cloned from *Solanum bulbocastanum* confers broad spectrum resistance to potato late blight. *Proc. Natl. Acad. Sci. U.S.A.* 100(16):9128-9133.

Soo Cho J, Wilson JT, DiGiulio DC, Vardy JA, Choi W (1997) Implementation of natural attenuation at a JP-4 jet fuel release after active remediation. *Biodegradation.* 8:265-273.

Stoger E, Ma JKC, Fischer R, Christou P (2005) Sowing the seeds of success: pharmaceutical proteinsfrom plants. *Curr. Opin. Biotechnol.* 16:167-173

Susarla S, Medina VF, McCutcheon SC (2002) Phytoremediation: an ecological solution to organic chemical contamination. *Ecol. Eng.* 18:647-658.

Sutherland TD, Horne I, Harcourt RL, Russel RJ, Oakeshott JG (2002) Isolation and characterization of a *Mycobacterium* strain that metabolizes the insecticide endosulfan. *J. Appl. Microbiol.* 93:380-389.

Sutherland TD, Horne I, Lacey MJ, Harcourt RL, Russell RJ, Oakeshott JG (2000) Enrichment of an endosulfan-degrading mixed bacterial culture. *Applied and Environmental Microbiology.* 66:2822-2828.

Svetlozar Velizarov, Joao G Crespo, Maria A Reis (2004) Removal of inorganic anions from drinking water supplies by membrane bio/processes. *Reviews in Environmental Science & Biotechnology.* 3: 361-380.

Tabak HH, Govind R (1997) Protocol for determining bioavailability and biokinetics of organic pollutants in dispersed, compacted and intact soil systems to enhance in situ bioremediation. *Journal of Industrial Microbiology & Biotechnology.* 18:330-339.

Terry C Hazen1, Henry H Tabak (2005) Developments in bioremediation of soils and sediments polluted with metals and radionuclides: 2. Field research on bioremediation of metals and radionuclides. *Reviews in Environmental Science and Bio/Technology.* 4:157-183.

Van Hamme JD, Singh A, Ward OP (2003) Recent advances in petroleum microbiology. *Microbiol. Mol. Biol. Rev.* 67:503-549.

Vasil IK (1996) Phosphinothricin-resistant crops. In: Duke SO (ed) Herbicide-resistant crops. CRC Press, Boca Raton, pp 85-91

Vasil IK (2003) The science and politics of plant biotechnology - a personal perspective. *Nat. Biotechnol.* 21:849-851

Vasil IK (ed) (1999) Advances in cellular and molecular biology of plants. *Molecular improvement of cereal crops.* Kluwer, Dordrecht, vol 5

Vasil IK, Bean S, Zhao J, McCluskey P, Lookhart G, Zhao H, Altpeter F, Vasil V (2001) Evaluation of baking properties and gluten protein composition of field grown transgenic wheat lines expressing high molecular weight glutenin gene 1Ax1. *J. Plant Physiol.* 158:521-528

Venkata S Mohan, Takuro Kisa, Takeru Ohkuma, Robert A Kanaly, Yoshihisa Shimizu(2006) Bioremediation technologies for treatment of PAH-contaminated soil and strategiesto enhance process efficiency. *Rev. Environ. Sci. Biotechnol.* 5:347-374.

Vishvesh K Bhupathiraju, Paula Krauter, Hoi-Ying N Holman, Mark E Conrad, Paul F. (2002) Assessment of in-situ bioremediation at a refinery waste-contaminated site and an aviation gasoline contaminated site. *Biodegradation.* 13:79-90.

Weir KM, Sutherland TD, Horne I, Russell RJ, Oakeshott JG (2006) A Single Monooxygenase, Ese, Is Involved in the Metabolism of the Organochlorides Endosulfan and Endosulfate in an *Arthrobacter* sp. *Appl. Environ. Microbiol.* 72:3524-3530.

Whitton BA, Say PJ, Jupp BP (1982) Accumulation of zinc, cadmium and lead by the aquatic liverwort Scapania. *Environmental Pollution.* 3:299-316.

Wolfenbarger LL, Phifer PR (2000) The ecological risks and benefits of genetically engineered plants. *Science.* 290: 2088-2093.

Wolters V (2000) Invertebrate control of soil organic matter stability. *Biol. Fert. Soils.* 31:1-19.

Zaharie Moldovan, Romeo Chira, Alfredo C Alder (2009) Environmental exposure of pharmaceuticals and musk fragrances in the Somes River before and after upgrading the municipal wastewater treatment plant Cluj-Napoca, Romania. *Environ. Sci. Pollut. Res.* 16 (Suppl 1):S46-S54.

Zhao zhe Hua, Yan Chen, Guo cheng Du, Jian Chen (2004) Effects of biosurfactants produced by *Candida antarctica* on the biodegradation of petroleum compounds. World *Journal of Microbiology & Biotechnology.* 20:25-29.

INDEX

A

absorption spectroscopy, 191
acclimatization, 88, 89, 90, 111, 196
accounting, 144
acetic acid, 129, 196
acetylcholinesterase, 237
acid, x, 8, 10, 11, 16, 19, 27, 31, 32, 33, 73, 92, 95, 107, 117, 119, 127, 129, 182, 189, 192, 193, 196, 228, 232, 254
activation, x, 6, 180, 182
active oxygen, x, 179, 180
adaptation, 51, 60, 61, 62, 63, 64, 66, 72, 75, 76, 112, 244
adenine, 12, 94
adenosine, 3, 53, 94
adenosine triphosphate, 53
adhesion, 236
adsorption, 240, 241, 245
agar, 127, 128
age, 4, 14, 23, 74, 141, 150, 248
agriculture, vii, ix, 23, 112, 122, 143, 144, 223, 224, 229, 233, 234, 249, 251
alcohols, 233
aldehydes, 9
aldolase, 6
alfalfa, 9, 26, 35, 192
algae, 237, 243
Algeria, 139
alkaloids, 94
allele, viii, xi, 41, 42, 45, 46, 49, 50, 52, 54, 60, 62, 64, 65, 67, 69, 73, 76, 203, 205, 206, 207, 211
allergens, 232
alterations, 7, 9, 11, 33, 226, 247
alternative, ix, xii, 13, 52, 56, 58, 63, 70, 71, 112, 113, 122, 129, 133, 143, 144, 145, 151, 152, 156, 194, 224, 226, 236, 237, 244
aluminium, 40, 240
American Indians, 42
amino acids, 4, 18
ammonia, 241, 242
ammonium, 141, 241, 251
amylase, 53, 79, 228, 234
aneuploidy, 12
angiogenesis, 201
anhydrase, 2, 6, 18
animals, 2, 12, 13, 22, 29, 232, 235
annealing, 133, 134, 136
ANOVA, 196
anthocyanin, 29
antibiotic, 237, 242, 251, 254
antibody, 201
antioxidant, 4, 9, 22, 23, 25, 27, 28, 32, 36, 96, 191
apoptosis, 15, 22, 40
apples, 204, 205, 206, 209, 211
applications, 21, 55, 157, 204, 226, 228, 232, 234, 242, 249, 250, 251, 252
aptitude, 138
Arabidopsis thaliana, 9, 23, 24, 26, 28, 38, 39, 253
ARC, 2, 12, 19
aromatic hydrocarbons, 237, 244
ascorbic acid, x, 10, 179, 181, 184, 192
ash, 17, 28, 31, 33, 35, 139, 155
Asia, 92, 111
assessment, 5, 7, 15, 17, 25, 27, 29, 32, 34, 38, 136, 147, 158, 159, 191, 230, 235
assimilation, 7, 20, 21, 22, 26, 36, 158
ATP, 9, 53
Australia, 129, 142
authors, 5, 9, 10, 11, 12, 14, 15, 16, 17, 46, 58, 59, 62, 65, 66, 68, 69, 70, 194
availability, viii, 3, 9, 21, 30, 38, 81, 146, 238, 244, 245

B

bacillus, 247
backcross, 14, 67, 69
background, 17, 30, 47, 49, 50, 55, 61, 62, 63, 66, 68, 70, 76, 77, 197, 239, 248
bacteria, 203, 228, 229, 237, 241, 242, 244, 246, 247, 248
bacterium, 249
Bangladesh, 116
barley, 6, 8, 13, 17, 32, 33, 38, 76
barriers, 20, 34, 230
behavior, 141, 146, 195, 200, 211
Beijing, 221
Belgium, 18
beneficial effect, ix, 225
bicarbonate, 144, 158
binding, 3, 8, 9, 10, 12, 18, 19, 21, 38, 39, 190
bioaccumulation, 2, 20, 23, 31, 34, 240, 243, 245
bioassay, 13, 29, 253
bioavailability, 3, 20, 22, 33, 34, 244, 251, 255
biochemistry, vii, 32, 34, 246
bioconversion, 233, 241
biodegradability, 239
biodegradation, 236, 239, 244, 245, 246, 247, 248, 251, 253, 256
biodiesel, 233
biodiversity, 225, 233, 234, 253
bioenergy, 233
bioindicators, 27
bioinformatics, 233, 249
biokinetics, 255
biological control, 234
biological processes, 225, 236
biological systems, 242
biomarkers, 34, 237
biomass, 4, 11, 26, 65, 78, 155, 239, 242, 248, 254
biomonitoring, 17
bioremediation, xii, 224, 225, 227, 233, 236, 239, 243, 244, 245, 246, 247, 248, 249, 250, 251, 252, 253, 255, 256
biosafety, 234, 236
biosurfactant, 247
biosynthesis, 5, 9, 25, 34, 36, 159
biosynthetic pathways, 5
biotechnology, vii, viii, xi, 121, 140, 223, 224, 225, 227, 232, 234, 235, 249, 251, 254, 255
biotic, 8
breeding, vii, xi, 42, 43, 45, 46, 51, 55, 59, 60, 61, 62, 63, 64, 66, 67, 69, 70, 73, 78, 120, 129, 137, 138, 140, 141, 213, 214, 219, 221, 227, 228, 229, 234, 251, 253
Britain, 51, 75
buffer, 132, 136
burn, 147, 151, 156
by-products, xi, 223

C

Ca^{2+}, 144, 146
cabbage, 29
cadmium, vii, 1, 2, 11, 22, 23, 25, 26, 27, 28, 29, 30, 31, 32, 33, 34, 35, 36, 37, 38, 39, 40, 256
caffeine, 240
calcium, 3, 27, 36, 40, 104, 112, 144
cambium, 82, 93, 94
Canada, 42, 43, 44, 45, 50
carbohydrate, viii, 41, 49, 53, 54, 72, 75, 79, 229
carbohydrates, vii, 1, 6, 47, 49, 53, 65
carbon, vii, 1, 4, 22, 37, 53, 241, 242, 244, 246, 248
carbon dioxide, 248
carcinogenesis, 39, 238
carrier, 10, 32, 206, 239, 248
case study, 33, 34
casting, 244
catalysis, 238
cation, 10, 12, 19, 20, 37, 40, 145
cDNA, 10, 11, 14, 27, 29, 211
cell, vii, 1, 3, 5, 7, 8, 10, 11, 12, 13, 15, 16, 17, 19, 22, 26, 29, 30, 40, 129, 197, 199, 200, 237, 238, 247, 248, 254
cell culture, 30, 238
cell cycle, 16, 17, 22
cell death, 5, 8, 10, 26
cell line, 12, 13, 40, 238
cell lines, 12, 13, 40, 238
cell membranes, 7
cellular homeostasis, 19
centromere, 15
chain transfer, 21
chemical properties, 157
chemotaxis, 244
China, 2, 32, 40, 115, 117, 119, 156, 194, 213, 214, 220, 221, 223
chlorine, 90
chloroform, 129
chlorophyll, vii, x, 1, 5, 6, 7, 25, 30, 33, 38, 147, 150, 158, 159, 160, 179, 181, 182, 191
chloroplast, 6, 7, 10, 14, 22, 30, 33
chromatography, 115
chromium, 28, 242, 254
chromosome, x, 12, 14, 15, 16, 33, 34, 36, 67, 68, 69, 129, 130, 131, 138, 140, 193, 195, 197
classes, 11, 28, 48, 58, 69, 239

classification, 77, 141
cloning, 10, 32, 211, 229
clusters, 120, 197
CO2, 7, 22, 25, 158, 236
coagulation, 240
coal, 17, 32, 33, 233
coding, 23, 226, 234
codominant, 133, 136
coefficient of variation, 16
colonization, 60, 180, 181, 182, 183, 184, 189
community, 233, 252, 253
compatibility, 66, 83, 90, 93, 101, 102, 103, 117, 205
complement, 9, 56, 70, 197
complexity, 20, 50, 54, 68, 246
components, 9, 11, 49, 52, 53, 54, 73, 79, 214, 215, 231, 233, 234, 236, 243
composition, 9, 19, 34, 49, 54, 68, 70, 71, 72, 74, 77, 79, 106, 112, 114, 117, 119, 125, 126, 127, 138, 145, 146, 148, 152, 157, 228, 248, 255
compost, 86, 244, 250
composting, 239, 244, 248, 252
compounds, 4, 9, 10, 12, 15, 49, 54, 146, 228, 236, 237, 238, 239, 240, 241, 244, 246, 247, 251, 256
concentration, 3, 11, 13, 16, 17, 21, 22, 23, 41, 46, 49, 53, 54, 68, 69, 70, 73, 107, 110, 113, 126, 128, 133, 144, 145, 148, 150, 191, 197, 239
concrete, 50, 67, 235
conditioning, 3, 22, 52, 112
conductance, 147, 150
Congress, iv, 29, 34, 112, 117, 119
consensus, 232, 234
conservation, vii, viii, 4, 64, 74, 121, 125, 126, 131, 137, 141, 158, 194, 225, 231
construction, ix, 122, 131, 132, 133, 137
consumers, x, 19, 20, 21, 41, 42, 52, 55, 56, 203, 204, 231, 232
consumption, 20, 42, 48, 232, 239
contaminant, 236, 239, 242, 245, 248
contaminated soils, 3, 14, 20, 31, 39, 230, 239, 245, 246, 252
contamination, 2, 4, 22, 31, 55, 94, 225, 239, 240, 242, 245, 247, 251, 252, 254, 255
contradiction, 68
control, 9, 10, 17, 20, 23, 24, 56, 59, 70, 72, 87, 89, 91, 92, 112, 113, 125, 146, 148, 152, 159, 181, 182, 183, 184, 185, 186, 187, 188, 189, 191, 201, 203, 225, 228, 229, 234, 235, 237, 239, 243, 254, 256
conversion, 8, 41, 46, 52, 145, 233, 236
cooling, 128, 129, 180, 192
copper, 2, 27, 28, 35, 36, 82, 240, 250
corn, viii, 41, 42, 43, 44, 45, 46, 47, 48, 49, 50, 51, 52, 53, 54, 55, 57, 58, 59, 60, 61, 62, 63, 64, 65, 66, 67, 68, 69, 70, 71, 72, 73, 74, 75, 76, 77, 78, 79, 227
corolla, 97, 107, 108
correlation, xi, 16, 50, 54, 58, 189, 213, 214, 215, 216, 218, 219, 220
correlation coefficient, 215
correlations, xi, 50, 53, 62, 213, 214, 216, 218, 220, 221
costs, 55, 92, 232, 238, 239
cotton, 227, 229
cotyledon, 87
covalent bond, 238
covalent bonding, 238
covering, 224, 227
crop production, vii, 115, 144
crops, vii, viii, ix, 1, 3, 4, 35, 56, 78, 81, 82, 90, 92, 95, 106, 107, 110, 115, 118, 143, 144, 145, 151, 154, 156, 157, 158, 160, 211, 224, 226, 227, 228, 229, 230, 231, 232, 233, 234, 235, 241, 243, 249, 250, 251, 254, 255
crude oil, 247
cryopreservation, 128, 139
crystals, 128
cultivation, vii, ix, x, 42, 55, 82, 83, 90, 95, 105, 107, 111, 115, 120, 141, 157, 179, 180, 195, 203, 204, 230, 231, 232
cultural practices, 146
culture, x, 90, 91, 111, 112, 113, 114, 115, 116, 122, 124, 125, 127, 128, 129, 137, 138, 140, 141, 157, 159, 180, 191, 192, 193, 194, 195, 196, 197, 198, 199, 200, 201, 226, 247, 255
culture media, 125, 127, 140, 196
customers, 55
cyanamide, 112
cyanide, 237, 241, 250, 251, 252
cycles, 64, 133, 136, 155
cycling, 20
cyclophosphamide, 240
cytochrome, 237, 250
cytokinesis, 16
cytology, 129
cytometry, viii, 1, 16, 22, 23, 24, 34
cytoplasm, 10, 15, 197
cytoskeleton, 11
cytotoxicity, 2, 18, 29, 38, 39

D

defence, 8, 9, 22, 25, 38, 191
defense, 192

deficiency, 5, 6, 7, 32, 37, 113, 114, 120
definition, 225, 236
degradation, vii, 1, 5, 7, 11, 17, 27, 122, 150, 159, 225, 237, 239, 242, 244, 246, 247, 248, 250, 251, 252, 253, 254
degradation process, 225
dehydration, 128
denitrification, 241
denitrifying, 241
density, 7, 214
dental caries, 228
Department of Agriculture, 44
Department of the Interior, 31
derivatives, 42, 190
desiccation, 129, 139
destruction, 7, 245
detection, 11, 12, 13, 30, 35, 71, 114, 158
developed countries, 82, 224
developing countries, 82, 83, 122, 131, 137, 224, 227, 232, 235, 250
differentiation, 14, 22, 50, 139, 201
diffusion, 3, 19
digestion, 205, 240, 253
dioxin, 225
diploid, 130, 197
disaster, 249
discharges, 240
discipline, 225
discs, 91, 192
disinfection, 127
disorder, 6, 114, 116
dispersion, 16
displacement, 8
disposition, 225
dissociation, 6
distilled water, 127
distribution, vii, 1, 18, 21, 26, 27, 30, 34, 35, 38, 65, 113, 122, 131, 192, 240, 245, 246
divergence, 64
diversity, 33, 70, 76, 138, 139, 194, 228, 233, 244
division, x, 7, 15, 40, 193, 194, 197
DNA, vii, viii, 1, 11, 12, 13, 14, 15, 16, 17, 21, 22, 24, 25, 26, 28, 29, 30, 31, 32, 33, 35, 36, 38, 40, 64, 67, 69, 70, 121, 124, 129, 131, 132, 133, 135, 136, 137, 138, 139, 226, 227, 233, 237
DNA damage, 13, 15, 16, 25, 29, 31, 32, 40
DNA polymerase, 133, 136
DNA repair, 12, 26
DNA sequencing, ix, 70, 122, 137
DNA strand breaks, 31
dominance, 59, 66, 194, 219
donors, 61, 62, 65, 75, 241
draft, 25, 160
drainage, 144, 145, 155, 156, 242
drinking water, 255
drought, ix, 122, 131, 159, 180, 190, 191
dry matter, 95, 157
drying, ix, 47, 90
duration, 4, 20, 214

E

early warning, 254
ears, 30, 42, 46, 51, 55, 56, 57, 58, 59, 70, 73, 78, 79
earthworms, 39, 244, 250, 255
eating, 41, 48, 49, 52, 54, 58, 67, 68, 69, 70, 71
ecology, 224, 225
economic development, 240
economics, 118
ecosystem, 122, 237, 243
Education, 200
effluent, 145, 158, 243, 254
effluents, 159, 241
elaboration, 124
electric current, 89
electrical conductivity, 53, 144, 155
electricity, 144
electrolyte, 50, 52
electron, 5, 6, 7, 8, 11, 19, 22, 30, 37, 239, 241
electrons, 8
electrophoresis, 131, 132, 136
ELISA, 237
elongation, x, 93, 117, 125, 126, 127, 128, 133, 193, 195, 196, 197
elucidation, 224
e-mail, 193
embryo, 54
embryogenesis, 138
emission, 239
encapsulation, 128
encoding, 10, 14, 227, 228, 229, 246
endangered species, viii, 121, 125, 129
endocrine, 225
endoderm, 3
endosperm, viii, 41, 42, 45, 47, 49, 50, 53, 54, 65, 69, 71, 72, 74, 75, 77, 78, 79
endotoxins, 77
energy, 5, 53, 112, 227, 240
environment, vii, viii, x, xi, 1, 2, 4, 17, 19, 34, 50, 52, 59, 66, 77, 81, 91, 111, 112, 125, 131, 146, 151, 152, 156, 159, 203, 204, 215, 216, 220, 223, 225, 227, 230, 231, 232, 233, 235, 236, 237, 238, 241, 243, 244, 245, 246, 249
environment control, 111

environmental conditions, viii, ix, 21, 76, 81, 93, 124, 131, 143, 144, 146, 155, 156, 242, 249
environmental effects, 66, 232
environmental impact, 235
environmental protection, xi, 223, 224, 240
Environmental Protection Agency, vii, 1, 2, 230
enzymatic activity, 147
enzymes, vii, xi, 1, 5, 6, 8, 9, 10, 11, 21, 22, 25, 36, 37, 119, 184, 190, 191, 192, 223, 225, 230, 237
EPA, vii, 1, 113, 230
epicotyl, 94
epidermis, 3
epistasis, 52
ERA, 23
erosion, 122
EST, 14, 139
estimating, 61, 79
ethanol, 90, 127, 129, 196
Euro, 75, 93
Europe, 2, 36, 46, 50, 56, 61, 63, 75, 83, 92, 194, 227, 235
European Commission, 24, 25
European Union, 232
evaporation, 145, 146, 242, 251
evolution, 6, 66, 73, 75, 78, 90, 141, 228, 233
excision, 86
excitation, 8, 16
exclusion, 148, 152
exopolysaccharides, 247
experimental condition, 23, 155
experimental design, 60
expertise, 87
explosives, 228, 236
exposure, 8, 9, 10, 12, 13, 14, 15, 17, 19, 20, 21, 23, 26, 31, 32, 35, 56, 148, 237, 240, 243, 256
expressed sequence tag, 14
external environment, 58
extinction, 191
extraction, 131, 132, 145

F

family, x, 19, 29, 38, 122, 129, 131, 136, 193, 246
farmers, 55, 60, 83, 106, 224, 227, 229, 233
fatty acids, 248, 253
fermentation, 113
fertility, 129, 141
fertilization, 208
fertilizers, 2, 3, 30, 111, 114, 192
field crops, 92
field trials, 146, 248
filtration, 240, 242
financing, 249
firms, 235
first generation, xi, 55, 57, 223, 227, 231, 232, 235
fish, 242, 244, 246
fitness, 65, 66, 230
flavonoids, 201
flavor, 41, 64, 73
floating, 25, 36
flora, 233
flotation, 242
fluidized bed, 243
fluorescence, vii, 1, 5, 16, 25, 30, 31, 33, 147, 150, 158, 159, 160
focusing, 13, 235
food, vii, 1, 2, 18, 19, 20, 21, 23, 27, 33, 34, 40, 122, 125, 144, 225, 226, 227, 228, 229, 230, 232, 234, 235, 241, 243, 249, 250
food production, 229
food safety, 230
forest management, 139
forests, 194
fossil, 2, 131
fragility, 122
fragments, 13, 14, 15, 28
France, 18, 30, 31, 50, 56, 103, 155
freedom, 128, 216
freezing, 141, 159, 228, 248
freshwater, 27, 34, 35
fructose, 6, 48
fruits, xi, 99, 100, 104, 107, 108, 110, 119, 138, 141, 147, 203, 207
FTIR, 245
FTIR spectroscopy, 245
fuel, 239, 255
fungi, 52, 115, 190, 191, 192, 203, 229, 237, 242, 244, 246
fungus, ix, 52, 107, 110, 179, 180, 184, 191, 192, 237

G

gasoline, 248, 256
gel, 132, 135, 136
gene, viii, xi, 8, 10, 11, 12, 13, 14, 22, 25, 29, 30, 31, 33, 35, 36, 38, 41, 44, 46, 47, 48, 56, 64, 67, 68, 70, 72, 74, 76, 78, 79, 137, 139, 141, 194, 223, 226, 227, 228, 229, 230, 231, 233, 234, 244, 246, 252, 254, 255
gene expression, 8, 10, 11, 22, 25, 36
gene pool, 137, 194
gene silencing, 226, 230

gene targeting, 226
gene therapy, 233
general knowledge, 197
generation, xi, 16, 57, 59, 67, 70, 71, 78, 137, 219, 223, 229, 232, 233, 234, 235
genes, viii, 9, 10, 11, 14, 22, 23, 29, 38, 41, 47, 49, 51, 52, 59, 62, 66, 70, 71, 73, 141, 224, 225, 226, 227, 228, 229, 231, 234, 237, 246, 250, 253, 254
genetic diversity, 230
genetic factors, 96
genetic linkage, 14
genetic marker, 12, 23
genetics, 13, 50, 51, 60, 67, 72, 78, 221, 227, 246
genome, 11, 12, 23, 29, 70, 78, 132, 141
genomic regions, 132
genomics, 225
genotype, 51, 58, 59, 62, 64, 65, 66, 68, 70, 74, 76, 77, 109, 118, 133, 144, 146, 148, 151, 156, 205, 215, 220
Germany, 28, 30
germination, 14, 46, 51, 52, 53, 65, 66, 73, 75, 76, 79, 125, 127, 141, 184, 194, 201
gibberellin, 52, 107
global climate change, ix
global markets, 249
globalization, 250
glucose, 48
glutathione, 9, 25, 33, 38
glycerol, 129
glycoproteins, 211
government, iv, 122, 232, 233, 249
granules, 19, 21
grass, 32, 141
Greece, 81, 83, 92, 93, 102, 103, 104, 110, 112, 115, 117, 118, 119
groundwater, 37, 236, 239, 240, 248, 252
grouping, 61
groups, 7, 8, 45, 64, 151, 205, 227, 232, 233, 235, 241, 243, 245
growth rate, 75, 147, 224
guanine, 12
Guatemala, 83

H

half-life, 2, 244
harmful effects, 249
Hawaii, 47
hazards, 232, 233, 242
healing, 88, 89
health, 2, 16, 18, 25, 26, 39, 45, 204, 225, 231, 232, 235, 237, 240, 245, 249, 252
health effects, 26, 204
heat, 21, 22, 58, 95, 180, 191, 192, 237
heat transfer, 237
heavy metals, 23, 28, 29, 31, 38, 40, 228, 240, 245, 252
height, xi, 85, 90, 108, 146, 152, 213, 214, 219, 221
hemisphere, 45
herbicide, 226, 229, 231, 250, 251
heritability, xi, 59, 73, 213, 214, 215, 216, 218, 220
hermaphrodite, 96
heterogeneity, 15
heterosis, 59, 61, 63, 64, 75, 77
histidine, 18, 37
histogram, 16
homeostasis, 3, 10, 18, 19, 23, 27, 35
Honda, 34, 95, 116
hormonal control, 107
hormone, x, 114, 193, 195, 197, 199, 200
host, x, 56, 179, 180, 190
humidity, 87, 88, 116, 146, 156, 196
humus, 196
hybrid, 43, 44, 45, 46, 47, 51, 54, 59, 61, 62, 64, 73, 90, 91, 95, 96, 103, 107, 109, 110, 113, 118, 148, 155, 191, 238
hydrazine, 192
hydrocarbons, 228, 247, 248, 250, 253
hydrogen, 240, 241, 245
hydrogen cyanide, 241
hydrophobicity, 247
hydrophyte, 25
hydroponics, 33
hypersensitivity, 10
hypocotyl, 86, 87, 90, 92, 94, 106, 117, 119
hypothesis, 45, 49, 53, 66, 68, 130, 233

I

ideal, 12, 235, 240
identification, 62, 74, 124, 131, 139, 237, 254
imbalances, 25
imbibition, 53
immersion, 128
immobilization, 251
immunization, 250
implementation, 247
in situ hybridization, 15
in vitro, 90, 122, 125, 126, 127, 128, 129, 137, 138, 140, 141, 194, 195, 197, 199, 200
in vivo, 17, 35, 36
inbreeding, 42, 43
inclusion, 125

incompatibility, x, 93, 106, 111, 203, 204, 205, 211
India, 110, 160
Indians, 42
indices, 32, 221
indigenous, 225, 231, 248, 253
indirect effect, 5, 231
inducer, 19, 237
induction, x, 11, 15, 26, 27, 36, 96, 193, 195, 196, 197, 198, 199, 200, 237
industry, vii, viii, ix, xii, 2, 30, 41, 44, 45, 48, 58, 143, 224, 236, 239, 241, 247, 248
infection, 97, 101, 105, 106, 109, 112, 191, 226
infectious disease, 203
inheritance, 51, 59, 61, 136, 226
inhibition, 4, 6, 7, 8, 10, 12, 30, 32, 35, 147
initiation, 125, 127, 128, 197
injuries, 128, 151, 152, 154, 156
injury, iv, ix, 22, 52, 79, 143, 147, 148, 151, 155, 156, 158, 234
inoculation, x, 179, 180, 181, 183, 184, 191, 237, 248
inositol, 228
insecticide, 55, 59, 255
insects, 55, 56, 206, 227, 231, 234
insertion, 83, 85, 86, 87, 226
instability, 12, 13, 34, 35
integration, xi, 12, 223, 226
intellectual property, 249
interaction, 10, 22, 30, 37, 38, 49, 50, 51, 54, 57, 58, 59, 66, 70, 76, 93, 94, 109, 144
interactions, 3, 6, 16, 31, 50, 64, 66, 190, 243
intervention, 12, 122, 233
intuition, 61
invertebrates, 20, 33, 35, 39, 238, 244, 246
ion transport, 25
ions, ix, 3, 5, 6, 8, 10, 18, 19, 22, 30, 94, 143, 146, 148, 151, 158, 236, 240, 242
Iran, 27, 139
iron, 2, 7, 36, 37, 38, 112, 120, 232, 240, 245
iron transport, 38
Islam, 116
isoflavonoid, 192
isolation, 46, 48, 132, 139
isopods, 20, 22
isozymes, 10, 73, 140, 192
Italy, 101, 102, 103, 131, 139, 140

J

Japan, 2, 46, 91, 92, 103, 110, 113, 114, 115, 116, 117, 118, 179, 180, 190, 191, 192, 203, 204, 205, 206, 210, 211, 253
Jordan, 247

K

K^+, 144, 146
karyotype, 125, 130, 137, 140, 195, 196, 200
karyotyping, viii, 121, 129, 137
kinetics, 26, 39, 246
Korea, 31, 92, 112, 113, 114, 117, 118, 253
Kuwait, 247

L

labor, 57, 128, 146, 232
labour, 85, 86, 87, 90, 91
land, viii, xii, 37, 42, 45, 81, 82, 122, 144, 224, 227, 239, 244, 248
landscape, 144, 145, 146, 147, 148, 150, 155, 156, 157, 158, 159, 160
Landscape, 78, 160
landscapes, ix, 143, 151, 152, 155, 156, 158
Latin America, 60
leaching, 155, 156
leakage, 50, 52, 53, 77
Lepidoptera, 73, 74, 77, 78
lesions, 4, 22, 82, 95
liberation, 241
life cycle, 40
ligand, 18, 19, 23
light transmission, 150
likelihood, 230, 231, 232
line, 46, 52, 57, 61, 64, 69, 129, 198, 199, 207, 210, 214, 216, 219
linkage, 64, 129
lipid peroxidation, 9, 27, 38
lipids, 7, 8, 35, 96, 247
liquid chromatography, 238
liquid phase, 240
localization, 8, 39
locus, 14, 66, 67
longevity, 74, 77, 220
low temperatures, viii, 81, 95, 96, 97, 101, 102, 103, 105, 106, 107, 108, 109, 111, 128, 248, 253
lung cancer, 13, 40
lymphocytes, 25
lysine, 228

M

macronutrients, 127

magnesium, 7, 114, 144
maintenance, xi, 10, 128, 129, 145, 223
maintenance tasks, 128
management, viii, ix, 20, 27, 56, 71, 81, 113, 125, 139, 143, 144, 147, 156, 157, 158, 159, 203, 225, 229, 234, 240
Manju, 252
manufacturer, 88
manufacturing, 30, 238
mapping, 14, 33, 67, 68, 70, 78, 131
marine environment, 238
marker genes, 230
market, 42, 43, 46, 55, 56, 91, 93, 106, 116, 219, 230, 232, 233, 235, 249
markets, 48, 56, 92, 224, 249
MAS, 69, 70
measurement, 16, 35, 54, 144, 145, 150, 237
measures, 15, 94
media, x, 126, 145, 157, 158, 193, 197, 228, 247
Mediterranean, 35, 55, 63, 111, 112, 122, 131
Mediterranean climate, 122
melon, viii, 81, 91, 95, 96, 101, 102, 103, 109, 110, 111, 112, 113, 114, 115, 116, 117, 118, 119
melting, 133, 136
membranes, 3, 7, 9, 10, 132, 240
menadione, 36
mercury, 2, 30, 35, 227
meristem, 15, 28, 29, 35, 85, 86, 87
MES, 126
mesophyll, 6, 7, 19
metabolic dysfunction, 147
metabolic pathways, 4, 10
metabolism, vii, 1, 4, 5, 6, 7, 10, 11, 22, 25, 29, 33, 36, 37, 38, 40, 118, 147, 229, 237, 239, 241, 250
metabolizing, 119, 191
metal salts, 28, 250
metals, vii, 1, 2, 3, 4, 5, 7, 8, 12, 13, 16, 17, 18, 19, 20, 21, 23, 24, 27, 29, 31, 33, 35, 37, 39, 230, 237, 245, 251, 255
methanol, 129
Mexico, 45, 46, 79, 83, 233, 251
Mg^{2+}, 3, 5, 6, 144, 146
mice, 31, 201
microbial communities, 248
microbial community, 243, 248
micronucleus, 14, 15, 23, 29, 36
micronutrients, 2, 3, 7, 19, 126, 127
microorganism, 54, 252
microsatellites, vii, 1, 23, 34, 131, 136
microscope, 196
microscopy, 40, 201, 248
microtome, 196
migration, 15
milligrams, 16
mitochondria, 8, 10, 26
mitosis, 16, 129, 197
mitotic index, 12
model, vii, 1, 10, 12, 14, 18, 20, 21, 22, 23, 25, 28, 38, 78, 118, 215, 220, 242, 245, 247, 250
model system, 23
models, 4, 12, 13, 31, 77, 230, 236
modernity, 225
moisture, 52, 54, 77, 103, 145
moisture content, 54, 145
molasses, 242
molecular biology, xi, 223, 255
molecular weight, 19, 21, 132, 228, 245, 255
molecules, 5, 8, 9, 22, 30, 236, 244
Montenegro, 116
Morocco, viii, 121, 122, 123, 131, 138, 139, 141
morphogenesis, 127
morphology, 4, 40, 73, 117, 119, 129, 140, 141, 159, 197, 200
mortality, 35, 56
mortality rate, 35
mosaic, 74, 75, 95, 111, 114
mother cell, 23
movement, 3, 53, 89, 90, 118, 233
multinational companies, 224
multiplication, 125, 126, 127, 131, 140
mung bean, 180, 191
mutagen, 12, 31
mutagenesis, 8, 12, 64
mutant, viii, 9, 10, 11, 30, 41, 42, 46, 48, 49, 50, 51, 60, 65, 66, 67, 70, 72, 112, 251
mutation, 31, 43, 45, 47, 49, 60, 64, 66, 68, 77, 78, 200
mutation rate, 31, 66
mycelium, 184
mycorrhiza, 192

N

Na^{+}, ix, 143, 144, 146, 148, 151, 152, 154
NaCl, 146, 150, 151, 152, 155, 160
naphthalene, 237, 246
national borders, 249
National Research Council, 233
native population, 194
native species, 233
natural selection, 64, 75
necrosis, 125, 128, 141, 147, 150
negative relation, 70
nematode, 106, 109, 113

Netherlands, 27, 29, 32, 34, 35, 50, 111, 157, 159, 200
network, 9, 31, 137
neurotoxicity, 244, 246
New England, 42, 133
next generation, 216, 232, 251
nickel, 13, 35, 37
nitric oxide, 36
nitrification, 241
nitrogen, 4, 27, 74, 104, 118, 128, 129, 132, 181, 240, 242
nitrogen gas, 242
NMR, 38
nodes, xi, 127, 213, 219
nodules, 197
North America, 34, 45, 46, 73, 115
nostalgia, 48
novelty, 235
nuclei, 15, 16, 33
nucleic acid, 8, 12
nucleus, 11, 15, 29, 32
nutraceutical, 124
nutrient imbalance, 147
nutrients, 3, 5, 7, 10, 22, 31, 65, 85, 89, 94, 104, 107, 140, 156, 239, 241, 243, 244, 248, 249
nutrition, 65, 113, 114, 116, 118, 156, 158

O

objectives, 43, 61, 226, 228
observations, 66, 239
occlusion, 82
oceans, 144
OECD, 7, 34, 235
oil, viii, 121, 122, 124, 220, 228, 247, 248
oil production, 124
oil spill, 248
oils, 236, 249
oilseed, 7, 25, 26
optimization, 236
order, ix, 52, 57, 70, 85, 91, 94, 97, 109, 133, 146, 148, 228, 237, 245
organ, 23, 24, 228
organelles, 10, 21
organic chemicals, 240, 244
organic compounds, 239, 241, 245
organic matter, 2, 3, 22, 241, 244, 256
organism, 16, 21, 56, 231, 233, 243, 246
osmosis, 240, 254
oxidation, 8, 10, 236, 243
oxidative damage, 8
oxidative stress, vii, 1, 4, 5, 8, 10, 15, 16, 22, 25, 31, 36, 37
oxygen, 6, 8, 22, 25, 40, 116, 191, 242, 248
oxygen consumption, 116

P

palm oil, 248, 253
parameter, 7, 23, 147, 150
parameters, vii, 1, 16, 20, 26, 215, 242, 246
parents, 61, 63, 67, 71, 72, 219, 220
pathogens, viii, 81, 82, 83, 90, 109, 111, 112, 117, 238, 240
pathways, 11, 18, 20, 22, 23
PCP, 244, 246
PCR, viii, 121, 131, 132, 133, 136, 137, 141, 142, 205, 229, 244, 246
peat, 155, 156
pedigree, 61
peptides, 18, 19, 39
perceptions, 224
peripheral blood, 25
permeability, 25, 27
pesticide, 204, 233, 245, 250
pests, 56, 61, 63, 66, 70, 72, 227, 229, 233, 234, 243
pH, 3, 20, 22, 39, 107, 108, 126, 127, 128, 195, 236, 242
pharmaceuticals, 227, 234, 249, 256
phenol, 190, 245, 246, 251
phenotype, 11, 47, 66, 67, 74
phloem, 93, 94
phosphorous, 241
phosphorus, x, 30, 179, 180, 240, 250
phosphorylation, 8, 11
photosynthesis, vii, 1, 4, 5, 6, 7, 8, 10, 25, 28, 31, 32, 37, 118, 147, 150, 190, 192
physical properties, 70
physiology, vii, ix, 147, 158, 246
phytoplankton, 20
phytoremediation, 11, 18, 23, 33, 40, 230, 237, 244, 252
pistil, 205pith, 86, 94
plasma, 3, 10, 30, 230
plasma membrane, 3, 10, 30
plasma proteins, 230
plasticity, 31, 126
Pliocene, 131
ploidy, 16, 29, 34, 129, 141
point mutation, vii, 1, 12, 13, 14
politics, 234, 255
pollen, xi, 23, 140, 141, 203, 205, 206, 208, 209, 231, 232
pollination, 41, 48, 49, 64, 205, 207, 210, 211, 214, 221

pollinators, 204, 206, 211
pollutants, xii, 2, 22, 224, 225, 236, 237, 240, 243, 244, 246, 249, 255
pollution, vii, ix, xii, 1, 2, 4, 12, 13, 23, 28, 29, 32, 224, 225, 227, 230, 231, 237, 240, 242, 243, 244, 247, 248, 254
polonium, 33
polyacrylamide, 136
polychlorinated biphenyls (PCBs), 236
polycyclic aromatic hydrocarbon, 17, 236, 246, 250, 252
polymer, 245, 251
polymorphism, 12, 14, 24, 70, 73, 136
polymorphisms, 13, 26, 133
polyploidy, 129
poor, ix, 4, 53, 65, 122, 123, 125, 143, 146, 156, 224, 234
population, ix, 13, 14, 16, 39, 49, 51, 54, 60, 65, 67, 68, 69, 71, 74, 78, 122, 123, 124, 139, 143, 144, 216, 220, 231, 233, 236, 244, 245
population density, 74
population growth, 144
Portugal, 1, 34
positive correlation, xi, 213, 214, 216, 219
potassium, 7, 113, 144
potato, ix, 96, 111, 229, 230, 255
poverty, 122, 224, 234
poverty alleviation, 122
preference, 52, 69, 77, 96, 220
pressure, 13, 56, 60, 122, 234
prevention, 12, 113, 114, 249
private sector, 45
probe, 15, 67, 225
producers, 20, 48, 83, 87, 91, 231, 232
production technology, 250
productivity, ix, 4, 159, 221, 232, 238
program, 43, 44, 45, 49, 51, 54, 59, 64, 67, 133, 136
proliferation, x, 22, 29, 127, 193, 196, 200, 201, 244, 245
promoter, 237, 254
propagation, vii, viii, x, 91, 111, 114, 121, 122, 125, 127, 137, 138, 139, 141, 155, 193, 194, 195, 199
protective role, 29
proteinase, 234
proteins, 7, 8, 10, 11, 18, 19, 21, 22, 23, 225, 229, 230, 231, 233, 238, 252
proteomics, 22, 225
protocol, viii, 121, 129, 132, 137, 246
purification, xi, 115, 223
pyrimidine, 229

Q

quantum yields, 5

R

race, 45, 55, 114
rain, 3, 16, 32
rainfall, 145
range, 2, 32, 70, 131, 132, 133, 136, 145, 146, 151, 152, 154, 196, 218, 226, 227, 229, 230, 231, 234, 238, 243
reactive oxygen, 8, 25, 26, 28, 29, 30, 36
reading, 145, 150
reason, 58, 91, 226, 232
recessive allele, viii, 41
recombination, 12, 24, 29, 70, 227, 237, 254
recovery, 226, 240, 250
redistribution, 18
regeneration, viii, 113, 121, 127, 138, 194, 195, 197, 200, 201, 226, 227
region, 15, 35, 65, 67, 68, 72, 156, 229, 242
regulation, 11, 22, 36, 65, 66, 85, 232, 235
regulators, 96, 113, 114, 125, 196, 200
regulatory bodies, 233
regulatory requirements, 232
rehabilitation, 122, 248
rejection, 56, 66
relationship, 7, 15, 21, 32, 53, 54, 57, 65, 68, 93, 116, 141, 145, 180, 184, 190, 221, 243
relationships, ix, 13, 45, 71, 76, 77, 122, 129, 131, 137, 140, 219, 239
relatives, 230, 231
relevance, vii, 1, 39, 58
remediation, xii, 224, 225, 227, 236, 239, 242, 244, 245, 250, 255
repair, 11, 12, 22, 30, 31
replication, 13, 22, 30, 181, 215
reserves, 53, 72, 75
residues, 19, 32, 108, 228, 238, 244, 252
resolution, x, 70, 179, 180
resources, 4, 125, 131, 139, 146, 228
respiration, 4, 8, 147
retention, 95, 108, 113
returns, ix
rhizome, 194
rice, 2, 13, 17, 30, 32, 74, 82, 226, 228, 232, 252
rings, 237
risk, 12, 19, 23, 53, 62, 225, 227, 230, 231, 232, 233, 234, 235
risk assessment, 12, 225, 230, 232, 233, 235
RNA, 230

RNAi, 226
Romania, 256
room temperature, 196

S

safety, 4, 227, 230, 234, 235, 236, 250
salinity, vii, ix, 20, 143, 144, 145, 146, 147, 148, 149, 150, 151, 152, 155, 156, 157, 158, 159, 160, 242
salt, ix, 116, 143, 144, 146, 147, 148, 150, 151, 152, 154, 155, 156, 157, 158, 159, 160, 243
salts, 94, 144, 145, 146, 155, 240, 241
sample, 16, 42, 144, 145
sampling, 182, 221, 238
saturation, 132, 156
scale system, 233
scavengers, 8, 9
scientific progress, 227
scores, 53, 54
search, 60
second generation, 55, 57, 58, 72, 227, 231, 232
secretion, 14, 32, 33
security, 122, 125, 225, 235, 250
sediment, 249
sedimentation, 242
sediments, 236, 243, 245, 248, 249, 251, 253, 255
seed, 28, 42, 43, 44, 46, 48, 50, 51, 52, 53, 71, 74, 77, 79, 90, 91, 92, 114, 122, 127, 139, 141, 214, 226, 228, 230, 231, 237
seedlings, viii, 13, 24, 26, 27, 29, 30, 32, 33, 35, 36, 81, 83, 85, 86, 87, 88, 89, 90, 91, 92, 93, 94, 95, 106, 107, 109, 111, 116, 117, 119, 127, 156, 191, 214
segregation, 135, 197
selenium, 10, 29, 40, 242, 249
senescence, 4, 5, 9
sensitivity, 6, 14, 133
sequencing, 70, 71, 233
sequestration proteins, 21
severity, x, 179, 182, 184
sewage, xi, 2, 30, 37, 223, 240, 241
sex, 96, 113, 115, 116, 119
shade, 86, 87, 152, 156, 194
shape, 7, 88, 95, 96, 107, 108, 194, 206
shelter, 210
shoot, ix, x, 7, 10, 30, 57, 82, 85, 86, 90, 94, 127, 129, 140, 141, 142, 152, 190, 193, 194, 195, 196, 197, 198, 199, 200, 201
shortage, 96
shrimp, 21, 32, 39
shrubs, 144, 147, 155, 156, 157, 158, 159
signal transduction, 11, 25
signalling, ix, 8
signs, 93, 122
silica, 144
Singapore, 248
single test, 50
SiO2, 144
sludge, 2, 3, 25, 30, 36, 37, 239, 240, 243, 248, 252
social problems, 122
sodium, ix, 3, 26, 127, 143, 144, 157, 158
soil particles, 243, 245
soil pollution, 225
solid phase, 244
solid waste, 24, 25, 28, 35, 225
solubility, 33, 145, 241, 246
solvents, 191, 228, 236
somatic cell, 129, 130
somatic mutations, 29
sorption, 244
South Africa, 140, 254
soybean, 26, 180, 182, 184, 192, 227
soybeans, 232, 253, 254
space, 85, 91, 125, 128, 137
Spain, 46, 50, 51, 63, 64, 76, 93, 103, 110, 121, 133, 137, 139, 140
speciation, 245, 251
231, 233, 234, 236, 237, 242, 243, 254
spectrum, xi, 56, 223, 227, 230, 255
speed, 11, 146
spindle, 124
stability, 34, 64, 66, 74, 195, 242, 245, 256
stabilization, 123
stages, 37, 50, 57, 58, 65, 77, 106, 226
standard deviation, 215
standard error, 153, 154, 183, 218
standards, 51, 58, 191, 236
starch, 46, 48, 49, 50, 53, 68, 70, 77, 79, 228
starch biosynthesis, 70
sterile, 90, 248
stimulus, 22, 118, 119
stock, 43, 51, 114, 116, 118, 120, 151
storage, 19, 20, 21, 23, 27, 53, 64, 128, 192, 228, 247
stormwater, 145
strain, 245, 247, 251, 255
strategies, vii, viii, ix, xii, 1, 5, 18, 21, 23, 74, 121, 126, 137, 159, 224, 225, 226, 229, 234, 237, 239, 240, 242, 244, 245, 249, 250, 253
strength, 128, 228, 242, 253
stress, ix, x, 4, 5, 7, 8, 9, 11, 15, 16, 22, 23, 25, 26, 28, 29, 30, 32, 33, 34, 35, 36, 38, 40, 52, 58, 64, 65, 66, 71, 90, 93, 116, 124, 143, 147,

148, 150, 151, 153, 155, 156, 157, 158, 159, 160, 179, 180, 181, 182, 183, 184, 190, 191
strong interaction, 5
structural protein, 230
subsistence, 224
substitution, 5, 141
substrates, 54, 146, 157, 241, 244
subsurface flow, 243
success rate, 83, 85, 86, 89, 91, 92, 94, 195
sucrose, 49, 53, 54, 68, 69, 127, 128, 129, 195
sugar, 5, 32, 41, 47, 48, 49, 53, 54, 67, 69, 73, 77, 110, 116, 118, 119
sugar beet, 5, 32
sugarcane, 221
summer, ix, 97, 101, 105, 108, 152, 179, 180, 194
Sun, 32, 113
supply, 9, 53, 152
suppression, 56
surfactant, 16, 32
survival, 22, 23, 32, 65, 88, 90, 117, 119, 123, 129, 244, 245
survival rate, 65, 88, 117, 119
susceptibility, 59, 60, 61, 65, 82
suspensions, 16
sustainability, xi, 223, 241
Switzerland, 31
symbiosis, x, 179, 180, 184, 191
symptom, 182
symptoms, 4, 5, 18, 22, 106, 128, 147, 155, 181, 182, 184
synthesis, x, 5, 22, 28, 39, 53, 71, 72, 77, 94, 107, 150, 179, 180

T

tactics, 56
Taiwan, 110, 112, 117
tanks, 243
targets, 5, 11, 61
technological developments, 235
telophase, 24, 36
temperature, ix, 3, 20, 22, 38, 66, 73, 83, 89, 90, 93, 94, 96, 107, 111, 112, 114, 115, 116, 117, 118, 119, 136, 146, 179, 180, 181, 182, 183, 184, 190, 191, 192, 214
testcross, 67
therapeutics, 225, 230
thermostability, 180, 192
threat, 19, 233, 239, 245
threshold, 17, 146
thresholds, 146, 152
thymine, 12
timber, 138
time frame, 246
timing, 147
tissue, x, 3, 4, 16, 20, 65, 90, 93, 111, 112, 116, 125, 128, 129, 132, 139, 140, 154, 157, 193, 194, 195, 196, 197, 198, 200, 226, 238, 254
tobacco, 15, 19, 29, 39, 95, 111, 116, 140, 190, 228
toxic effect, 12, 40, 245
toxic metals, 7, 237, 243, 245
toxic substances, 25, 245
toxicity, vii, 1, 2, 3, 4, 5, 6, 7, 8, 11, 16, 20, 21, 23, 25, 26, 27, 28, 30, 31, 32, 33, 34, 35, 36, 37, 38, 39, 230, 236, 237, 241, 244, 245, 246, 252
toxicology, 11, 13, 18, 36
toxicology studies, 11
tracks, 208
trade, 91, 250
traits, xi, 12, 14, 28, 49, 51, 54, 58, 61, 65, 68, 69, 70, 73, 79, 111, 116, 131, 213, 214, 215, 216, 217, 219, 220, 226, 228, 231, 234
transactions, 249
transcription, 11, 22, 53
transcripts, 11, 14, 22, 249
transduction, 11
transformation, 5, 225, 226, 227, 229
transformations, 245
transgene, 230, 233
translocation, 7, 17, 94, 119
transpiration, 4, 7, 89, 116, 192
transplantation, 86, 87, 88, 125, 138
transport, ix, 3, 5, 7, 9, 10, 11, 14, 18, 19, 22, 23, 27, 30, 36, 37, 38, 39, 112, 143, 148, 236, 245, 247, 252
transportation, 93
trees, 82, 141, 144, 147, 155, 156, 157, 159, 182, 192, 194, 195, 206, 209, 210
triploid, 116
trust, 236
tryptophan, 228
tumours, 12, 228

U

UK, 138, 140, 141, 211
ultrastructure, 6
UN, xi, 83, 223, 254
uncertainty, 234, 246
UNESCO, 122
uniform, viii, 81, 86, 91, 92, 106, 125
United Kingdom, 50, 220
United Nations, 139
United States, 44, 56, 144, 151, 160, 194, 233

urban population, 144
USDA, 157
UV, 133, 251

V

vaccine, 238
vacuole, 18, 19
Valencia, 116
validation, 69, 238
values, x, xi, 69, 145, 147, 151, 179, 181, 182, 184, 213, 216
vapor, 116
variability, 13, 18, 52, 64, 66, 124, 131, 136, 137, 220, 246
variables, 54, 224, 245
variance, 60, 67, 68, 196, 214, 215, 216, 218
variation, xi, 13, 31, 50, 51, 54, 59, 66, 67, 68, 70, 72, 76, 77, 78, 90, 120, 131, 133, 140, 146, 213, 215, 216, 249
vascular bundle, 82, 86, 87, 92, 94
vascular system, 3, 89
vegetable oil, 247
vegetables, ix, 2, 82, 83, 90, 91, 93, 111, 112, 113, 114, 115, 116, 117, 118
vertebrates, 238
vested interests, 232
virus infection, 230
viruses, 94, 203, 229, 254
vitamin A, 232
vitamin C, 107
vitamins, 126, 127
volatilization, 228, 243, 245

W

waste treatment, 225
waste water, 253
wastewater, xii, 26, 36, 144, 145, 152, 157, 224, 225, 240, 241, 247, 251, 253, 254, 256
water policy, 24
water quality, ix, 143, 157, 240, 242, 243, 253
water resources, 236, 249
water supplies, ix, 143, 151, 158
wetlands, 38, 242
wheat, 6, 13, 14, 28, 33, 34, 35, 158, 228, 229, 247, 255
wild type, 49, 50, 62, 64, 66
winter, 97, 101, 105, 107, 108, 113, 152, 214
wood, 122, 124
woodland, x, 193, 194
work study, 124, 131

X

xylem, 3, 93, 94, 106, 114, 115, 117, 119

Y

yeast, 9, 10, 27
yield, viii, ix, 61, 62, 63, 65, 72, 75, 79, 81, 82, 83, 92, 94, 97, 104, 107, 108, 109, 110, 111, 112, 113, 114, 116, 118, 119, 124, 132, 143, 147, 155, 157, 190, 206, 214, 220, 221, 224, 227, 238, 239

Z

zinc, 3, 27, 28, 30, 31, 40, 256